T0207371

Lecture Notes of the Institute for Computer Sciences, Social Informatics and Telecommunications Engineering 512

The LNICST series publishes ICST's conferences, symposia and workshops.

LNICST reports state-of-the-art results in areas related to the scope of the Institute.

The type of material published includes

- Proceedings (published in time for the respective event)
- Other edited monographs (such as project reports or invited volumes)

LNICST topics span the following areas:

- General Computer Science
- E-Economy
- E-Medicine
- Knowledge Management
- Multimedia
- Operations, Management and Policy
- Social Informatics
- Systems

Yifan Chen · Dezhong Yao · Tadashi Nakano
Editors

Bio-inspired Information and Communications Technologies

14th EAI International Conference, BICT 2023
Okinawa, Japan, April 11–12, 2023
Proceedings

Editors
Yifan Chen
University of Electronic Science
and Technology of China
Sichuan, China

Dezhong Yao
University of Electronic Science
and Technology of China
Sichuan, China

Tadashi Nakano
Osaka Metropolitan University
Osaka, Japan

ISSN 1867-8211 ISSN 1867-822X (electronic)
Lecture Notes of the Institute for Computer Sciences, Social Informatics
and Telecommunications Engineering
ISBN 978-3-031-43134-0 ISBN 978-3-031-43135-7 (eBook)
https://doi.org/10.1007/978-3-031-43135-7

This Springer imprint is published by the registered company Springer Nature Switzerland AG
The registered company address is: Gewerbestrasse 11, 6330 Cham, Switzerland

Paper in this product is recyclable.

Preface

We are delighted to introduce the proceedings of the 14th EAI International Conference on Bio-inspired Information and Communications Technologies (BICT 2023). Looking back, the first edition of this conference was held as BIONETICS (International Conference on Bio-Inspired Models of Network, Information and Computing Systems) in 2006 with the goal of facilitating research and development of bio-inspired ICT. The following editions of the conference since then have been held almost annually, providing a world-leading venue for researchers and practitioners to discuss recent results on bio-inspired ICT. Due to the safety concerns and travel restrictions caused by COVID-19, however, the last two editions of BICT took place online as a virtual interactive conference in 2020 and 2021, respectively. For the first time after the COVID-19 pandemic, this year, we had BICT 2023 as a face-to-face conference in the beautiful city of Okinawa, Japan from April 11 to 12 in 2023.

The technical program of BICT 2023 consisted of two keynote talks and 30 paper presentations. On day 1, William C. Tang of National Taiwan University delivered his keynote talk on "Subnetwork Communications within the Hippocampus and High-Efficiency Computing" where a novel neuromorphic computing paradigm for low-power artificial intelligence was presented. Day 1 also included three technical sessions. Day 2 began with another keynote by Winston K. G. Seah of Victoria University of Wellington, New Zealand. Entitled "Quantum Internet", he overviewed the emerging area of quantum communications and networking, touching on how designing quantum internet protocols requires a major paradigm shift and presents new challenges to network design. Day 2 also included one technical session and one online session.

This year, we received 63 paper submissions, and accepted 25 papers. Additionally, we accepted 7 extended abstracts for oral presentations. We appreciate our Program Committee members for their hard work in reviewing papers carefully and rigorously. With our congratulations to the authors of accepted papers, the BICT 2023 conference proceedings consists of 25 high-quality papers.

The organization of the BICT 2023 conference proceedings relied on the contributions by Organizing Committee members as well as PC members. It was our privilege to work with these respected colleagues. Last but not least, special thanks go to the EAI, particularly Mikita Yelnitski, for helping us organize BICT 2023 and publish these proceedings successfully.

Yifan Chen
Dezhong Yao
Tadashi Nakano

Conference Organization

Steering Committee

Imrich Chlamtac	University of Trento, Italy
Jun Suzuki	University of Massachusetts, Boston, USA
Tadashi Nakano	Osaka Metropolitan University, Japan

Organizing Committee

General Chairs

Yifan Chen	University of Electronic Science and Technology of China, China
Dezhong Yao	Sichuan Research Institute of Brain Science and Brain-inspired Intelligence, China

TPC Chairs and Co-chairs

Chan-Byoung Chae	Yonsei University, South Korea
Mengjie Zhang	Victoria University of Wellington, New Zealand
Weisi Guo	Cranfield University, UK

Sponsorship and Exhibit Chair

Qiang Liu	University of Electronic Science and Technology of China, China

Local Chair

Yue Sun	Chengdu University of Technology, China

Workshops Chair

Yansha Deng	King's College London, UK

Publicity and Social Media Chairs

Nan Yang	Australian National University, Australia
Muhammad Upal Mahfuz	University of Wisconsin-Green Bay, USA
Qammer H. Abbasi	University of Glasgow, UK

Publications Chair

Lin Lin	Tongji University, China

Web Chair

Shaolong Shi	University of Electronic Science and Technology of China, China

Tutorials Chair

Adam Noel	University of Warwick, UK

Technical Program Committee

Yifan Chen	University of Electronic Science and Technology of China, China
Dezhong Yao	Sichuan Research Institute of Brain Science and Brain-Inspired Intelligence, China
Chan-Byoung Chae	Yonsei University, South Korea
Mengjie Zhang	Victoria University of Wellington, New Zealand
Weisi Guo	Cranfield University, UK
Shaolong Shi	University of Electronic Science and Technology of China, China
Nan Yang	Australian National University, Australia
Muhammad Upal Mahfuz	University of Wisconsin-Green Bay, USA
Qammer H. Abbasi	University of Glasgow, UK
Yansha Deng	King's College London, UK
Qiang Liu	University of Electronic Science and Technology of China, China
Lin Lin	Tongji University, China
Adam Noel	University of Warwick, UK
Yue Sun	Chengdu University of Technology, China
Zhen Cheng	Zhejiang University of Technology
Wang Chen	Tongji University, China

Contents

Electromagnetic-Induced Calcium Signal with Network Coding for Molecular Communications

Mengnan Su[1,2,3] and Peng He[1,2,3](✉)

[1] School of Communication and Information Engineering, Chongqing University of Posts and Telecommunications, Chongqing, China
s200101066@stu.cqupt.edu.cn

[2] Advanced Network and Intelligent Connection Technology, Key Laboratory of Chongqing Education Commission of China, Chongqing, China

[3] Chongqing Key Laboratory of Ubiquitous Sensing and Networking, Chongqing, China
hepeng@cqupt.edu.cn

Abstract. Molecular communication (MC) has become a new communication technology between nano-scale devices due to its biocompatibility and low energy consumption. Calcium signaling gradually becomes a hot research topic as a typical case of molecular communication in biological cells, but the system performance of Ca^{2+} signal-based molecular communication is low because the intracellular Ca^{2+} concentration decays with time and space. In this work, we firstly introduce a hybrid communication scheme based on electromagnetic and molecular communication to investigate the mechanism of action of cytoplasmic calcium ions induced by alternating fields. Secondly, the Ca^{2+} signal is analyzed in a multimodal analysis using an external electromagnetic device that emits electromagnetic waves as a control wave to drive the transmitter. In addition, we propose a network coding scheme based on Ca^{2+} signal frequency and a high-efficiency communication system with XOR logic gates. The proposed network coding communication system reduces the number of information exchanges and has a higher communication efficiency compared to conventional communication.

Keywords: Molecular communication · Network Coding · Electromagnetic-induced · Calcium Oscillation

---

This work was partially supported by Natural Science Foundation of China (Grant 61901070, 61801065, 61771082, 61871062, U20A20157 and 62061007), in part by the Science and Technology Research Program of Chongqing Municipal Education Commission (Grant KJQN202000603 and KJQN201900611), in part by the Natural Science Foundation of Chongqing (Grant CSTB2022NSCQ-MSX0468, cstc2020jcyjzdxmX0024 and cstc2021jcyjmsxmX0892) and in part by University Innovation Research Group of Chongqing (Grant CxQT20017), in part by Youth Innovation Group Support Program of ICE Discipline of CQUPT (SCIE-QN-2022-04).

Y. Chen et al. (Eds.): BICT 2023, LNICST 512, pp. 1–10, 2023.
https://doi.org/10.1007/978-3-031-43135-7_1

1 Introduction

With the development of interdisciplinary fields such as synthetic biology and nanotechnology, the construction of a heterogeneous and ubiquitous Internet of nano-things (IoNTs) is becoming a reality in vivo. However, bio-nanomachines must overcome power and functional limitations to achieve a large number of complex biomedical applications, such as bio-detection and tissue repair [1]. Currently, it is difficult for nanolevel devices to wireless communication due to the size and energy consumption limitations of the device unit (e.g., antenna). To solve this challenge, a novel communication technology between nanoscale devices, i.e., molecular communication (MC) [2], is proposed.

Molecular communication, as an innovative communication technology, uses molecular or chemical signals as carriers to communicate between biological nanomachines [3]. Information molecules are small in size usually in the nanoscale and can be biological compounds or synthetic compounds such as proteins, nanoparticles, etc. The MC channels are water or gas environments where tiny molecules can freely pass. The transmitter encodes messages onto the molecules and releases them into the molecular communication channel. After the receiver detects the message molecule, the receiver reacts biochemically with the molecule to decode the message and perform the corresponding function [4]. The biocompatibility and low energy consumption of molecular communication are promising for applications in the human body, such as target detection, organ repair, etc [5].

Many works have now investigated calcium (Ca^{2+}) signaling as typical of MC in bio-cellular networks and have identified Ca^{2+} signaling as a potential physical mechanism to research [6]. Ca^{2+} signaling plays an significant role in a variety of physiological processes, including fertilization and prominence transfer [7], and it is important to understand its conduction process. From a biological point of view, Ca^{2+} signaling is prevalent in cells as intracellular second messenger. From a communication engineering point of view, the multimodal waveforms of calcium signals are suitable for information encoding and for transmitting information in biological cell networks [8]. Due to the intracellular Ca^{2+} concentration decays with time space, the system performance of MC based on Ca^{2+} signals is low. However, communication between nanomachines can be established by nanomechanical, acoustic, chemical, and electromagnetic (EM) communication methods, in addition to using molecular communication.

Molecular communication and electromagnetic communication are envisioned as the most promising paradigms in vivo, and a great deal of research has been conducted in both of them. MC is a novel biomimetic communication technology that offers the possibility of building IoNTs in vivo. However, existing MC methods exhibit slow and unstable properties in biological environments due to the decay and interactions of biochemical molecules, which limits their application. Therefore, many studies have considered combining molecular communication with other communication paradigms to enhance molecular communication performance. In addition, the communication efficiency of molecular communication systems is a problem to be solved. Although the communication distance can be extended by relay nodes, it will also the complexity of information transmission between nodes.

Many previous works have investigated the molecular communication of calcium signals. In [9], a model of Ca^{2+} oscillations inside biological cells is developed and combined with a biological perspective to describe Ca^{2+} signal generation. In [10], the mechanism of the effect of alternating current field on cytoplasmic calcium concentration was proposed, and the effect of electric field strength on Ca^{2+} signal was analyzed. [11] introduces a hybrid communication scheme based on electromagnetic nano-communication and molecular communication, and explores some open problems and challenges. In addition, many works have begun to investigate efficient molecular communication networks to build molecular computers of the future. In [12], an XOR gate is constructed using details of the interactions of certain types of two-reactant reaction diffusion. In [13], frequency coding based on Ca^{2+} signals is investigated and it is shown how noise degrades the performance of channel switching. [14] proposes the use of biological cells to develop logic gates that are implemented by controlling the intercellular Ca2+ signal through an external input signal.

In this work, we propose a hybrid communication scheme based on electromagnetic communication and molecular communication, and theoretically investigate the mechanism of action of cytoplasmic calcium ions induced by alternating fields. Then, we use an external electromagnetic device that emits electromagnetic waves as a control wave to drive the transmitter, oscillatory forces can be applied to each free electrolyte and affect changes in the biological cell. In addition, we introduce an MC network coding scheme based on Ca^{2+} signal frequencies to generate different Ca^{2+} signals for coding by an external electromagnetic device. The exchange of information between cells is then done efficiently using XOR logic gates as a way to improve communication efficiency with the aim of expanding and implementing more applications.

The rest of paper is presented below. Section 2 describes the Ca^{2+} signal generation mechanism for electromagnetic enablement, Sect. 3 describes Ca^{2+} signal-based network coding, and Sect. 4 summarizes the full paper and presents future related work.

2 Electromagnetic-Induced Ca^{2+} Signaling Mechanism

In this section, we describe the mechanism of Ca^{2+} signal oscillation induced by external electromagnetic devices. The mechanism consists of a phase of membrane potential change induced by electromagnetic potential coupling and a phase of potential-induced calcium signal oscillation, as shown in Fig. 1.

2.1 Electromagnetic-Induced Membrane Potential Change Phase

It has been shown that low-frequency electric and magnetic fields can influence the activity of biological cells, which is mainly caused by forced vibrations of all free ions by external oscillatory fields. The function of the biological cell and the internal electrochemical balance will be disrupted by the coherent vibration of this charge [15]. In addition, the free ions (Na, Ca^{2+}, etc.) present around the

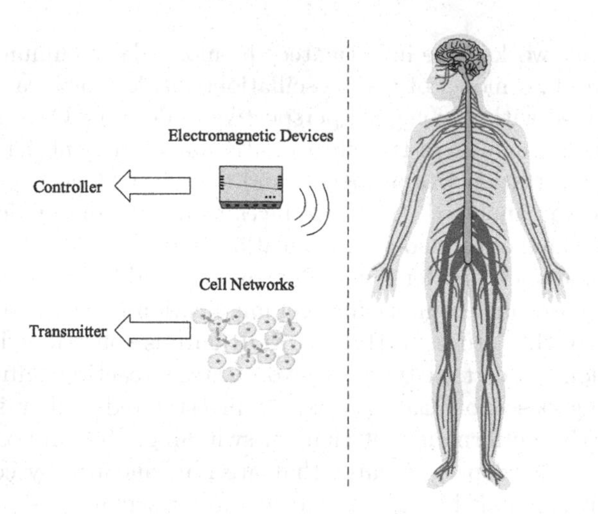

Fig. 1. System model.

cell membrane also play an essential role in the intercellular signaling process. Thus an oscillating external electric or magnetic field will apply an oscillating force for the free ions crossing the plasma membrane, and this force induces coherent forced vibrations that can be combined with the random motion of the ions [16].

The total potential difference in the cell membrane is caused by the movement of all available ions, which is facilitated by three different types of forces, as follows [15]:

$$m\frac{d^2l}{dt^2} - F_3 + F_1 = F_2 \tag{1}$$

$$F_1 = -m\omega^2 l \tag{2}$$

$$F_2 = E_m z q_e \sin(2\pi f t) \tag{3}$$

$$F_3 = -\lambda\frac{dl}{dt} \tag{4}$$

where F_1 is the restoring force generated by the electrochemical gradient, which is determined mainly by the ion mass m, the angular frequency of self-sustained oscillations ω and the distance travelled l by the ions. F_2 denotes the external force brought about by electromagnetism, where E_m is the electromagnetic intensity, f is the alternating frequency, z is the chemical valence of the ion, and $q_e = 1.6 * 10^{-19}C$. F_3 indicates the attenuation force and λ is the attenuation coefficient.

In addition, the cell membrane potential V_m is associated with the cell membrane thickness s, effector force F and charge q, i.e., the membrane potential can be expressed as:

$$V_m = F \cdot \frac{s}{q} \tag{5}$$

Here, we regard the influence of electromagnetism on SOC channels in cell membranes caused by a force.

$$F = \frac{1}{4\pi\varepsilon\varepsilon_0} \cdot \frac{q \cdot zq_e}{r^2} \tag{6}$$

where ε_0 and ε are the vacuum permittivity and relative permittivity, respectively. $q = 1.6 * q_e$, r indicates the distance between the free charge and the SOC channel. Therefore, it can be solved by the difference operation:

$$\partial V_m = \frac{1}{2\pi\varepsilon\varepsilon_0} \cdot \frac{q \cdot zq_e}{r^3} \partial r \tag{7}$$

Assuming $\partial x = \partial r$, i.e. the position of the ion is the initial origin, and the membrane potential can be expressed as:

$$\partial V_m = \frac{1}{2\pi\varepsilon\varepsilon_0} \cdot \frac{q \cdot zq_e}{r^3} \cdot \frac{s}{q} \cdot \frac{E_0 zq_e}{\lambda} \sin(2\pi vt)\partial t \tag{8}$$

2.2 Potential-Induced Ca^{2+} Oscillation Phase

The cell undergoes an intracellular signaling cascade in response to an external stimulus. The intracellular calcium reservoir (ER) opens calcium channels and then Ca^{2+} begins to diffuse into the cytoplasm generating a large Ca^{2+} concentration gradient. However, excessive Ca^{2+} concentration may lead to apoptosis, so cells maintain Ca^{2+} concentration homeostasis through internal ion regulatory mechanisms. The constant oscillation of Ca^{2+} generates signals within the cell and transmits them to adjacent cells via intercellular channels. Here, we use a widely accepted Ca^{2+} oscillation model to describe the generation of Ca^{2+} signals [17]:

$$\frac{dx}{dt} = K_1 - K_2 + K_3 + n_1 y - n_2 x + I_G \tag{9}$$

$$\frac{dy}{dt} = K_2 - K_3 - n_1 y \tag{10}$$

where x denotes the cytoplasmic Ca^{2+} concentration, y is Ca^{2+} concentration in the internal calcium pool, K_1 indicates the influx of extracellular Ca^{2+} into the cell through different classes of channels. K_2 and K_3 for Ca^{2+} exchange between cytoplasm and internal stores. Here, K_1 is mainly composed of non-electromagnetic and electromagnetic induced components, which can be expressed as follows:

$$K_1 = w_1 + w_2(t) \tag{11}$$

$$C_m \frac{dV_m}{dt} = w_2(t) \tag{12}$$

where w_1 indicates a non-EM-induced increase in Ca^{2+}, i.e., a constant influx of Ca^{2+} stimuli into the cell interior. $w_2(t)$ is induced by electromagnetism leading to an increase in Ca^{2+} concentration, which is mainly associated with changes in the membrane potential V_m and C_m the capacitance of the membrane. The specific parameter settings for the two stages are shown in [10,15–17].

Figure 2 indicates that the Ca^{2+} signal shows different calcium waves with the electric field intensity, including pulsed calcium signal and sinusoidal calcium signal. Figure 3 demonstrates that the Ca^{2+} signal varies with alternating frequency and that the resulting sinusoidal calcium signal has a higher frequency (Fig. 3b). Simulation results show that it is feasible to use an external electromagnetic device as an input device to control Ca^{2+} signal generation. And the generated Ca^{2+} signals exhibit multimodal characteristics, i.e., pulsed calcium signals and sinusoidal calcium signals. Based on these findings, we generated Ca^{2+} signals of different frequencies by external electromagnetic devices and used Ca^{2+} signals for molecular coding to enhance intercellular communication efficiency.

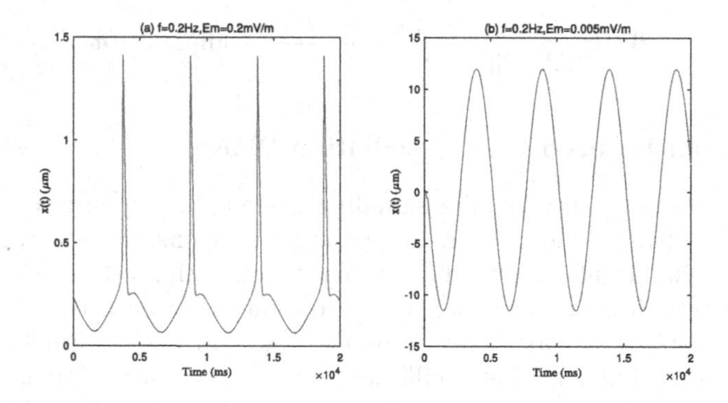

Fig. 2. Variation of Ca^{2+} waveform with electric field intensity.

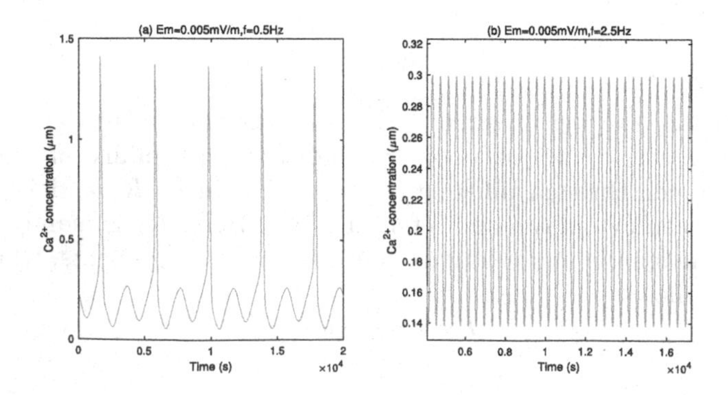

Fig. 3. Variation of Ca^{2+} waveform with alternating frequency.

3 Network Coding

In this section, we introduce a molecular communication network coding system based on multimodal Ca^{2+} signals and investigate the use of logic gates to achieve efficient communication between cells.

3.1 Intercellular Communication Model

Logic gate circuits constructed by biological cells will enable a new molecular computing technology that can rely on Boolean logic gates for intercellular communication [18]. A logic circuit consisting of Boolean logic gates can be used to reconfigure the computation of metacell operations. A fundamental function of synthetic logic circuits is communication, including communication between cells or between groups of cells [14]. Therefore, constructing logic gate circuits in intercellular molecular communication network species can not only accomplish efficient communication but also enable reconfigurability of logic operations.

The Ca^{2+} signaling molecular communication system is a short-term intercellular communicatio that uses Ca^{2+} as signaling between cellular gap junctions. To improve the system communication efficiency, we propose a Ca^{2+} signaling network encoding mechanism to accomplish efficient communication between cells or between cell populations through the XOR gate. Figure 4 briefly illustrates the two communication modes between cells, including the traditional relay communication mode (Fig. 4a) and the network coded communication mode (Fig. 4b). The network model includes two communicating cells, N1 and N2, and a relay cell, R, and cells N1 and N2 can only communicate through the relay cell. Without the use of network coding, the information of exchange cells N1 and N2 needs to be transmitted four times, i.e., regular relay communication. The information exchange realized by network coding only needs to be transmitted 3 times, i.e. N1 to R for A and N2 to R for B, which has higher communication efficiency. The cell R passes the information molecule to both N1 and N2 after the logic gate operation, and N1 and N2 then perform the logic operation to complete the information exchange, i.e., network coded communication.

3.2 Intercellular Communication

In Sect. 3, the Ca^{2+} signals induced by external electromagnetic devices exhibit multimodal characteristics (i.e., pulsed calcium signals and sinusoidal calcium signals). Here, we consider coding using the Ca^{2+} waveform frequency, and the cell treats the pulsed Ca^{2+} signal as releasing the A molecule (bit "1") and the sinusoidal Ca^{2+} signal as releasing the B molecule (bit "0"). This makes it possible to construct a combination of logic gates using Ca^{2+} frequency coding exactly to perform logic gate operations. It is assumed that every biological cell can have XOR logic gate and the cell can know the molecule it sends. In addition, we define the XOR logic gate operation rules:

$$A \oplus B = A \tag{13}$$

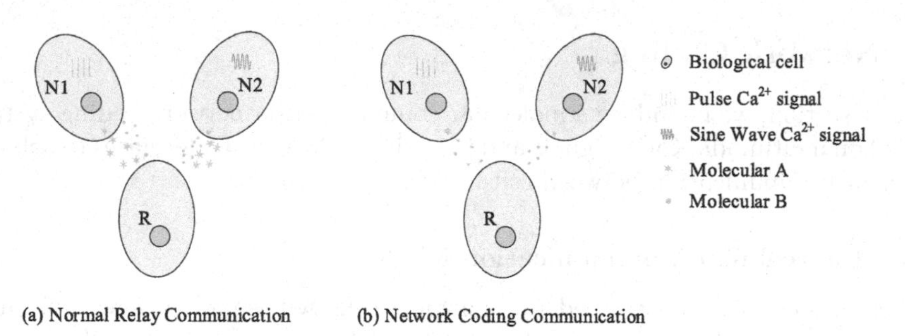

Fig. 4. Intercellular communication patterns.

$$B \oplus B = B \tag{14}$$

$$A \oplus A = B \tag{15}$$

The coding rules follow the difference as A and the same as B. The specific network code table is shown in Table I. After the relay cell R performs the heterogeneous operation and releases the information molecules to N1 and N2 for heterogeneous operation, N1 and N2 realize the intercellular communication.

Table 1. Molecular communication codes.

Bits transmitted(N1, N2)	N1 releases	N2 releases	Concentration of R	R releases
(0,0)	B	B	Low	B
(0,1)	B	A	High	A
(1,0)	A	B	High	A
(1,1)	A	A	Low	B

Figure 5 illustrates the process of intercellular communication based on the XOR gate. At first cells N1 and N2 use Ca^{2+} signaling generated by external electric fields to transmit molecular information chains. Relay cell R undergoes XOR operation after receiving the information chain from cells N1 and N2, as in Fig. 5a. The relay cell R then generates a new information link according to the molecular communication coding rules in Table 1 and simultaneously transmits it to the sending cell. After receiving the information transmitted by R, cells N1 and N2 perform XOR operations with their own information chains to complete the information exchange between cells, i.e., Fig. 5b. The proposed molecular communication coding system has better communication efficiency compared to conventional relay communication where four transmissions are required to complete the information exchange.

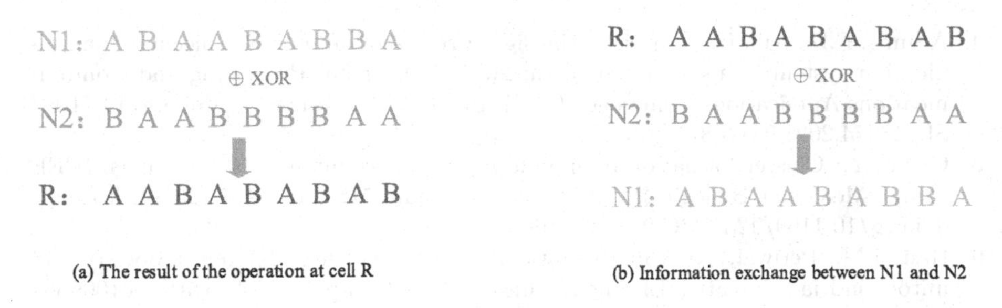

Fig. 5. Intercellular communication between N1 and N2.

4 Conclusion

In this paper, we introduce a hybrid molecular communication scheme based on alternating electric fields and propose the use of molecular communication codes to enhance intercellular communication efficiency. First, we theoretically investigated the alternating field-induced membrane potential changes in cells and explored the mechanism of potential-induced cytoplasmic calcium ion oscillations. In addition, a molecular communication coding system based on Ca^{2+} signal frequency was proposed using the multimodal characteristics of calcium signals induced by electric fields. We also demonstrate the rules of the molecular communication coding system, which is able to improve the efficiency of intercellular communication compared to the codeless communication mode.

In future work, we will continue to investigate the coding system for long-range communication between cell populations. We also consider the coding efficiency between coding rules, anti-interference and other issues, and combine with the actual biological cell network scenario to find a more efficient calcium signal molecular communication system through indepth research, aiming to build a real nano-Internet of things.

References

1. Bi, D., Almpanis, A., Noel, A., Deng, Y., Schober, R.: A survey of molecular communication in cell biology: establishing a new hierarchy for interdisciplinary applications. IEEE Commun. Surv. Tutorials **23**(3), 1494–1545 (2021). https://doi.org/10.1109/COMST.2021.3066117
2. Nakano, T., et al.: Random cell motion enhances the capacity of cell-cell communication. IEEE Trans. Molecular Biolog. Multi-Scale Commun. **5**(2), 158–162 (2019). https://doi.org/10.1109/TMBMC.2020.2983909
3. Barros, M.T., Dey, S.: Feed-forward and feedback control in astrocytes for Ca^{2+}-based molecular communications nanonetworks. IEEE/ACM Trans. Comput. Biol. Bioinform. **17**(4), 1174–1186 (2020). https://doi.org/10.1109/TCBB.2018.2887222

4. Atamiş, F.B., Işık, İ.: Analysis of the signal reception in mobile molecular communication system by using antenna. In: 2021 29th Signal Processing and Communications Applications Conference (SIU), pp. 1–4 (2021). https://doi.org/10.1109/SIU53274.2021.9477881
5. Okaie, Y.: Cluster formation by mobile molecular communication systems. IEEE Trans. Molecular Biological Multi-Scale Commun. **5**(2), 153–157 (2019). https://doi.org/10.1109/TMBMC.2020.2981662
6. Han, J.M., Periwal, V.: A mathematical model of calcium dynamics: obesity and mitochondria-associated ER membranes. PLoS Comput. Biol. **15**(8), e1006661 (2019)
7. He, P., Pi, B., Liu, Q.: Calcium signaling in mobile molecular communication networks: from a multimedia view. IEEE Access **7**, 164825–164834 (2019). https://doi.org/10.1109/ACCESS.2019.2953074
8. He, P., Su, M., Cui, Y., Wu, D., Wang, R.: Epidemic-like calcium signaling in mobile molecular communication networks. IEEE Trans. Nanobiosci. **21**(3), 425–438 (2022). https://doi.org/10.1109/TNB.2022.3155644
9. Barros, M.T.: Ca^{2+}-signaling-based molecular communication systems: design and future research directions. Nano Commun. Netw. **11**, 103–113 (2017)
10. MingYan, L., Kun, S., Xu, Z., et al.: Mechanism for alternating electric fields induced-effects on cytosolic calcium. Chin. Phys. Lett. **26**(1), 017102 (2009)
11. Yang, K., Bi, D., Deng, Y., et al.: A comprehensive survey on hybrid communication in context of molecular communication and terahertz communication for body-centric nanonetworks[J]. IEEE Trans. Molecular Biological Multi-Scale Commun. **6**(2), 107–133 (2020)
12. Adamatzky, A., Costello, B.D.L.: Experimental logical gates in a reaction-diffusion medium: the XOR gate and beyond. Phys. Rev. E **66**(4), 046112 (2002)
13. He, P., Nakano, T., Mao, Y., et al.: Stochastic channel switching of frequency-encoded signals in molecular communication networks. IEEE Commun. Lett. **22**(2), 332–335 (2017)
14. Barros, M.T., Doan, P., Kandhavelu, M., et al.: Engineering calcium signaling of astrocytes for neural-molecular computing logic gates. Sci. Rep. **11**(1), 1–10 (2021)
15. Panagopoulos, D.J., Karabarbounis, A., Margaritis, L.H.: Mechanism for action of electromagnetic fields on cells. Biochem. Biophys. Res. Commun. **298**(1), 95–102 (2002)
16. Barnes, F.S.: Interaction of DC and ELF electric fields with biological materials and systems. In: Polk, C., Postow, E. (eds.) CRC Handbook of Biological Effects of Electromagnetic Fields, pp. 27–96. CRC Press, Boca Raton, FL (1996)
17. Goldbeter, A., Dupont, G., Berridge, M.J.: Minimal model for signal-induced Ca^{2+} oscillations and for their frequency encoding through protein phosphorylation. Proc. Natl. Acad. Sci. **87**(4), 1461–1465 (1990)
18. Mao, C., LaBean, T.H., Reif, J.H., et al.: Logical computation using algorithmic self-assembly of DNA triple-crossover molecules. Nature **407**(6803), 493–496 (2000)

Smart Farm Teaching Aids Based on STEM Concepts

Hsin-Te Wu[1](✉) and Kuo-Chun Tseng[2]

[1] Department of Computer Science and Information Engineering, National Taitung University, Taitung, Taiwan
wuhsinte@nttu.edu.tw

[2] Department of Computer Science and Information Engineering, National I-lan University, Yilan, Taiwan

Abstract. In view of the above-mentioned problems, this paper is based on STEM education and constructs teaching aids for smart farms, allowing students to practice the teaching aids developed by this paper in the field. The teaching aids of this paper are mainly based on the detection of farm and honeycomb status. Determine whether there are any abnormalities between the farm and the activity status of bees, such as: crop growth, bee reproduction, etc. Students can increase their interest in IT practical learning through teaching aid assembly and program operation. In addition, students should correct the teaching aid parameters during actual operation. Improving the recognition accuracy will further arouse students' interest in artificial intelligence theory learning and achieve STEM education concepts. This paper will mainly use traditional artificial intelligence practical teaching methods and the innovative teaching methods proposed by this paper for learning comparison. The teaching method will be evaluated through the T test method through front and back questionnaires, teaching evaluation and student achievement scores. Help show whether the teaching method of this paper has achieved the expected goal. This article aims to cultivate more talents through Artificial Intelligence of Things (AIoT) teaching aids, analyzing data using t-tests in Statistical Product and Service Solutions (SPSS). Therefore, the experimental results prove that the STEM 4.0 approach proposed in this study can enhance students' learning performance and willingness.

Keywords: Artificial Intelligence · Smart Farming · Interdisciplinary Teaching · STEM · Digital Transformation

1 Introduction

Concerning the teaching of Artificial Intelligence (AI) and multimedia theories, most students tend to be less willing to learn as they lack motivation or have learning disabilities. The possible reasons for this situation are their incomprehension of the concepts and definitions and limited computational skills. In the long term, students become afraid of mathematics and even choose to give up

Y. Chen et al. (Eds.): BICT 2023, LNICST 512, pp. 11–21, 2023.
https://doi.org/10.1007/978-3-031-43135-7_2

when encountering basic calculations. When students lack the ability for mathematical logic and proof and can only memorize formulas by route, they are likely to forget those formulas as time passes, making them fail to calculate. On the other hand, the complex mathematical theories in AI and multimedia studies can sometimes lower students' learning motivation; additionally, they might give up the entire course because these mathematical theories are the application foundation of these sectors. If learners want to be more creative in developing relevant applications, they must possess the inferential capacity to tackle mathematical theories and comprehend the full process.

This article focuses on smart agriculture and aims to cultivate interdisciplinary talents. Firstly, the study lists the relevant questions of smart agriculture for students to choose the ones they are interested in, allowing students to discuss the issues and solutions. Next, this research will create smart agriculture teaching aids for students to assemble, write programs, and derive mathematical reasoning for each module, expecting to achieve the goals of Science, Technology, Engineering, and Mathematics (STEM). By focusing on smart agriculture, the study will provide students with relevant teaching aids; the features of this article are: 1. The equipment utilizes Artificial Intelligence of Things (AIoT) sensors to detect the environmental factors in the farm, such as temperatures or humidity; 2. The system can identify various animal sounds during different periods, such as farm animals or bees, by wavelet transformation to process signals; furthermore, the equipment then analyzes the pitches, eigenfrequencies, and frequency distributions to build the condition eigenvalues of various farm animals or bees, helping AI identify the signals; 3. The system uses You Only Look Once (YOLO) to recognize objects, primarily checking whether there are hornets in beehives; meanwhile, as the high recognition rate of YOLO, the technique can be applied in animal husbandry to detect whether a pig is crippled or in other smart agriculture environments. Additionally, this research starts by creating the model of the proposed teaching aids and further asks the students to train the model and develop the recognition rate. This process allows students to encounter challenges during the practice and encourages them to discuss with teachers and understand the theories, stimulating students' learning motivation toward the mathematical theories in multimedia and AI.

2 Related Works

Reference [2] utilizes 3-D Computer-Aided Design (CAD) to create a model, design, and print 3-D products to improve students' spatial visualization, creativity, and problem-solving capacity. Applying 3-D CAD software and 3-D printing can influence creativity needs' perception when students resolve the issues they might encounter in their STEM careers. Reference [5] points out that the lower numbers of female participants in STEM courses has always been a popular research topic, and the research has found that the scarcity of female students enrolled does not always show a uniform distribution in STEM studies. In some subjects, such as computer, communications, electric, and electrical engineering, the number of female students does not increase but declines instead.

The experiment result of Reference [6] reveals that many pieces of research focused on discussing and evaluating the appropriateness and effectiveness of Virtual Reality (VR) supported teaching applications. Moreover, some studies point out the learning outcomes, positive viewpoints of user experience, and the improvement of perceived availability; yet, only limited research measures students' learning performance. The current scope review aims to encourage instructional designers to develop innovative VR applications or integrate existing methods into their teaching. On the other hand, Reference [1] mentions that Japanese high school students in STEM subjects expect to work in scientific and technical sectors but lack the motivation for learning English as a second language (L2), weakening their current L2 learning capability and the ability to communicate globally during their career in the future. Finally, the summary of the research states that it seems to have a close correlation between students' English learning motivation and gaining high scores in standardized English language tests, obtaining employment, and striving for promotions in occupational hierarchies; however, these external motivations in L2 learning is never stronger than the positive image of using English directly in the future.

While setting STEM as the basic foundation, this study aims to arouse students' interests in mathematic theories by practicing experiments. In Reference [8], the research states that the graduates who adopted STEM education in the United States tend to have higher work capacity and ability; meanwhile, students who participate in STEM education have better wages and job positions. Therefore, some universities further changed their educational entrance examination methods by including STEM scores in the rating scale, hoping to increase students' interest in practicing STEM. Reference [7] argues that although the United States has introduced STEM education proactively, the country's mathematical ranking still falls to 35; the reason is probably due to the interference of parents in selecting STEM courses. Hence, apart from implementing STEM education, governments and schools also need to provide educational propaganda and win support from parents. By integrating and summarizing the best practice papers that focused on identifying, sharing, and evaluating experiences and items, the articles in Reference [3] intend to foster diversity and gender equality in STEM subjects in academia and the professions. The experiment in Reference [4] anticipates filling the educational gap by evaluating the quality of teachers' predictions (labelled expert prediction). In the experiment, 43 elementary teachers predicted students' step-by-step actions whilst solving Ill-defined Problems (IDPs) through a Light Path Task (LPT); next, the experiment compared the quality with machine predictions executed by sequential pattern mining techniques. In the meantime, the experiment also collected students' lines of action data from 501 fifth- and sixth-grade students aged between 11 and 12.

Based on the concept of STEM 4.0, our research empowers students to team up and explore more problems and solutions independently. Summarizing the above literature reviews completes the research and teaching of this article, and the developed smart agriculture teaching aids allow students to extend to more topics.

3 The Proposed Scheme

3.1 Research Method

Setting STEM 4.0 education as the foundation, the course in this study separates students into controlled and experiment groups, and the course process is shown in Fig. 1. Firstly, we explore various topics for students to choose from, and the students will discuss the questions of the topic accordingly. As the course agenda is about farms, students need to acquire interdisciplinary knowledge of agriculture and stock farming. On the other hand, teachers will conduct the traditional didactic teaching to students in the controlled group and teaching aid assembling and programming to the other group. In the experiment group, the teacher will provide a 30-minute discussion time in each class; when students encounter questions, teachers can offer theoretical teaching, encouraging students to discover issues through practical experiments and stimulating a higher interest in learning theories and knowledge.

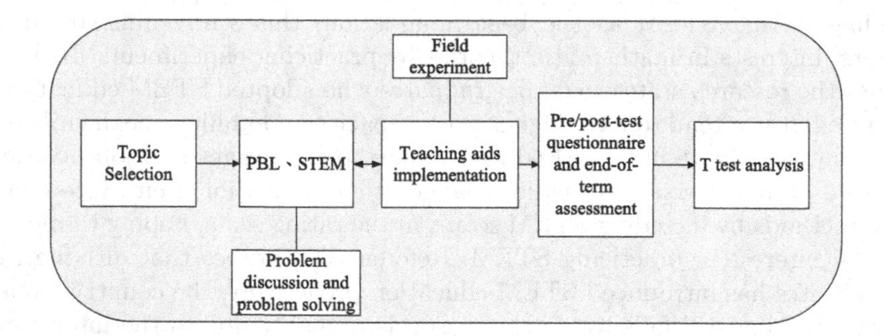

Fig. 1. Teaching Mode.

We will conduct pre-, post-, and final evaluation questionnaires to analyze the course effectiveness of the experiment; next, a t-test will be utilized to examine the performance, verifying whether the teaching model proposed in this study can achieve the expected teaching goal. Moreover, we will use a questionnaire to understand students' practicing interest in those teaching aids and check if they fully absorb the theories. Apart from confirming the performance of teaching aids, we will also take students to visit and do actual experiments in the field, testing those teaching aids' usability. Finally, the t-test results from all questionnaires allow us to learn the difference between the two groups and check students' satisfaction with the course design.

3.2 The Design of Teaching Aids

The structure of designed teaching aids is demonstrated in Fig. 2. To enhance the preciseness, we start the process by sorting the teaching aids. Next, we separate the teaching aids' sound and image and use YOLO to recognize objects; to increase the accuracy, we have collected videos directly from farms to separate the sound and video. After collecting enough data, we implement feature extraction, training, and testing on the images. In the meantime, using wavelet transformation as the foundation, we recognize the sound by its frequency, eigenvalue, and pitch period. Finally, as we aim to detect future environmental factors in this research, we utilize the Internet of Things (IoT) to obtain data on current environmental factors and judge if there are abnormal conditions. Additionally, this article employs machine learning to process and recognize the conditions of the ongoing topics, hoping to improve the model's accuracy and data ownership.

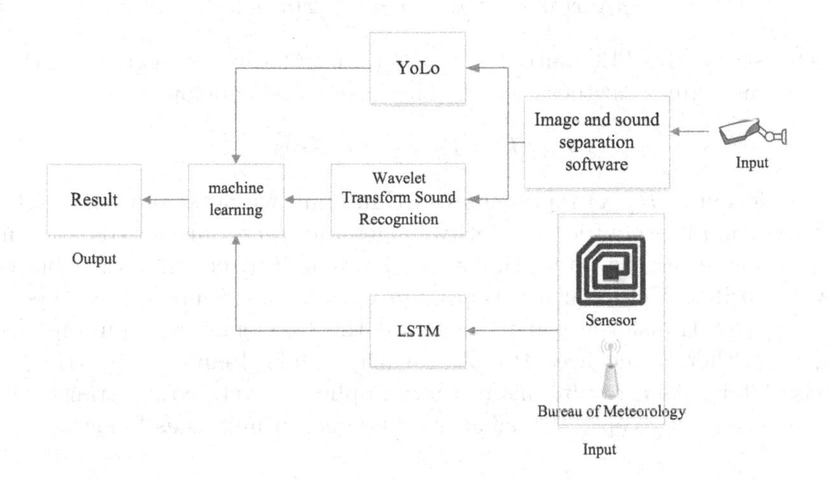

Fig. 2. Schematic diagram of teaching aids.

3.3 The Design of Sound Recognition Teaching Aids

The sound recognition technique in this research utilizes the frequency domain. Firstly, based on the auditory model to divide input signals into 24 sub-band signals, the decomposition tree structure consists of six hierarchies. As the highest frequency will not be over 8 KHz; thus, according to the Nyquist Sampling Rate, we set the sampling frequency to be 16 HKz, which is 0 to 16 KHz in the human audibility range. Afterward, we use wavelet transformation to construct 24 sub-band signals, $X[n]$, based on the human auditory range; after gaining the audio frame, we can get $x[i]$, where i represents each different frame. Each wavelet packet transformation of the frame can further build 24 sub-band signals, $w_{j,m}(k)$, where j means the hierarchy of the tree structure, m represents the 24 sub-band signals, respectively, and k shows the length of the sub-band signal; therefore, we select 24 sub-bands for further steps. Meanwhile, the function of the Teager Energy Operator (TEO) is to strengthen stable or semi-stable signals and attenuate other unstable signals; it is an extremely effective nonlinear algorithm that extracts signal energy through mechanical and physical methods. There are two ways of expression in continuous and discrete signals; the TEO calculation in continuous signals can be shown as the first- and second-order functions of the signal $x[i]$, and the formation is as follows:

$$\varphi_c\left[x[i]\right] = \left(\frac{dx[i]}{dt}\right)^2 - x[i]\frac{d^2x[i]}{dt} \tag{1}$$

On the other hand, the differential is used to replace derivative operations in discrete signals, and the formula is illustrated below:

$$\varphi_D\left[x[n]\right] = x^2[n] - x[n+1]x[n-1] \tag{2}$$

In this study, the TEO output is $t_{j,m}(k)$, and MASK means to smooth the waveform and reduce sawtooth waves. The formula is as below:

$$T_{j,m}(k) = t_{j,m}(k) * H_j(k) \tag{3}$$

In the formula, $H_j(k)$ represents the Hamming window, and the length is $256/2^j$. As sound recognition primarily judges the eigenvalues of sound; hence, our study measures the pitch, timbre, and sonorant (frequency distribution). Firstly, we utilize Mel-frequency Cepstrum to calculate timbre. Next, because we use wavelet transformation to calculate the frequency distribution of the sound; thus, there is no need to obtain each sound's frequency by triangular bandpass filters. As a result, this research applies discrete cosine transform to derive Mel-frequency cepstral coefficients, and the formula is as follows:

$$C_m^i = \sum_{j=0}^{J-1} \cos\left(m\frac{\pi}{J}(k+0.5)\right)\log_{10}(T_{j,m}(k)) \tag{4}$$

where C_m^i represents Mel-frequency cepstral coefficients, and the calculation of differential cepstral coefficient is shown below:

$$\Delta C_m(t) = \frac{\partial C_m(t)}{\partial t} = \frac{\sum_{\tau=-M}^{M} \tau \cdot C_m(t+\tau)}{\sum_{\tau=-M}^{M} \tau^2} = \frac{\sum_{\tau=1}^{M} \tau \cdot (C_m(t+\tau) - C_m(t-\tau))}{2 \cdot \sum_{\tau=1}^{M} \tau^2} \tag{5}$$

In the formula, C_m^i and $\Delta C_m(t)$ mean the sound's timbre eigenvalues. Next, we need to calculate the pitch period of the sound, and the formula is as follows:

$$P_{j,m} = \prod_{j=0}^{n-1} X[j] \tag{6}$$

Table 1. Post-analysis

	Number	Minimum	Maximum	Sum	Average Deviation	Standard	Variation
@1 I possess the capacity for practical execution	35	2.0	4.0	95.0	2.714	0.5725	0.328
@2 I possess teamwork abilities	35	1.0	4.0	89.0	2.543	0.6572	.432
@3 I possess communication skills, including group communication in discussing coursework and presentation reports	35	2.0	4.0	87.0	2.486	0.5621	0.316
@4 I possess data collection capacity	35	1.0	4.0	90.0	2.571	0.6081	0.370
@5 I possess problem-solving abilities	35	2.0	4.0	92.0	2.629	0.6897	0.476
@6 I possess innovation capability	35	2.0	4.0	96.0	2.743	0.6572	0.432
@7 I possess a logical thinking capacity	35	2.0	4.0	92.0	2.629	0.5470	0.299
@8 I possess the ability to analyze problems	35	2.0	3.0	90.0	2.571	0.5021	0.252
@9 I possess employability	35	1.0	4.0	95.0	2.714	0.6217	0.387
@10 Overall speaking, this course benefits the improvement of professional expertise	35	2.0	3.0	90.0	2.571	0.5021	0.252
Valid N (Listwise)	35						

This article plans to utilize wavelet transformation to convert frequencies first; then, combined with deep learning, we use the timbre, pitch, and sonorant to recognize the sound, enhancing the accuracy of sound recognition.

4 Experiment Results

This article evaluates the effectiveness of the proposed method by pre- and post-questionnaire and the final teaching assessment. Table 1 demonstrates the post-analysis result, Table 2 presents the pre-analysis result, and Table 3 shows the integrated analysis. Furthermore, a t-test is conducted to check the significance of the research, and the data in Table 4 has proved that our proposed approach has statistical significance, showing the feasibility of our study.

Table 2. Pre-analysis

	Number	Minimum	Maximum	Sum	Average Deviation	Standard	Variation
@1 I possess the capacity for practical execution	35	2.0	5.0	131.0	3.743	0.7413	0.550
@2 I possess teamwork abilities	35	3.0	5.0	146.0	4.171	0.5681	0.323
@3 I possess communication skills, including group communication in discussing coursework and presentation reports	35	3.0	5.0	147.0	4.200	0.6325	0.400
@4 I possess data collection capacity	35	4.0	5.0	152.0	4.343	0.4816	0.232
@5 I possess problem-solving abilities	35	4.0	5.0	153.0	4.371	0.4902	0.240
@6 I possess innovation capability	35	3.0	5.0	137.0	3.914	0.6122	0.375
@7 I possess a logical thinking capacity	35	3.0	5.0	133.0	3.800	0.6325	0.400
@8 I possess the ability to analyze problems	35	3.0	5.0	137.0	3.914	0.6585	0.434
@9 I possess employability	35	3.0	5.0	131.0	3.743	0.6108	0.373
@10 I anticipate gaining sufficient support from this course	35	4.0	5.0	157.0	4.486	0.5071	0.257
Valid N (Listwise)	35						

Table 3. Integrated Analysis

	Test	Number	Average	Standard Deviation	Standard Error of the Mean
@1 I possess the capacity for practical execution	1.0	35	3.743	0.7413	0.1253
	2.0	35	2.714	0.5725	0.0968
@2 I possess teamwork abilities	1.0	35	4.171	0.5681	0.0960
	2.0	35	2.543	0.6572	0.1111
@3 I possess communication skills, including group communication in discussing coursework and presentation reports	1.0	35	4.200	0.6325	0.1069
	2.0	35	2.486	0.5621	0.0950
@4 I possess data collection capacity	1.0	35	4.343	0.4816	0.0814
	2.0	35	2.571	0.6081	0.1028
@5 I possess problem-solving abilities	1.0	35	4.371	0.4902	0.0829
	2.0	35	2.629	0.6897	0.1166
@6 I possess innovation capability	1.0	35	3.914	0.6122	0.1035
	2.0	35	2.743	0.6572	0.1111
@7 I possess a logical thinking capacity	1.0	35	3.800	0.6325	0.1069
	2.0	35	2.629	0.5470	0.0925
@8 I possess the ability to analyze problems	1.0	35	3.914	0.6585	0.1113
	2.0	35	2.571	0.5021	0.0849
@9 I possess employability	1.0	35	3.743	0.6108	0.1032
	2.0	35	2.714	0.6217	0.1051
@10 I anticipate gaining sufficient support from this course	1.0	35	4.486	0.5071	0.0857
	2.0	35	2.571	0.5021	0.0849
Valid N (Listwise)	35				

Table 4. Integrated Analysis

	T	Degrees of Freedom	Significance (two-tailed)
@1 I possess the capacity for practical execution	6.497	68	0.000
@2 I possess teamwork abilities	11.091	68	0.000
@3 I possess communication skills, including group communication in discussing coursework and presentation reports	11.986	68	0.000
@4 I possess data collection capacity	13.511	68	0.000
@5 I possess problem-solving abilities	12.186	68	0.000
@6 I possess innovation capability	7.716	68	0.000
@7 I possess a logical thinking capacity	8.288	68	0.000
@8 I possess the ability to analyze problems	9.594	68	0.000
@9 I possess employability	6.982	68	0.000
@10 I anticipate gaining sufficient	15.870	68	0.000

A pre- and post-questionnaire was conducted before and after the course. We also operated the t-test in SPSS to analyze data. The analytic result shows that our data is an F-distribution, and the obtained value of F means statistical significance if the F-value is lower than 0.05; in other words, the statistical result proves that the teaching model and aids can enrich students' learning performance. The pretest reveals that the top three personalities among the participants are innovation, hands-on execution ability, and employability, presenting that students expect to focus on personal development like innovation and practical execution capacity. On the other hand, the posttest shows that the top three abilities students expect are: this course benefits the improvement of professional expertise, I possess problem-solving abilities, and I possess data collection capacity. Finally, the integrated analysis indicates that the average difference between the pre- and post-questionnaire is two to four, demonstrating the discrepancy between the experimental and control groups. The result also illustrates that students possessed expectations toward the course content before experimenting, and the class improved students' learning performance. Furthermore, the t-test analysis illustrates that participants significantly improved each evaluation item after the class, particularly the three sub-statements in question four: I possess data collection capacity, problem-solving abilities, and communication skills, including group communication in discussing coursework and presentation reports, present high significance. This result proves that the study can indeed enhance students' performance. The t value of the integrated analysis is 15.87, showing a high significance. As a result, the analyses indicate that apart from enriching students' learning capacity, they have also improved their teamwork and practical abilities through participating in the course.

5 Conclusion

Universities have been promoting interdisciplinary education; consequently, apart from acquiring professional skills in the major subject, students are required to gain additional knowledge in other sectors. This project focuses on STEM 4.0 education and encourages students to develop self-directed learning, cultivating their problem-seeking and problem-solving abilities. By taking the STEM 4.0 concepts as the key concept, this project creates an initial model, attracting students to collect, train, and test data spontaneously, guiding students to understand relevant mathematical theories during the process and increasing their interest in learning the background knowledge. Aside from developing teaching aids, the course is designed to foster AI and smart agriculture talents, benefiting talent cultivation in the technology industry. The experiment result has proven that students possess high acceptance toward the designed course, which is feasible to boost students' learning enthusiasm for mathematics. Meanwhile, we will also discuss blockchain technology to record students' full learning performance and special topic conditions, enhancing the analysis of students' learning progress and special topic performance.

Acknowledgments. This work was supported in part by the Ministry of Science and Technology of Taiwan, R.O.C., under Contracts NSTC 109-2622-E-197-012, NSTC 110-2622-E-197-015, NSTC 111-2410-H-143-023, NSTC 111-2222-E-197-001-MY2 and NSTC 111-2410-H-197-002.

References

1. Apple, M.T., Falout, J., Hill, G.: The relationship between future career self images and English achievement test scores of Japanese stem students. IEEE Trans. Prof. Commun. **63**(4), 372–385 (2020). https://doi.org/10.1109/TPC.2020.3029662
2. Bicer, A., Nite, S.B., Capraro, R.M., Barroso, L.R., Capraro, M.M., Lee, Y.: Moving from stem to steam: the effects of informal stem learning on students' creativity and problem solving skills with 3D printing. In: 2017 IEEE Frontiers in Education Conference (FIE), pp. 1–6 (2017). https://doi.org/10.1109/FIE.2017.8190545
3. González-González, C.S., García-Holgado, A., Peixoto, A.: Diversity and equity in stem: second part. IEEE Revista Iberoamericana de Tecnologias del Aprendizaje **15**(4), 314–316 (2020). https://doi.org/10.1109/RITA.2020.3033221
4. Norm Lien, Y.C., Wu, W.J., Lu, Y.L.: How well do teachers predict students' actions in solving an ill-defined problem in stem education: a solution using sequential pattern mining. IEEE Access **8**, 134976–134986 (2020). https://doi.org/10.1109/ACCESS.2020.3010168
5. Olmedo-Torre, N., Sánchez Carracedo, F., Salán Ballesteros, M.N., López, D., Perez-Poch, A., López-Beltrán, M.: Do female motives for enrolling vary according to stem profile? IEEE Trans. Educ. **61**(4), 289–297 (2018). https://doi.org/10.1109/TE.2018.2820643
6. Pellas, N., Dengel, A., Christopoulos, A.: A scoping review of immersive virtual reality in stem education. IEEE Trans. Learn. Technol. **13**(4), 748–761 (2020). https://doi.org/10.1109/TLT.2020.3019405
7. Rozek, C.S., Svoboda, R.C., Harackiewicz, J.M., Hulleman, C.S., Hyde, J.S.: Utility-value intervention with parents increases students' stem preparation and career pursuit. Proc. Natl. Acad. Sci. USA **114**(5), 909–914 (2017). https://doi.org/10.1073/pnas.1607386114. https://app.dimensions.ai/details/publication/pub.1074198524. https://europepmc.org/articles/pmc5293025?pdf=render
8. Sjoquist, D., Winters, J.: State merit-aid programs and college major: a focus on stem. J. Labor Econ. **33** (2015). https://doi.org/10.1086/681108

Reinforcement Learning for Multifocal Tumour Targeting

Yi Hao[1], Zhijing Wang[2], Minghao Liu[1], Yifan Chen[1,3(✉)], and Yue Sun[1,4(✉)]

[1] School of Life Science and Technology, University of Electronic Science and Technology of China, Chengdu, China
yifan.chen@uestc.edu.cn, sunyuestc90@126.com

[2] Glasgow College, University of Electronic Science and Technology of China, Chengdu, China

[3] Yangtze Delta Region Institute (Huzhou), University of Electronic Science and Technology of China, Chengdu, China

[4] School of Mechanical and Electrical Engineering, Chengdu University of Technology, Chengdu, China

Abstract. This paper implements a reinforcement learning (RL) targeting strategy for multifocal tumour lesions in the framework of computational nanobiosensing (CONA). Multi-tumours are promoted by the metastatic interaction between the surrounding tissues and the tumour suppressor. Nanorobots, regarded as computing agents, aim to search the multi-tumour lesions within the complicated vessel network. The Biological information gradient fields (BGFs) indicate the formation of the tumour microenvironment regulated by the nearby vessel network. By using reinforced learning and applying the knowledge of BGFs, this work achieves a higher tumour targeting efficiency than the previous work. The Markov and BGFs rewards are included in the total RL reward, in which the Markov reward is utilized for training nanorobots to find the path and avoid colliding with vessel walls, allowing them to learn the vascular network's topology, whereas the knowledge of BGFs incentive benefits faster convergence of the searching process. Therefore, this method enables the discovery of the path planning for the multi-tumour in a heterogeneous vessel network by combining viable vessel path planning with BGFs information.

Keywords: Reinforcement Learning · Biological Gradient Field · Markov Rewards · Multifocal Tumours

1 Introduction

Tumour is the primary disease that threatens humans health worldwide [1]. According to [1], more than fifty thousand people may have died because of cancer in 2020. Detecting tumours early and precisely delivering medication treatment could significantly reduce fatalities. However, this is a great challenge

Y. Chen et al. (Eds.): BICT 2023, LNICST 512, pp. 22–30, 2023.
https://doi.org/10.1007/978-3-031-43135-7_3

for traditional medical imaging techniques such as MRI and CT for early-stage tumours detection due to the limited facility resolution and acquired information.

In emerging nanotechnology, nanoparticles are used to enhance image effects based on physiochemical properties of *in vivo* environment, such as temperature, optical characteristic, tissue elasticity [2]. However, injected nanoparticles can only rely on human system circulation without external control, which results in low efficiency with 0.7% achieving tumour targeting [3]. This outcome is unsatisfactory in a practical situation, so overcoming the previous hurdle is crucial to moving forward nanomedicines into practice. Computational nanoparticle-mediated drug delivery has great potential in cancer diagnoses [4]. Therefore, we propose a computational nanobiosensing (CONA) framework as a "smart" strategy for targeting tumours, in which external manipulable nanorobots replace nanoparticles [5,6]. According to [7], pH and oxygen tension in the human body, nutrition distribution, enzymatic activity may become heterogeneous. The changes in biological gradient fields (BGFs) resulting from these factors can therefore be used to define the location of tumours, and the visualized tumour-triggered BGFs can be considered "objective functions". In these objective functions, the high-risk tissue is the domain, the targeted tumour sites are the optimal solutions, and the nanoparticles are the computing agents [6]. By this means, tumour targeting can be performed under the guidance of an external steering field to manipulate the internal nanorobots. Besides, a previous study showed improved efficiency and feasibility [5]. Conclusively, the knowledge of BGFs is essential in the CONA framework for tumour detection.

Indeed, the knowledge of BGFs is helpful to *in vivo* tumour detection; nonetheless, the vessel walls are failed to be considered [8]. As a result, the agents that adopt CONA strategies may adhere to the vessel walls during the searching process. Combining the BGFs with human vasculature during the tumour targeting process to avoid the obstacle caused by blood vessel walls can overcome the challenge mentioned above. The previous work mainly contributes to the detection of the single cancer [8], while we focus on the multimodal bio-detection scenario in this paper. Different from the single tumour, multifocal tumours originated from a specific cellular cone and migrated to other lesions in the metastatic interaction process [9]. Based on our previous work in [8], reinforcement learning (RL) achieved considerable results with sphere BGFs function, which has only one optimal solution in the global domain. The current work focuses on the BGFs having optimal solutions using the RL algorithm with Q-learning, which means that nanorobots can search multifocal tumours in more complex BGFs with higher accuracy.

This paper is organized as follows. Section 2 describes the model of vascular network and BGFs. In Sect. 3, methodology including the Markov decision process, RL algorithm, and multifocal tumour search strategies are introduced. Following that, simulations with different scenarios are presented to illustrate the performance of the smart strategy in a more complex environment, and the results are shown in Sect. 4. Finally, Sect. 5 shows the main outcomes and the conclusion is drawn.

2 Models of Vascular Network and BGFs

2.1 Tumour Vascular Network

Due to the high demand for oxygen and nutrition during tumour growth, the vascular network around the tumour, high-density interconnected, is more complicated than normal vascular networks [10].

Additionally, tumour vascular networks have many unique properties, such as the tortuous vessels and wide range of avascular spaces in tumours, which are modelled by the fractal dimension [10,11]. Moreover, the percolation structure indicates the similarity with the local growth process [11]. Therefore, the invasion percolation algorithm is used to represent the growth process of the tumour vascular network.

When using the invasion percolation algorithm, first randomly assign uniformly distributed intensity values to each lattice point. Next, from any location as the starting point, the network gradually occupies the minimum lattice point adjacent to the current location and iteratively grows until the desired lattice occupancy is reached. The blood vessels are interconnected to all the occupied lattice points, while the blood flows in from the initial entry point and flows out from the specified outlet point. Then the network is pruned, and only the blood vessels of the non-zero blood flow part are retained to obtain the required tumour vascular network.

The occupancy of the grid indicates the fractal dimension of tumour vessels. According to [11], the lattice occupancy corresponding to fractal dimensions 1.6, 1.8, 1.9 and 2.0 are 40%, 60%, 80% and 10%, respectively. An example of a tumour vascular network with 77.00% occupancy is shown in Fig. 1(a), and the distance between two adjacent vessels is 50 µm.

For the multi-tumour vascular network, as shown in Fig. 1(b), the length per grid side is 100 µm, which represents the distance between blood vessels of healthy tissue, and the adjacent area with neovascularization is represented as three small squares.

2.2 Biological Gradient Fields

As mentioned above, oxygen tension and pH in the tumour microenvironment are heterogeneous. Besides, the distribution of glucose and other nutrients such as growth factors may be uneven or deficient [7]. These passive physical properties of tumour tissue like blood flow velocity and vascular network structure can be utilized to generate BGFs.

In the RL computation, BGFs in the high-risk tissues are transferred into representative objective functions used to evaluate the tumour detection performance, such as convergence, accuracy, and robustness of touch computing. Previously published studies have primarily focused on the experimental observation of the changes in the tumour microenvironment; however, it lacks the proper quantitative BGFs models. Therefore, this paper focuses on representative objective functions shown in Eq. (1), which could evocatively represent the

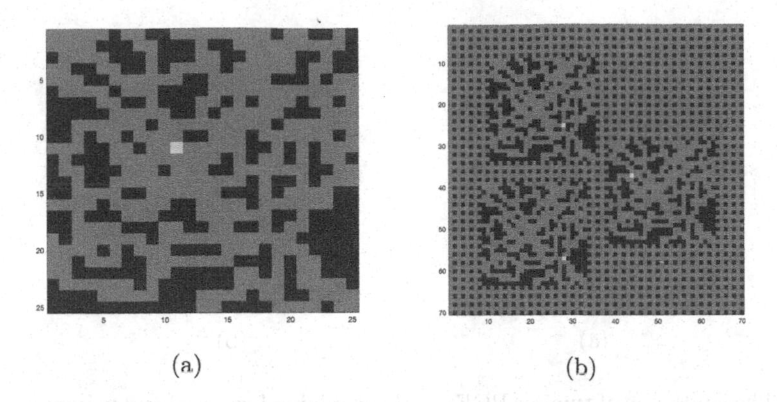

Fig. 1. Vessel Framework Model (a) Simulated tumour vascular network using invasion percolation algorithm. The level of occupancy is 0.77 and the distance of tumour vascular is set to 50 μm. The light blue area represents the vascular spaces around the tumour, and the dark blue area indicates the vessels. (b) Represents the three tumour vascular network regions generated by invasive infiltration techniques, and the scope of the search space is 70 × 70. (Color figure online)

presence and absence of random fluctuations in the BGFs around tumours [9]. Besides, they are consistent with the qualitative observation results in the existing literature and can represent the BGFs in different change modes, which is used to verify the effectiveness of *in vivo* computing strategy in different BGFs. The expression of BGFs is shown in Eq. (2), and the corresponding landscape is shown in Fig. 2. It is worth noting that this BGF landscape is different from the traditional optimization problems in the concept test function [9]. Since traditional optimization problems require comprehensive features in the functions, while this model only focuses on the biological characteristics of high-risk tumour areas, thus, this paper proposed in vivo computing concept has a significant difference from the traditional optimization problem. The detail of this model is illustrated and discussed in Sect. 4.

$$U_T(x, y) = U_{T1}(x, y) + U_{T2}(x, y) + U_{T3}(x, y) \tag{1}$$

$$U_{T1}(x, y) = 1 - exp((-(x - 56)^2 - (y - 27)^2)/300) \tag{1-1}$$

$$U_{T2}(x, y) = 1 - exp((-(x - 36)^2 - (y - 43)^2)/300) \tag{1-2}$$

$$U_{T3}(x, y) = 1 - exp((-(x - 24)^2 - (y - 27)^2)/300) \tag{1-3}$$

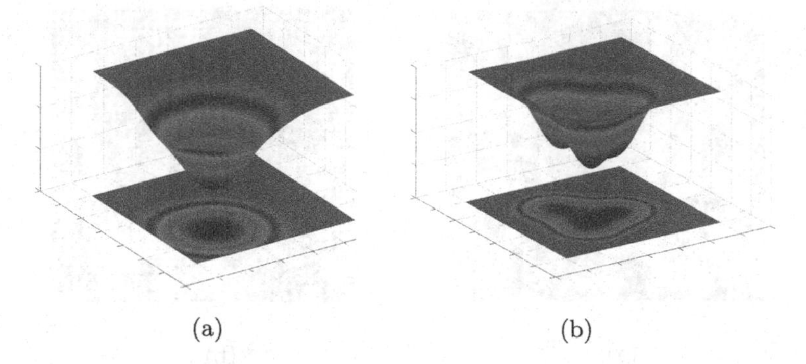

Fig. 2. The landscape of tumour BGFs, and the minimal values 0 and maximum values 1 are the range of the functions, and the tumour is located in the optimal minimal values. (a) The landscape of singular tumour with coordinates (70,70). (b) The landscape of three tumours, with location at coordinates $(56, 27), (36, 43)$, and $(24, 27)$.

3 Multifocal Tumour Searching Model

The searching scenario is a 70×70 grid-like vascular network with three tumours, where the nanorobots find the shortest path to the tumour locations. This optimisation problem can be mapped into a reward function as part of a Markov decision process (MDP), which the RL algorithm can solve.

In conventional mathematical optimization, continuous niche technology can effectively solve multi-solution computing problems. As an overall training process, the tumours are examined one by one [12]. Firstly, we use an optimization algorithm to find the optimal global solution. Secondly, the objective function is modified in the optimal solution location to update the knowledge of BGFs. Thirdly, re-run the optimization algorithm with updated BGFs to find the next tumour. Repeat the above process until all tumours are found. It is worth noting that Q-learning is adopted for each step of the optimization algorithm.

3.1 Markov Decision Process

The optimization problem, as mentioned before, can be regarded as an MDP, which is defined through the tuple (S, A, R) with state-space S, action space A, and reward function R [13].

The state at mission time t in the vascular network is given by $s_t = (p_t, p_t) \in \mathbb{R}^2$, which is the location of the nanorobot. There are five actions that are allowed nanorobots to take:

$$A = \{up, down, left, right, idle\} \tag{2}$$

If the nanorobot collides with the wall of a vessel, then it goes to idle mode; otherwise, it takes a step forward in one of the four directions specified in (2).

The reward function maps the state-actions to a real-valued reward, i.e., $S \times A \rightarrow R$. The mission goals, R, consist of the following components:

- Markov reward, which is utilized to avoid colliding with the vessel wall during the tumour searching process.

$$r_{Markov} = \begin{cases} 2, & \text{one tumour found} \\ 3, & \text{two tumours found} \\ 4, & \text{three tumours found} \\ 0.01, & \text{tumour not found \& vessel wall not reached} \\ -1, & \text{vessel wall reached} \end{cases} \tag{3}$$

- BGFs reward, which is used to speed up the convergence and is defined by (1).

To sum up, the total reward, divided into two parts, the Markov reward and the BGFs reward of the multifocal tumours, is defined as:

$$r = r_{Markov} + \beta \times r_{BGFs} \tag{4}$$

where $\beta(0 \leqslant \beta \leqslant 100)$ regulates the weight of the Markov reward and BGFs reward in the model.

3.2 Q-Learning

Q-learning is one of the basic algorithms of reinforcement learning [14]. A model-free learning method allows agents to select optimal actions using experienced action sequences in a Markov environment. The interaction between the agent and the multi-tumour environment can be regarded as an MDP, a critical assumption in Q-learning. The agent could perform the tasks in iterations, firstly observe the state $s_t \in S$ and then perform an action $a_t \in A$ at time t and subsequently receive a reward $r(s_t, a_t) \in R$ to the agent from the environment, and finally restart in a new state $s_t + 1$. The behavioural policy of the agent is to obtain the highest reward, which means finding the optimal paths from the starting point to the destination. A probabilistic policy $\pi(a|s)$ is a distribution over actions based on the state such that $\pi : S \times A \rightarrow R$. It reduces to $\pi(a|s)$ in the predictable situation, resulting in $\pi : S \rightarrow A$.

To learn the policy π, Q-Learning updates the state-action value function by iterating, given as:

$$Q^{\pi}(s, a) = \mathbb{E}_{\pi}[R_t | s_t = s, s_t = a] \tag{5}$$

which denotes an expectation of the discounted cumulative return R_t from the current state s_t up to a terminal state at time T given by

$$R_t = \sum_{k-t}^{T} \gamma^{k-t} r(s_k, a_k) \tag{6}$$

with $\gamma \in [0,1]$ being the discount factor, balancing the priority of current and future rewards. s_t and a_t are simplified to s and a, s_{t+1} and a_{t+1} are simplified to s' and a' for the sake of clarity. We use the Bellman equation to train our model [15]:

$$Q_{k+1}(s,a) = r_s^a + \gamma Q_k(s', \pi(s')) \tag{7}$$

where k is the number of steps and r_s^a is the same as r in (3). In practice, a learning rate of $0 \leq \alpha \leq 1$ is used to keep the agent from becoming stuck in a locally optional solution:

$$Q_{k+1}(s,a) = Q_k(s,a) + \alpha(r_s^a + \gamma Q_k(s', \pi(s) - Q_k(s,a)) \tag{8}$$

4 Simulation and Results

4.1 Simulation Setup

Each new iteration is conducted in a 70×70 discretized vascular network with $10\,\mu\text{m} \times 10\,\mu\text{m}$ in each grid cell size. Considering the limited lifespan of the nanorobot, we set the maximum exploration steps to 3000. During the simulation process, training steps are used to determine the performance of nanorobots in different scenarios.

4.2 Simulation Results

Figure 3(a) shows the trajectories of nanorobots, which are represented by the colour orange. As shown in the figure, the movement of nanorobots is coordinated towards the maximum-gradient direction estimated in the Markov reward and BGFs. The method of Q-learning can detect all three tumour centres with high accuracy and efficiency. It is shown that the simulation result is correlated with the tumour locations and the weight β, these factors will be introduced in the following paragraphs.

Q-learning is assessed regarding its tumour targeting efficiency with different multi-tumour locations considering the practical situation. Figure 3(b) presents different tumour locations and different nanorobot injection locations, which verifies the robustness of this tumour research strategy.

4.3 Parameter Optimization in Different Scenarios

According to Eq. (3), the value of the Markov reward is between -1 and 1. The maximum value of the function is set to 0, and the minimum value is set to -1. As a result, when the weight β is set to 1, the Markov and BGFs rewards will be of the same size. The number of training steps for the convergent model decreases as the weight β goes from -3 to -1.5. As shown in Fig. 4, the choice of β value is critical, and both much large and much small cannot achieve the goal. Taking [15,15], [44,46], and [70,70] tumour locations, for example, the training steps reached the low bound when β was equal to -1.5 because the reward was applied to avoid attaching to the vessel wall and the gradient reward was used to speed up the convergence. Furthermore, given the restricted resource of computing capacity, setting the incentive weight $\log \beta$ to -1.5 is the best option.

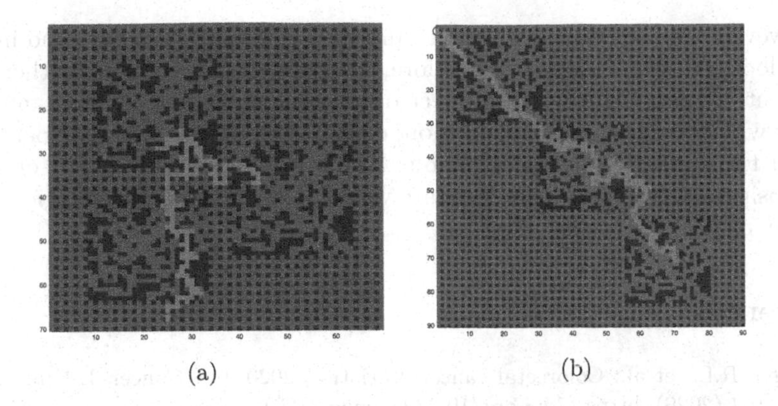

Fig. 3. The simulation demonstrated multifocal tumour targeting. (a) and (b) are simulations of the search process in the environment with different tumour locations. The yellow dots and red circles represent the tumour locations and injection point of the agent, and the trajectory of the agent is marked by orange. (Color figure online)

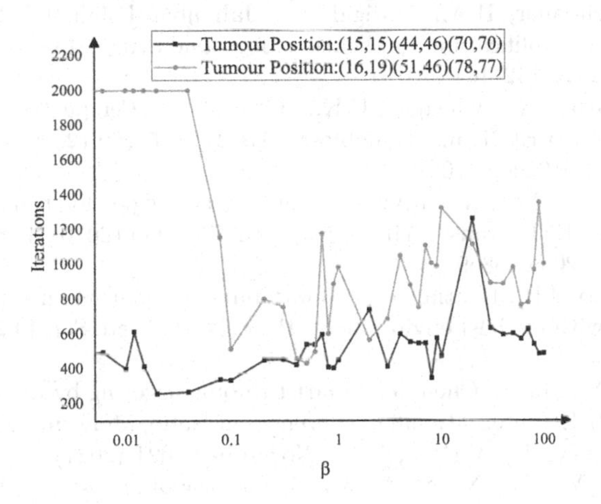

Fig. 4. Training steps (iteration) related to different β values, two different sets of three-tumour locations are tested.

5 Conclusion

This paper investigated the novel CONA strategy using a reinforcement learning algorithm to target multi-tumours. The main contributions of this paper are as follows:

- This paper introduces an RL algorithm for multi-tumour detection, which shows a better performance with fewer iterations than the brute-force searching strategy.
- The model uses the reward of BGFs in multifocal tumour targeting processes.
- For different scenarios, the nanorobots are injected in different locations to search multi-tumours simultaneously.

However, the algorithm is unstable, and its simulation results depend in part on the location of the tumour. Therefore, further research could use other reinforcement learning algorithms to select optimal actions, or focus on nanorobot swarms with RL algorithms. In addition, considering the complex microenvironment of the human body, multi-tumour BGFs may not be the result of direct superposition of BGFs, so more accurate vascular models and physiological environment models are needed.

References

1. Siegel, R.L., et al.: Colorectal cancer statistics, 2020. CA Cancer J. Clin. **70**(3), 145–164 (2020). https://doi.org/10.3322/caac.21601
2. Chen, H., Zhang, W., Zhu, G., Xie, J., Chen, X.: Rethinking cancer nanotheranostics. Nat. Rev. Mater. **2**(7), Art. no. 17024 (2017)
3. Wilhelm, S., Tavares, A.J., Dai, Q., Ohta, S., Chan, W.C.W.: Analysis of nanoparticle delivery to tumours. Nat. Rev. Mater. **1**(5), 16014 (2016)
4. Seidi, K., Neubauer, H.A., Moriggl, R., Jahanban-Esfahlan, R., Javaheri, T.: Tumour target amplification: implications for nano drug delivery systems. J. Control. Release **275**, 142–161 (2018)
5. Shi, S., Sharifi, N., Cheang, U.K., Chen, Y.: Perspective: computational nanobiosensing. IEEE Trans. Nanobiosci. **19**(2), 267–269 (2020). https://doi.org/10.1109/TNB.2019.2956470
6. Shi, S., Chen, Y., Yao, X.: In vivo computing strategies for tumour sensitization and targeting. IEEE Trans. Cybern. **52**(6), 4970–4980 (2020). https://doi.org/10.1109/TCYB.2020.3025859
7. Kwon, E.J., Lo, J.H., Bhatia, S.N.: Smart nanosystems: bio-inspired technologies that interact with the host environment. Proc. Natl. Acad. Sci. **112**(47), 201508522 (2015)
8. Liu, L., Sun, Y., Shi, S., Chen, Y.: Smart tumour targeting by reinforcement learning. In: 2021 IEEE International Conference on Nano/Molecular Medicine & Engineering (NANOMED), Virtual, 15–18 November 2021 (2021)
9. Shi, S., Chen, Y., Yao, X.: NGA-inspired nanorobots-assisted detection of multifocal cancer. IEEE Trans. Cybern. (2020). https://doi.org/10.1109/TCYB.2020.3024868
10. Gazit, Y., et al.: Fractal characteristics of tumour vascular architecture during tumour growth and regression. Microcirculation **4**(4), 395–402 (1997). https://doi.org/10.3109/10739689709146803
11. Baish, J.W., et al.: Role of tumour vascular architecture in nutrient and drug delivery: an invasion percolation-based network model. Microvasc. Res. **51**(3), 327–46 (1996). https://doi.org/10.1006/mvre.1996.0031
12. Beasley, D., Bull, D.R., Martin, R.R.: A sequential niche technique for multimodal function optimization. Evol. Comput. **1**, 101–125 (1993)
13. Kaelbling, L.P., Littman, M.L., Cassandra, A.R.: Planning and acting in partially observable stochastic domains. Artif. Intell. **101**(1–2), 99–134 (1998)
14. Watkins, C.: Technical note: Q-learning. Mach. Learn. **8** (1992)
15. Ford, R., Delbert, F.: A simple algorithm for finding maximal network flows and an application to the Hitchcock problem. Rand Corporation (1955)

Automatic Soil Testing Device for Agriculture

Vikranth Vakati, Mark Rosado, Nitin Bohra, and Douglas E. Dow(✉)

Computer Engineering, School of Engineering, Wentworth Institute of Technology, Boston, MA 02115, USA
{vakativ,dowd}@wit.edu

Abstract. Farm produce is essential to feed growing world population, even as the area of land available for agriculture decreases. Farmers tend to over apply water and fertilizer to maximize crop yield, since knowledge of soil conditions is insufficient for a more targeted application. The over application of water needlessly uses a scarce resource, especially in drier climates. The over application of fertilizer wastes the fertilizer, increases greenhouse gas (GHG) emissions, and degrades downstream water as algae increases and oxygen decreases. The purpose of our project was to develop modules for soil testing and transmitting of data to the hub computer that would be accessible by the farmer. The sensors and transmitter were developed and tested to be mounted on a stake that would be implanted in the soil of the field, and results transmitted to a hub computer that would provide a dashboard of results and control for the farmer to use in making decisions. Prototype modules were developed for soil nutrients and pH. Modules were tested for monitoring moisture and wireless data transmission. Such a system would provide soil condition information that the farmer could use to more apply appropriate amounts of water and fertilizer, and not over apply.

Keywords: soil · moisture · nitrogen · phosphorus · potassium · NPK sensor · agriculture · stake · internet of things · IoT · ESP-32 · ESP-NOW · farm monitor

1 Introduction

Agricultural farming accounts for 70% of freshwater usage across the globe. Of that amount, 40% is used for crops. According to the Organization for Economic Cooperation and Development (OECD), by the year 2050 water demand will increase by 55% due in part to rising populations [1]. The risk of water scarcity will increase and there may not be enough water to distribute adequately between the various stakeholders. The limited water supply may become polluted from the overapplication of fertilizers, and poor management of manure and sludge [2].

Farmers desire to ensure enough water and fertilizer for their plants, so they may tend to over apply. Over application of resources is not only wasteful of those resources but may cause much harm. When fertilizer is overapplied, or if rain occurs soon after application, eutrophication may occur and promote the growth of harmful bacteria and algae [3]. Some soil microorganisms consume and convert excess nitrogen into nitrous

Y. Chen et al. (Eds.): BICT 2023, LNICST 512, pp. 31–39, 2023.
https://doi.org/10.1007/978-3-031-43135-7_4

oxide. Nitrous oxide is a greenhouse gas (GHG) that has a warming potential almost 300 times higher than carbon dioxide [4]. Runoff of fertilizer into surface water, rivers, and lakes can increase overgrowths of algae, or algal blooms. These algal blooms are harmful because they produce toxins and decrease oxygen content in the water [5]. One such toxin is nitrate, which depletes oxygen levels and thus is a risk to health [2]. In the U.S.A., agriculture was estimated to be responsible for about 700 million metric tons of carbon dioxide, methane and nitrous oxide that were emitted into the atmosphere. About 50% of that amount was nitrous oxide, annually released by bacteria in soil and seawater after excess nitrogen fertilizer was applied [6].

The quality of the soil for plant growth is dependent on the criteria of soil parameters such as moisture, pH, temperature, and macronutrient content. Farmers currently attempt to monitor soil quality by shipping a soil sample to a testing laboratory. Soil labs recommend testing every 3–5 years, where the test results are analyzed, and recommendations made for fertilizers to improve the quality of the soil. Farmers may also use at home soil testing kits to determine moisture content and pH. However, these methods are either costly, complicated, or time-consuming [7].

The concept of a stationary soil sensing station has been reported [8]. This concept could be developed to provide the user with real-time data about the soil conditions. Such a device should have the ability to monitor soil moisture content, pH level, temperature, and macronutrient composition, which could be used to optimize application of water and fertilizer, and minimize over application [9]. Automatic transmission of data could be performed from a microcontroller-based soil testing edge node to a hub computer. The hub computer would store a copy of the recommended soil parameters, and act as a web server where the farmer can see summary data, analyze the data, and input control parameters on a dashboard.

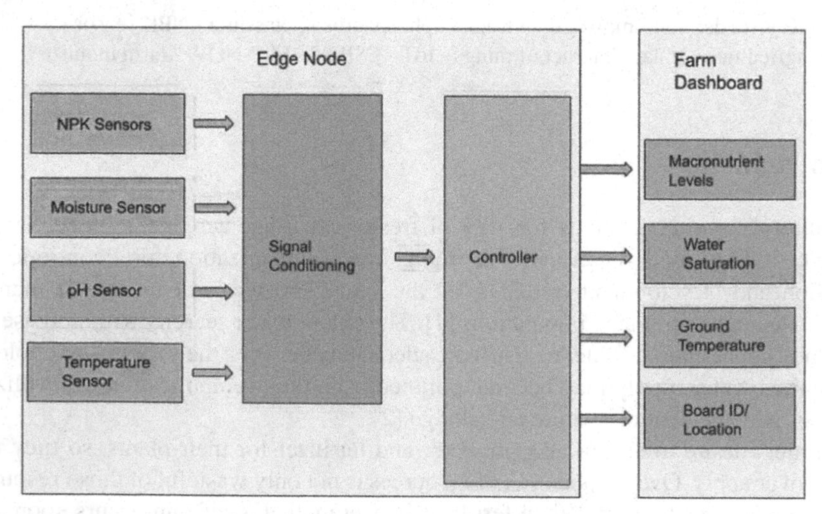

Fig. 1. Block diagram showing the edge node that would be mounted on a stake in the field, and the dashboard on a computer the office or home of the farmer. Communication between units is wireless.

The purpose of our project was to develop modules for soil testing and for transmitting of data to the hub computer that would by accessible by the farmer. The collected data of soil conditions could be analyzed to improve soil conditions in order to increase crop yield while conserving resources. This could result in a more optimal application of water and fertilizer, reducing the harm of overapplication.

2 Materials and Methods

The full design of the system had three units: an edge node mounted on stakes in the field, a hub computer and a web server with a dashboard accessible by the farmer (Fig. 1). One or more edge nodes would be mounted on stakes (Fig. 2) in the field to monitor soil and crop conditions. The edge nodes would wirelessly communicate to the hub computer. The farmer would interact with the web server through a dashboard.

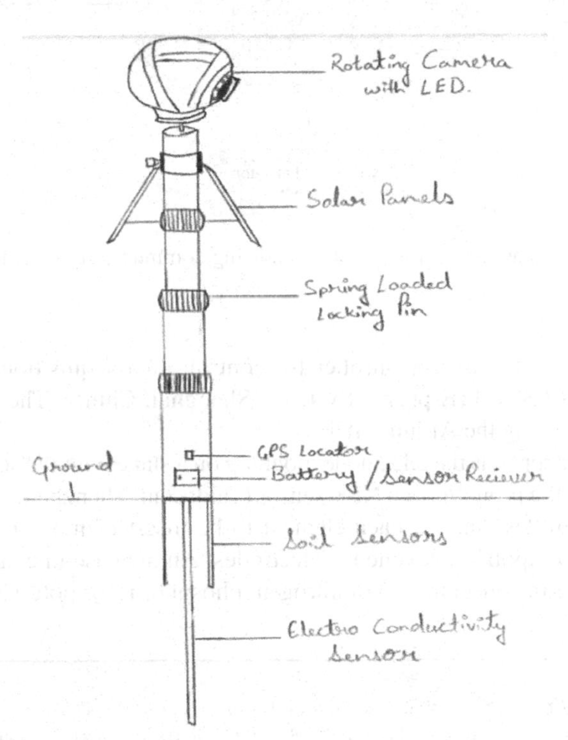

Fig. 2. The design concept has 3 below ground sensors and solar panels that would be externally mounted to the stake. All electronics and power supply for the edge node would be stored internally to stay protected from the weather. The prototype would be height adjustable, allowing for the solar panels to remain above the height of the crop and receive direct sunlight.

The edge node had sensors, including a water moisture sensor, a nitrogen, phosphorus and potassium (NPK) sensor, a pH sensor, a temperature sensor, and a camera. The edge node also had electronics for signal processing, microcontroller, wireless communication module and power management module. The hub computer had a corresponding communication module to receive that data from the edge nodes and transmit to the web server, which hosted a dashboard accessible by the farmer.

Modules of the design concept were implemented and tested as a prototype. The modules are seen in the system block diagram (Fig. 3).

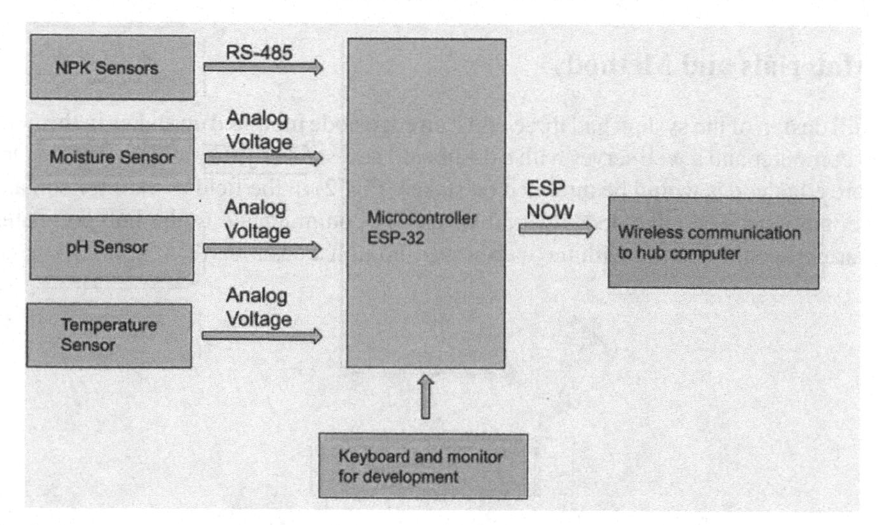

Fig. 3. System block diagram for the prototype showing communication methods between each component

In the prototype, the microcontroller for control, data acquisition and communication was the ESP-32 S3 (Espressif Systems, Shanghai, China). The ESP-32 S3 was programmed in C using the Arduino IDE.

The design concept has the edge node mounted on a stake with the soil sensor probes in the ground. For the prototype, an NPK sensor (Taidacent, Shenzhen, China) was used, which has 3 electrodes, one for each element to be inserted into soil. An alternating current voltage was applied to excite the electrodes. An increase in conductivity would signify an increase in concentration of nitrogen, phosphorus, or potassium.

```
//From NPK sensor datasheet
const byte nitro[] = {0x01,0x03, 0x00, 0x1e, 0x00, 0x01, 0xe4, 0x0c};
const byte phos[] = {0x01,0x03, 0x00, 0x1f, 0x00, 0x01, 0xb5, 0xcc};
const byte pota[] = {0x01,0x03, 0x00, 0x20, 0x00, 0x01, 0x85, 0xc0};
byte values[11];
```

Fig. 4. Inquiry frame for all elements of NPK sensor, used to interpret RS-485 into analog values for the ESP-32 to read data values from the response frame for each element.

To communicate from the NPK sensor to the microcontroller in the edge node, the half-duplex RS-485 serial communication was used. Figure 4 shows the inquiry frame for the NPK sensor elements. The inquiry frame was used in the software that ran on the microcontroller for serial communication over RS-485. The RS-485 serial communication sent and received the data one bit at a time. Figure 5 shows the software function used to read the phosphorus concentration from the NPK sensor.

```
byte phosphorous(){
  digitalWrite(DE,HIGH);
  digitalWrite(RE,HIGH);
  delay(10);
  if(mod.write(phos,sizeof(phos))==8){
    digitalWrite(DE,LOW);
    digitalWrite(RE,LOW);
    for(byte i=0;i<7;i++){
    // Serial.print(mod.read(),HEX);
    values[i] = mod.read();
    // Serial.print(values[i],HEX);
    }
    // Serial.println();
  }
  return values[4];
}
```

Fig. 5. Function to read phosphorus concentration.

A module was developed to measure moisture content in the soil. The water sensor (ICStation, Shenzhen, China) used in the prototype had 2 metal prongs and returned an analog value based on how resistive the soil was. Resistance decreased as water content increased. The analog voltage value was then converted by the microcontroller into a percentage to display on the dashboard.

The pH sensor (Atlas Scientific Environmental Robotics, Long Island City, New York, U.S.A) used in the prototype was made of a single probe consisting of a glass electrode, made from a special glass containing metal salts and a reference electrode, which had a potassium chloride wire suspended in a solution of potassium chloride. By measuring the potential difference between a known solution and the soil, the pH was returned as an analog voltage value to the microcontroller.

A)

```
void OnDataSent(const uint8_t *mac_addr, esp_now_send_status_t status) {
  Serial.print("\r\nLast Packet Send Status: \t");
  Serial.println(status == ESP_NOW_SEND_SUCCESS ? "Delivery Success" : "Delivery Fail");
  packet_value++;
}
```

B)

```
void OnDataRecv(const uint8_t *mac, const uint8_t *incomingData, int len) {
memcpy(&myData, incomingData, sizeof(myData));
```

Fig. 6. ESP-NOW software callback functions to A) send and B) receive data.

Wireless communication was achieved using the ESP-NOW protocol. For the user to view data remotely, a chain mesh network would be used to pass sensor data along a series of edge nodes until a computer hub was reached in the network. Then the data would be uploaded from the computer to a web server. To pass the data along the chain mesh network, the MAC address of the receiving edge node was altered as shown in Fig. 6 according to the successive edge node in the network. The data from each edge node was passed as a C program struct with a device ID so that the location the data was coming from would be known.

The dashboard of the prototype was hosted on a web server from an ESP-32 hub (Fig. 7). The dashboard was made using HTML to display the data from the edge nodes. As the data was uploaded to a web server hosted on the ESP-32 hub computer, it could be accessed remotely from any device with a network connection.

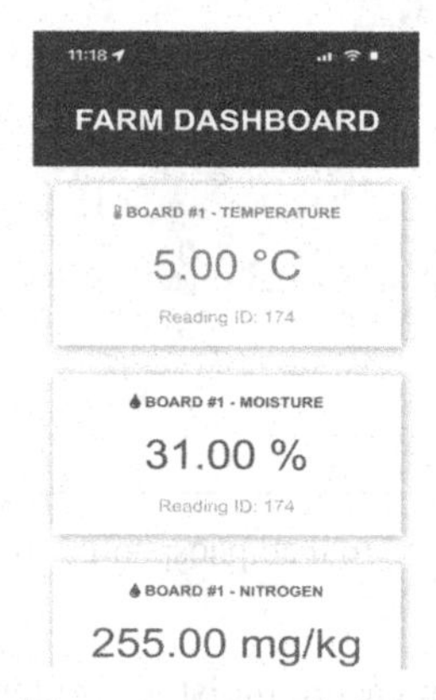

Fig. 7. Example of the dashboard of the prototype to display values from the sensor modules. This dashboard could be accessed on the phone or PC of the farmer. The displayed data was simulated on an edge node and then transmitted to the hub node, which published it to the web server.

Two power sources were used in the prototype: a 12-V source for the NPK sensor, and a 5-V source for all other modules.

3 Testing and Results

Two of the modules of the prototype were tested using the ESP-32 microcontroller: the soil moisture sensor and the wireless communication modules.

3.1 Soil Moisture

The moisture sensor was utilized to assess whether the module could classify soil between wet, good, and dry soils. First, the moisture sensor needed to be calibrated. The prototype module for soil moisture testing consisted of the moisture sensor and ESP-32 microcontroller. Results were displayed on an attached laptop computer display during testing.

The procedure to evaluate the moisture sensor was to progressively add water to a sample of soil, mix and measure. A plastic tumbler with a diameter of 7 cm was filled with loam soil, such that the bottom 13 cm of the tumbler was filled with loam. The loam was obtained at a consumer gardening store, and was used as the soil for this test procedure. The first measurement was made prior to any water being added, so the loam was considered dry. Thereafter, known amounts of water (multiples of 5 mL) were added and mixed with the loam, then moisture was measured again. Prior to each measurement, the sensor electrodes were wiped clean with a cloth, inserted into the freshly mixed loam, 30 s were allowed to pass prior to recording the moisture value to ensure the sensor reached a steady state. Two moisture sensors were used for each trial. The ICStation sensor used in the prototype and a commercial moisture sensor (Sonkir, MS02, Ashland City, Tennessee) were both used. The Sonkir sensor values were reported on a scale from 1 to 10, with 1 being dry and 10 being moist. The commercial sensor was used to help classify the soil as dry, good and wet.

Five samples of dry loam had this testing procedure done. The total number of trials was 30. The number of trials for one loam sample ranged from 2 to 10. The results of the measurements are shown in Fig. 8, a scatterplot of the raw reading on the microcontroller for the moisture sensor vs. the known amount of water mixed into the soil. The raw sensor values decreased as the amount of water increased. An equation for linear fit and a R^2 values were determined for the measurements in Fig. 8. The R^2 value was only 0.67, so the relation was not strongly linear.

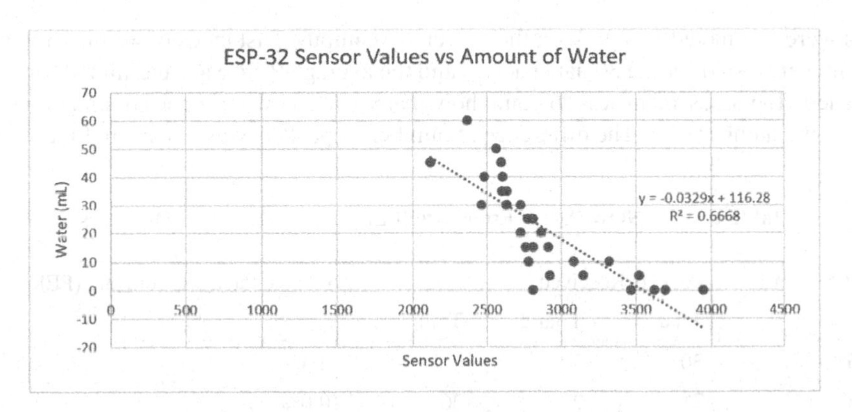

Fig. 8. Uncalibrated Sensor Readings vs. Amount of Water (mL).

A similar plot was made for the values (1–10) from the Sonkir sensor vs. the known amount of water mixed into the soil (Fig. 9). The Sonkir value increased as the known amount of water increased, but the relation also had a moderate R^2 value of 0.67. Based on this analysis, the relation between the moisture sensor readings and the Sonkir values could be determined and used for classification of soil moisture level.

3.2 Wireless Communication

ESP-NOW is a wireless communication protocol developed by Espressif. ESP-NOW is compatible with ESP-32 boards, inexpensive and may have a higher transmission range in comparison to some other protocols. The ESP-NOW communication protocol had a specified range of 76 m. To assess the effective range of ESP-NOW in a natural outdoor space, a test was conducted by sending 30 data packets at various distances and determining the packet error rate (PER). Each packet contained 250 bytes of data, which was the maximum packet size. For each distance, 30 data packets were sent. The

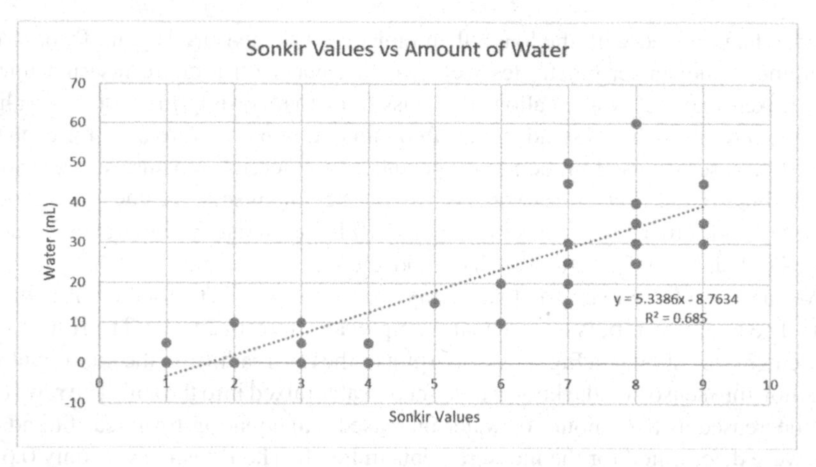

Fig. 9. Sonkir Readings vs Amount of Water (mL).

tests were conducted at a park on the university campus. Distances of 46 m, 76 m and 107 m were tested for the 30 data packets and the average PER was determined for each distance. The assessment was to count how many packets were received, and compare that to the number sent. The difference in number of packets was considered the error.

Table 1. ESP-NOW Packet Error Rate (PER) Tested at Various Distances

Distance (m)	Packets Received			Average Packet Error Rate (PER)
	Trial 1	Trial 2	Trial 3	
46	30	30	26	4.4%
76	28	27	26	10.0%
107	15	10	8	63.33%

Results from testing showed that the ESP-NOW protocol could transmit data over all three tested distances, but the PER of errors increased with increasing distance. Results are shown in Table 1. Performance on a farm may be no worse than the tested conditions, since testing was conducted in a large park in an urban environment. Since there were some errors at all the tested distances, the software should add an error detection and retransmission scheme.

4 Discussion and Future Directions

Prototype modules were developed for measuring in the soil the following: nitrogen, phosphorus and potassium (NPK), pH, temperature and moisture. Testing was done toward calibration of the moisture sensor with moisture content in the soil. A communication module was developed and tested using the ESP-NOW wireless protocol. This could be used in a mesh network to pass messages from the soil monitoring stakes back to a hub computer. A prototype hub was made with the ESP-32 microcontroller that hosted a website that displayed a dashboard of sensor values. The farmer would be able to view the website to make decisions about application of water and fertilizer.

The design and testing results appear promising. More development to complete the envisioned system and testing would be required toward full implementation. Development of such a system would help farmers monitor soil parameters and assist with the application of appropriate amounts of water and fertilizers, reducing the need of over application of these resources. Minimizing over application of fertilizer would reduce the amount of pollutants downstream.

References

1. Islam, S.M.F., Karim, Z.: World's Demand for Food and Water: The Consequences of Climate Change (2019). https://doi.org/10.5772/intechopen.85919. https://discovery.ebsco.com/linkprocessor/plink?id=dca6e9e7-b733-3e28-8f5f-369dd57fcc46
2. Powers, S.: Quantifying Cradle-to-Farm Gate Life-Cycle Impacts Associated with Fertilizer used for Corn, Soybean, and Stover Production. (2005). https://doi.org/10.2172/15016223. https://discovery.ebsco.com/linkprocessor/plink?id=78254792-ed1e-316d-920b-a26e573b0efa
3. Erisman, J.W., et al.: Nitrogen : too much of a vital resource: science brief (2015). https://discovery.ebsco.com/linkprocessor/plink?id=50583922-9386-3f35-9393-86bc617e5302
4. Sedlacek, C.J., Giguere, A.T., Pjevac, P.: Is Too Much Fertilizer a Problem? (2020). https://doi.org/10.3389/frym.2020.00063
5. Graneli, E.: Influence of anthropogenic nutrients on harmful algal blooms (2019)
6. Lehuger, S., Gabrielle, B., Laville, P., Lamboni, M., Loubet, B., Cellier, P.: Predicting and mitigating the net greenhouse gas emissions of crop rotations in Western Europe (2011). https://doi.org/10.1016/j.agrformet.2011.07.002. https://discovery.ebsco.com/linkprocessor/plink?id=c17a29d3-5766-3743-afa8-4734272c6541
7. Reshma, R., Sathiyavathi, V., Sindhu, T., Selvakumar, K., SaiRamesh, L.: IoT based Classification Techniques for Soil Content Analysis and Crop Yield Prediction (2020). https://doi.org/10.1109/I-SMAC49090.2020.9243600. https://discovery.ebsco.com/linkprocessor/plink?id=70dea23b-6858-39e7-8a01-8b02a61d3f21
8. Alasco, R., et al.: SoilMATTic: arduino-based automated soil nutrient and pH level analyzer using digital image processing and artificial neural network. Presented at the 2018 IEEE 10th International Conference on Humanoid, Nanotechnology, Information Technology, Communication and Control, Environment and Management (HNICEM) (2018)
9. Chauhdary, J.N., et al.: Impact Assessment of Precision Agriculture and Optimization of Fertigation for Corn Growth, vol. 57, pp. 993–1001 (2020). https://doi.org/10.21162/PAKJAS/20.9305

A Novel Visualization Method of Vessel Network for Tumour Targeting: A Vessel Matrix Approach

Mengsheng Zhai[1], Minghao Liu[2], Zhijing Wang[3], Yifan Chen[2,4(✉)], and Yue Sun[2,5(✉)]

[1] School of Optoelectronic Science and Engineering, University of Electronic Science and Technology of China, Chengdu, China

[2] School of Life Science and Technology, University of Electronic Science and Technology of China, Chengdu, China

[3] Glasgow College, University of Electronic Science and Technology of China, Chengdu, China

[4] Yangtze Delta Region Institute (Huzhou), University of Electronic Science and Technology of China, Chengdu, China

yifan.chen@uestc.edu.cn

[5] School of Mechanical and Electrical Engineering, Chengdu University of Technology, Chengdu, China

sunyuestc90@126.com

Abstract. When the tumour grows, the microvessel density in the surrounding area increases and exhibits irregular curvature, which shows a difference from the regular vascular network. Therefore, a model is needed to describe the vascular network around the tumour. However, the existing models can not provide a good representation of the vascular network. This paper proposes a Vessel Matrix Model (VMM), a visualization vascular network model which has the potential to resemble the complicated vessels networks around the tumour microenvironment. VMM is conducive to the works such as drug delivery and tumour search and can perform a tumour-targeting search by combining with the computational nanobiosensing (CONA) framework. CONA uses nanorobots as computing agents to learn the surrounding environment to regulate the path-planning to the tumour location through algorithms such as reinforcement learning. A CONA method is performed in searching for a tumour to verify the feasibility of this vascular network. In order to seek optimal routing in the vascular network, VMM provides distance reward and weight reward for the agents, where the rewards are determined by the distance of starting point to the tumour lesion and the gradient of BGF, respectively. Therefore, VMM enables the tumour search with the CONA method. By introducing different weights between the destination and weights rewards, it is found that targeting efficiency can be affected by branch rate and size of the network.

M. Zhai and M. Liu—These authors contributed to the work equally and should be regarded as co-first authors.

Y. Chen et al. (Eds.): BICT 2023, LNICST 512, pp. 40–49, 2023.
https://doi.org/10.1007/978-3-031-43135-7_5

Keywords: tumour targeting · Vessel Network Visualization · Reinforcement learning · Vessel Matrix Model · CONA

1 Introduction

Cancer is regarded as abnormal and uncountable cell growth due to the specific genetic accumulation, which is the primary factor that results in fatal death globally [1,2]. As a result, clinical diagnosis is needed to detect cancer at an early stage [3]. Nevertheless, traditional medical imaging technology, such as MRI and CT, has a significant challenge since the resolution and information acquired by these facilities are restricted.

With the advances in nanotechnology, nanomedicines, using nanoparticles, are considered a new solution to prevent and treat disease. For decades, researchers have been working on developing nanoparticles that can accurately detect tumours and deliver drugs to cancer lesions. However, just 0.7 percent of the nanoparticles' injected dose (ID) can be delivered to tumour targets without external guidance, and delivery efficiency has not increased significantly in the last decade [4]. Therefore, recent years have made immense progress on the externally controllable nanobots, which can be manipulated by an external magnetic field, in the *in vivo* environment [5]. Utilizing the manipulable nanorobots in tumour targeting and drug delivery, as a result, we propose a novel computational nanobiosensing (CONA) framework. In addition, the performance of autonomous nanorobots achieved considerable high targeting efficiency [6]. These types of nanorobots have ability to explore the *in vivo* environment using reinforcement learning [7]. Besides, the biological gradient field (BGF) is introduced to this model, and through the principles of cooperation and coordination, the sensing nanoparticles successfully find the target. Internal structural changes generated by *in vivo* physical, chemical, or biological perturbations in the peritumoral area allow these nanoparticles to perform target-directed motility [8].

In our previous work [10,11], the drug delivery process is mapped into Molecular Communication (MC). The vessel network is regarded as the communication channel from NPs released site to the tumour lesions. The impulse response of the blood vessel-based communication channel contributes to the calculation of the drug concentration at the received end, which indicates a quantified drug delivery system. Therefore, a visualization vessel model is needed for drug delivery and tumour targeting applications. Recently, there have been proved that natural blood vessels can be transformed into the form of matrices utilizing images, which lays the foundation for a more realistic simulation of blood vessels. Based on this, this paper proposes Vascular Matrix Model (VMM), a method for describing blood vessels using a matrix combined with CONA to guide agents to target tumours. VMM may continue to display more features of the vascular network in the future, such as the transition model from fractal to lattice.

This paper is organized as follows. Section 2 describes the vascular matrix model. In Sect. 3, methodology including the Markov decision process, the allocation of values is introduced. Following that, simulations based on this model are presented, and the results are shown in Sect. 4. Finally, Sect. 5 shows the major outcomes and the conclusion is drawn.

2 The Vessel Matrix Model

The VMM assumes blood vessels are equivalent to matrices. The idea is inspired from [12], where they have proved that it is feasible for the robot to accomplish route path-planning using a matrix.

Given the complexity of blood vessels in the human body, only one matrix can not well generalize the topography of blood vessels. We use the following matrices to simulate the blood vessels:

- H_{label} represents the state of blood vessel i and j. For each element,

$$x_{ij} = \begin{cases} 1, & \text{two blood vessels are connected} \\ 0, & \text{two blood vessels are blocked} \end{cases} \tag{1}$$

- H_{loc} represents the distance from blood vessels i and j. In the matrix, the elements indicate where the vessel is connected.
- H_{weight} represents the information of BGF for the branch of each blood vessel. Some vessels may have a high value of BGF, so their weight must be greater than others and more valuable for the agent.

It has been proved that it is achievable to extract real blood vessels into the form of matrices based on vessel images [13], which is shown in Fig. 1, revealing the mapping from the vessel to the matrix.

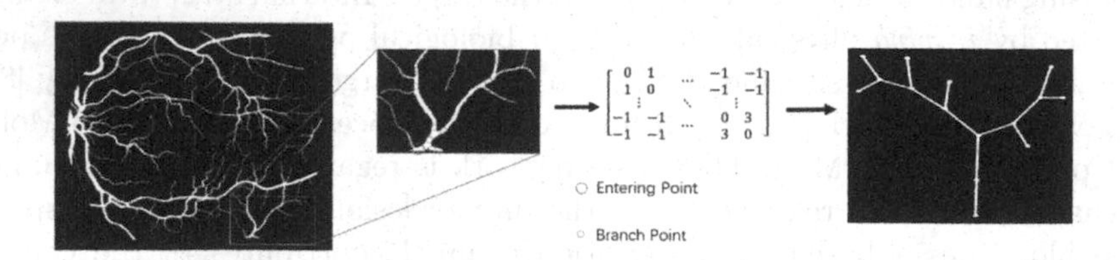

Fig. 1. Generating matrices from blood vessels images. The H_{label} matrix can be generated by transforming the blood vessels images to define the connectivity of the blood vessels.

We use the VMM model to reconstruct the blood vessel model corresponding to the medical image, in which two matrices H_{loc} and H_{weight} are used to describe the weights and the length of vessel branches. The result is shown in Fig. 2.

However, it lacks vessel datasets to train the model, so the following guidelines are used to generate the training datasets:

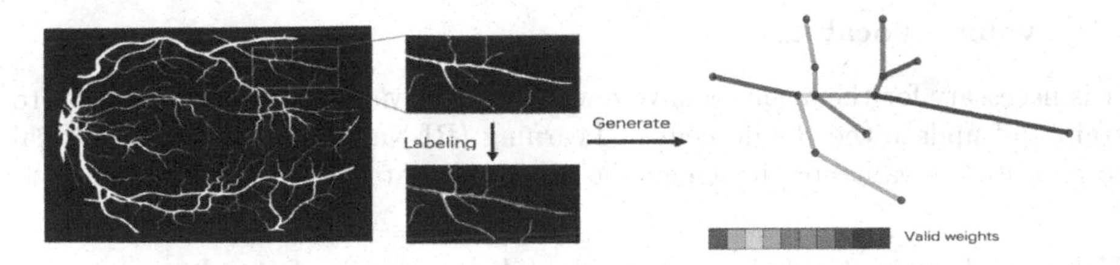

Fig. 2. Schematic diagram of the blood vessel model generated by the VMM model. Different weights of BGFs information are represented by different colors.

1. Select a size for the matrix H_{label} and initialize the matrix.
2. Initialize branch rate r which refers to the degree of branching of blood vessels.
3. Generate a random integer $r_{x_{ij}}$ and compare the $r_{x_{ij}}$ with the branch rate r. If $r_{x_{ij}}$ is greater than r, set the value to 1; otherwise, set it to 0.
4. Repeat step 3) until all points has been iterated and computed.
5. Initialize the matrix H_{loc} which has the same size of H_{label} and generate a random integer for each element x_{ij} in H_{loc} whose value in H_{label} is 1.
6. Initialize the matrix H_{weight} which has the same size of H_{label} and generate a random integer for each element x_{ij} in H_{weight} whose value in H_{label} is 1.

3 Tumour Searching Method

In the simulation process, the nanorobots detected the position of tumours in a grid-like vascular network of a given size. The core goal is to locate the tumor with a minimum exploring path. This problem is considered an optimization of the reward function, which uses the Markov decision process (MDP).

3.1 Markov Decision Process

The aforementioned problem is formulated as an MDP [14]. In this process, the definition (S, A, R) is represented separately as state space, action space, and reward function. The state at mission time t in the vascular network is given by $s_t = (p_t, p_t) \in \mathbb{R}^2$, which is the position of the nanorobots. Nanorobots are allowed to take the following actions:

$$A_{ij} = \{1, 2, 3, ..., n\} \tag{2}$$

where n is the number of branches from x_{ij}.

If the nanorobots collide with a vessel wall, they enter idle mode; otherwise, they move ahead in one of the four directions defined in the equation (2). The state-actions are mapped to a real-valued reward using the reward function, i.e., S × A → R.

3.2 Value Allocating

It is necessary for the agent to have reward, action, value, and states in order to train and update the Reinforcement Learning (RL) algorithm. So, it is critical to allocate the value for the agent, the rule of allocated value will be proposed.

Value of Destination. In order to describe the value of the destination for our agent, we use the Dijkstra algorithm to find the shortest distance from the destination. The calculated shortest path is stored in the $Dis(x_i)$ array, where x_i is the index of the distance array.

According to the Dijkstra algorithm [13], we follow these steps to find the shortest distances:

1. Initialize the matrix and select a point x_i from the matrix. The array Dis initializes with the distance the x_i contained.
2. The array r_{ix}, and select a point x_i from the matrix. The array Dis now contains some distances, and we select the shortest one x_j to continue our search.
3. Comparing the array Dis with the current distance from the point x_j which the distance has added the $Dis(x_j)$, and keeping the miniature data into the array Dis.
4. Repeat step 2) until all points have been iterated and computed

When the shortest distance is available, we have the following steps to compute the value of the destination.

$$V_{dst} = [Dis(X_n) - Dis(X_c)]V_0 \tag{3}$$

where $Dis(X_n)$ and $Dis(X_c)$ represent the distance from the current location and the next location to the destination, respectively. The last parameter V_o is the offset value for the V_{dst}.

Value of Weights in Vessel. By introducing H_{weight} and H_{loc}, the value of weights could be described in following equation.

$$V_{weight} = \begin{bmatrix} z_{1,1} & z_{1,2} & \cdots & z_{1,j} \\ z_{2,1} & z_{2,2} & \cdots & z_{2,j} \\ \vdots & \vdots & \ddots & \vdots \\ z_{i,1} & \cdots & \cdots & z_{i,j} \end{bmatrix}, \tag{4}$$

$$z_{i,j} = \frac{x_{i,j}}{y_{i,j}} \tag{5}$$

where $x_{i,j}$ and $y_{i,j}$ are elements at corresponding positions in matrix H_{weight} and matrix H_{loc}, respectively.

V_{weight} could be equivalent to the gradient of BGF. The greater V_{weight} is, the more valuable for the agent this area is.

Value in Total. Let V_{total} denote the value in total, which contain both V_{dst} and V_{weight}. In case of an imbalance value's negative impact on the agent, some biases are used to balance the values. Here we write the equation as the following equation:

$$V_{total} = aV_{dst} + bV_{weight}, \quad (6)$$

where the parameter a and b are the rates to weight the value for the parameter V_{dst} and V_{weight}.

3.3 Deep Q-Network

Due to the complexity of the vascular network, the number of states of the agent may be very large. Therefore, Q-learning in RL will result in the exponential growth of the Q-table, we adopt DQN to train agents to learn in CONA.

The DQN (Deep Q-Network) algorithm is a neural network architecture for model-free reinforcement learning [15]. It successfully realizes the end-to-end from perception to action and has been applied in gaming and navigation. The DQN algorithm introduces deep learning into reinforcement learning, in which the interaction between the nanorobots agent and the environment enables the agent to learn and optimize its behaviour. The learning process is evaluated by improving the Q-function iteratively. The computation and updating of the Q-function can be written as the following equation:

$$Q(s, a) \leftarrow Q(s, a) + \alpha[r + \gamma \max_{a'} Q(s', a') - Q(s, a)] \quad (7)$$

where s is the agent state which includes the overall situation of the whole vessel, a is the action performed by the agent, and r represents the reward to the agent. The constants α and γ are the learning rate and decaying rate, which control the convergence rate of the agent and the impact factor from the future. Practically, DQN's neural network (NN) weights should be updated by the gradient of its loss function:

$$L(\theta) = E[T_Q - Q(s, a; \theta)^2] \quad (8)$$

where T_Q is the optimization objective, which is calculated as follows:

$$T_Q = r + \gamma \max_{a'} Q(s', a'; \theta) \quad (9)$$

3.4 Target Network

The target network is used to generate a T_Q for the main NN so that the main NN could update its NN weights according to the gradient for a period of iterations, and we set it 100 times here. The loss is counted by the quality of the main NNQ and the T_Q. Such a measure can reduce the relevance between Q and T_Q, usable for increasing the stability of DQN.

3.5 Experience Replay

DQN proposes a buffer in which the agent would randomly select the situation to train itself. In our work, the environment information and the action generated by the agent are stored in a buffer. While training, the agent selects a batch of buffers to review the past situation to train itself. As a result, the lack of relevance with samples and its problem of non-static distribution is partially compensated. Hence, experience replay can improve the robustness of the NN.

4 Simulation and Results

4.1 Simulation Setup

According to the input state, DQN model calculates the reward of each vessel branch. As a result, the vessel branch with the highest reward will be selected. The action is invalid when the action calculated according to the reward is marked as impassable in the H_{label}. To correct this action, we propose two methods.

1. Return a false result when the output is invalid so that the agent can learn and act accordingly. However, massive training is needed in this method and hardly ensures efficiency.
2. Execute a random action when the action is invalid. Although this method cannot guarantee accuracy, it does not need massive training.

Therefore, the second method is utilized to test the proposed VMM model. In the simulation, we set the total test time to 50'000 times to eliminate the effect of random movement in different circumstances. Considering the limited lifespan of the nanorobot, we set the maximum detection time as 100. Sessions that do not reach the end within the time will be considered invalid.

We changed the vascular branch rate and the number of vascular branches in the VMM model to simulate different vascular environments, as shown in Fig. 2. Set the parameters in the simulation as follows: $\gamma = 0.9$, $V_0 = 60$.

4.2 Different Scenarios

The result of the different branches generating probabilities r and different sizes is compared in a single circumstance. As shown in Fig. 3, each circumstance has a good efficiency in finding the tumour. From the found times and found rate data, we notice that the agent's efficiency mainly depends on the number of branches and size of the vascular network.

According to equation (7), we know that the process of searching for tumours is affected by both reward V_{dst} and V_{weight}, so the parameter β = a/b is set to play a role in regulating the relative weight of the two rewards in the simulation. As the weight, β increases from 0.01 to 100, the number of found times and found rate do not show significant differences, which indicates that searching efficiency is less affected by β.

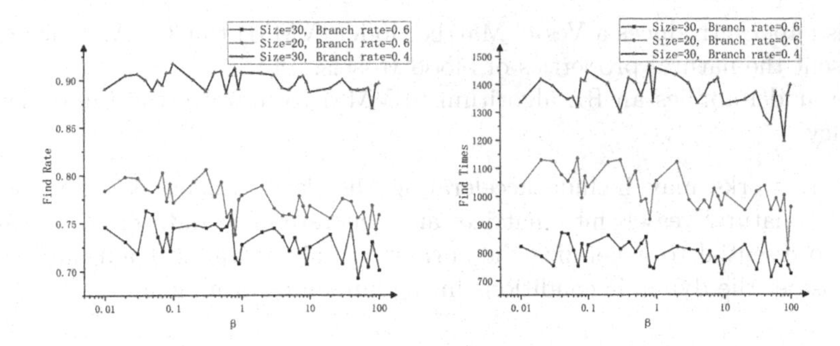

Fig. 3. Found times (a) and found rates (b) for different branch generate probabilities r and different sizes when model converged, the log of β is taken.

The fluctuation of reward can reflect the stability of the search process, and a huge reward indicates that the search process is greatly affected by random actions. As shown in Fig. 4, when the value of beta is too high, the standard deviation of the reward increases significantly, and the learning effect of the neural network is not good. When the value of β is too low, it is not in line with the actual situation in the human body, so the value of β should be considered by considering the above factors and making a compromise.

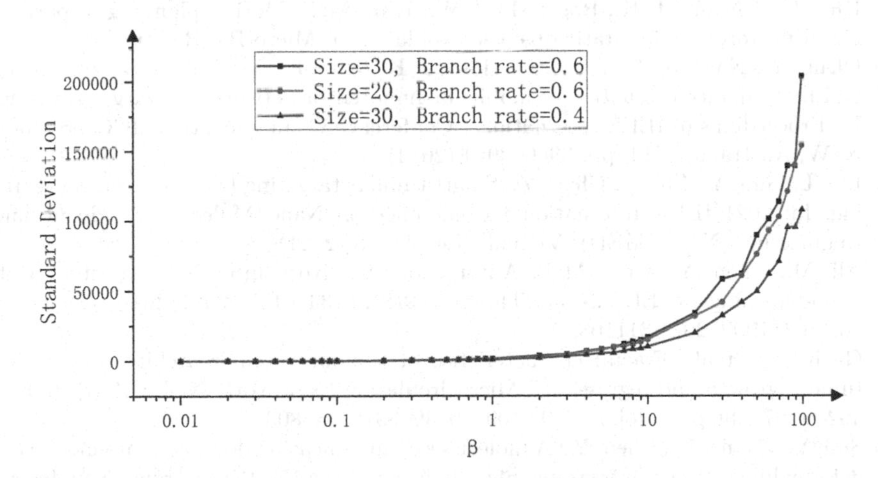

Fig. 4. The standard deviation of total reward when model converged, the log of β is taken.

5 Conclusion

We proposed a novel model for displaying blood vessels that can be used to expand tumour sensitization and tumour targeting. The major contributions are as follows:

– This paper introduces a Vessel Matrix Model (VMM) to CONA, which could present the natural properties of blood vessels.
– The model applies an RL algorithm to VMM to improve the targeting efficiency.

Future works may include accelerating the development of techniques to transform natural vessels into matrices and generate more datasets for training. It is also essential to generalize the current RL algorithm to train nanorobots and consider the dynamic conditions in the human environment.

References

1. Schulz, W.: Molecular Biology of Human Cancers: An Advanced Student's Textbook. Springer, Dordrecht, The Netherlands (2005)
2. Abubakar, I., Tillmann, T., Banerjee, A.: Global, regional, and national age-sex specific all-cause and cause-specific mortality for 240 causes of death, 1990–2013: A systematic analysis for the global burden of disease study 2013. Lancet **385**(9963), 117–171 (2015)
3. Bohunicky, B., Mousa, S.A.: Biosensors: the new wave in cancer diagnosis. Nanotechnol. Sci. Appl. **4**(1), 1–10 (2011)
4. Wilhelm, S., Tavares, A.J., Dai, Q., Ohta, S., Chan, W.C.W.: Analysis of nanoparticle delivery to tumours. Nat. Rev. Mater. **1**(5), 16014 (2016)
5. Kim, H., Cheang, U.K., Rogowski, L.W., Kim, M.J.: Motion planning of particle based microrobots for static obstacle avoidance. J. Micro-Bio Robot. (2018)
6. Okaie, Y., Nakano, T., Hara, T., Hosoda, K., Hiraoka, Y., Nishio, S.: "Modeling and performance evaluation of mobile bionanosensor networks for target tracking. In: Proceedings of IEEE International Conference Communication (ICC), Sydney, NSW, Australia, 2014, pp. 3969–3974 (2014)
7. Liu, L., Sun, Y., Shi, S., Chen, Y.: Smart tumour targeting by reinforcement learning. In: 2021 IEEE International Conference on Nano/Molecular Medicine and Engineering (NANOMED), Virtual, Nov. 15–18 (2021)
8. Ali, M., Chen, Y., Cree, M.J.: Autonomous in vivo computation in internet-of-nano-bio-things. IEEE Internet Things J. **9**(8), 6134–6147 (2021) https://doi.org/10.1109/JIOT.2021.3111089
9. Gazit, Y., et al.: Fractal characteristics of tumour vascular architecture during tumour growth and regression." Microcirculation (New York, N.Y. : 1994) **4**, 395–402 (1997). https://doi.org/10.3109/10739689709146803
10. Sun, Y., Zhang, R., Chen, Y.: A molecular communication detection method for the deformability of erythrocyte membrane in blood vessels. IEEE Trans. Nanobiosci. **20**(4), 387–395 (2021). https://doi.org/10.1109/TNB.2021.3064194
11. Wang, D., Sun, Y., Xiao, Y., Chen, Y.: An Optimal Strategy for Individualized Drug Delivery Therapy: A Molecular Communication Inspired Waveform Design Perspective, 2021 43rd Annual International Conference of the IEEE Engineering in Medicine & Biology Society (EMBC), pp. 866–869 (2021)https://doi.org/10.1109/EMBC46164.2021.9629560
12. McDougall, S.R., Anderson, A.R.A., Chaplain, M.A.J., Sherratt, J.A., et al.: Mathematical modelling of flow through vascular networks: Implications for tumour-induced angiogenesis and chemotherapy strategies[J]. Bull. Math. Biol. **64**, 673–702 (2022)

13. Wang, H., Yu, Y., Yuan, Q.: Application of dijkstra algorithm in robot path-planning. IEEE (2011)
14. Rahebi, J., HardalaÇ, F.: Retinal blood vessel segmentation with neural network by using gray-level co-occurrence matrix-based features. J. Med. Syst. **38**(8), 1–2 (2014)
15. Kaelbling, L.P., Littman, M.L., Cassandra, A.R.: Planning and acting in partially observable stochastic domains. Artif. Intell. **101**(1–2), 99–134 (1998)

Heterogeneous Group of Fish Response to *Escape Reaction*

Violet Mwaffo(✉)

United States Naval Academy, Annapolis, MD, USA
mwaffo@usna.edu

Abstract. The response of heterogeneous groups of fish including a few leaders, several followers, and a few fish initiating *escape reaction* is investigated. This alarm response is often observed in animal groups where exposure to strong stimuli such as a predator can force a few individuals to initiate sudden and abrupt turns to move to safer locations. In this work, a coupled stochastic process is leveraged to recreate this behavior and investigate their effects on the group collective dynamics. At the vicinity of a synchronized state, for small perturbations introduced by *startled* fish, a closed-form expression of the polarization order parameter is determined and shown effective in predicting group alignment. A numerical analysis suggests that a variation of the frequency and the amplitude of the jumps introduced by escaping fish can result in a transition to several states including an ordered state where individual align their heading direction, a disorganized state where they move in random direction, and two other states where the group split up resulting either into a change of leadership or individuals swimming away from the startled fish and therefore recovering their initial synchronized state. The findings from this work are in line with observations on fish groups exposed to a predator where initially a completely disordered state can be observed but groups tend to progressively recover a synchronized state.

Keywords: Bio-inspired systems · Collective dynamics · *Escape reaction* · Heterogeneous group

1 Introduction

Individual differences are often listed among important factors at the origin of collective behavior in biological groups. These differences can be characterized based on morphological or physiological traits, an individual position within the group, the knowledge of the environment, or individual social dominant statute [1–3]. The effects of individual traits on group response are either observed in nature through observations or tested in controlled laboratory environments [4,5]. In recent years, to address ethical issues about the use of animals in laboratory studies, *in-silico* experiments [6] have become popular. These computational study allow to reduce significantly the number of subjects by pretesting hypothesis in order to better

Supported by the United States Naval Academy.

Y. Chen et al. (Eds.): BICT 2023, LNICST 512, pp. 50–63, 2023.
https://doi.org/10.1007/978-3-031-43135-7_6

plan experimental studies with real life subjects. In addition, they have allowed to unravel some microscopic factors explaining the emergence of groups collective behavior [7–10] such as leaders-followers relationships and information flow within groups [11].

In biological groups, *escape reaction* [3], is classified among alarm responses originating from exposure to fear-inducing stimuli such as animal exposure to a predator [12]. This behavior has been observed in fish groups causing individuals to exhibit erratic or zig-zagging swimming [13] before maintaining a coordinated swimming behavior [14,15]. In the literature, computational models have been utilized to dissect fish kinematics during *fast-start swimming* [3,16–18] or to unravel group collective response during *escape reaction*. The authors in [19] working on group collective response have leveraged a mathematical model to observe a propagation wave following the initiation of *escape reaction*.

Different from most prior works [3,16–18] dissecting fish kinematics or the response of homogeneous groups, this work investigates the collective response of a heterogeneous group of fish to the observable *escape reaction* [20] characterized by sudden and fast-turns away from a strong stimulus source [21]. The heterogeneous group of fish includes a few individuals denoted *leaders* having a dominant statute [3] or a good knowledge of their environment [4] to guide other team members denoted *followers* implementing local updating rules to maintain alignment with other fish in the group [22,23]. In addition to *leaders* and *followers*, the group includes a few *startled* fish initiating sudden and fast-turning maneuvers after exposure to a strong stimulus. These bursts of activity are captured through a stochastic jump-diffusion process introduced in [24] and adapted here with a biased distribution to stir a *startled* toward a preferential heading direction.

Group response to small perturbations at the vicinity of a synchronized state is conducted allowing to establish a closed form expression for the polarization order parameter introduced in [25,26] to measure group of fish tendency to swim in the same direction. The closed form expression is validated against numerical simulations which in addition revealed unexplored forms of state transitions in addition to the traditional transition from a state of complete order to a disorganized state [25,26]. Further, as observed in the literature [3], the modeled *escape reaction* is shown capable to induce new form of leadership defined as the initiation of new directions of motion by a few individuals followed by other fish in the group [4].

Section 2 introduces the mathematical modeling framework recreating *escape reaction*. Section 3 analyzes group response to small perturbations introduced by *startled* fish leading to the derivation of a closed form expression for the polarization order parameter. Section 4 presents results from the numerical analysis and Sect. 5 discusses the main findings and concludes the work.

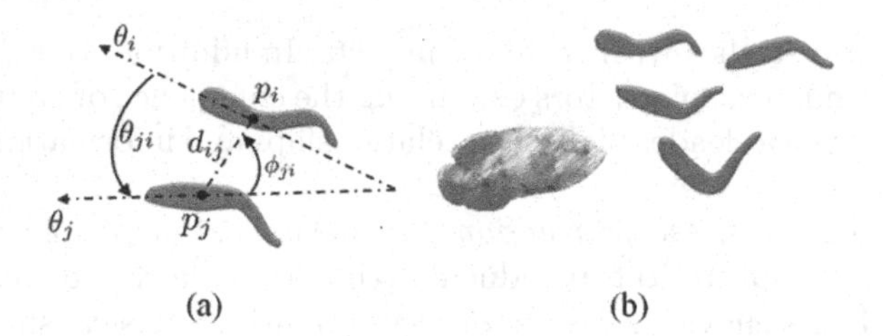

Fig. 1. Pair of fish i, j interaction (a) with positions p_i, p_j, heading angles θ_i, θ_j with respect to a fixed reference frame, their inter-distance d_{ij}, difference heading angle $\theta_{ji} = \theta_j - \theta_i$, and relative angular position of fish i with respect to fish j heading angle ϕ_{ji}; and initiation of *escape reaction* (b) by a group of fish exposed to a predator.

2 Fish Individual and Collective Behavior

2.1 Mathematical Modeling of the Heterogeneous Group

A group of N fish swimming in a 2D unbounded and non-periodic open water domain at a constant speed is considered. A fish i position is captured at any time instant τ by the 2D vector p_i of the position and the scalar θ_i representing the heading angle in a fixed frame as illustrated in Fig. 1(a). The group of fish includes a few *leaders* well aware of their environment [22] to move freely similar to dominant fish [3] and guide the rest of the group [9,27] towards set location such as migration route or foraging source. These fish are not coupled to the rest of the group which can interact with them creating a sort of potential difference forcing individuals implementing a social interaction function with closer neighbors to follow them. Another subset of fish is denoted as *startled* fish initiating *escape reaction* (see Fig. 1(b)) characterized by sudden and fast turns with frequency ι and intensity δ. The remaining fish are denoted as *followers* which concomitantly to *startled* fish update their heading angle through the predefined social interaction function.

A stochastic jump-diffusion process introduced in [24] to capture the fast turn rate $w(\tau)$ observed in the swimming locomotion of small fish species is adapted here to model fish exhibiting *escape reaction* in term of a dimensionless coupled system [28]:

$$\mathrm{d}w_i(\tau) = -\left(w_i(\tau) - w_i^\star(\tau)\right)\mathrm{d}\tau + \varsigma \mathrm{dW}_i(\tau) + \mathrm{dJ}_i(\tau), \tag{1a}$$

$$w_i^\star(\tau) = \sum_{j \in \mathcal{N}_i(\tau)} \frac{1}{|\mathcal{N}_i(\tau)|}\left[\kappa_v \sin\left(\theta_{ij}(\tau)\right) + \kappa_p \bar{d}_{ij}(\tau) \sin\left(\phi_{ij}(\tau)\right)\right], \tag{1b}$$

where ς is the noise intensity; $w^\star$ the interaction function with coupling gains κ_v and κ_p; $|\mathcal{N}_i(\tau)|$ the number of fish interacting with fish i; $\bar{d}_{ij}$ the inter-distance between pair of fish; $\mathrm{W}_i(\tau)$ is the Wiener process which is defined such that the increments $\mathrm{dW}_i(\tau)$ follow a normal distributed random variable with standard deviation $\sqrt{\mathrm{d}\tau}$; and $\mathrm{J}(\tau) = \sum_{j=1}^{\nu_i(\tau)} \delta_i Z_j$ is the jump diffusion process with $\nu_i(\tau)$

defining a counting process with parameter ι_i capturing the frequency of the fast turns while δ_i is a scaling coefficient of the jumps.

In the model above, the heterogeneous fish behavior is captured by a single parameter ς_i for the *informed fish*, three parameters ς_i, κ_{vi}, and κ_{pi} for *followers*, and by five parameters ς_i, κ_{vi}, κ_{pi}, ι_i and δ_i for the *startled* fish. The parameters of the jump-diffusion process in [24] have been calibrated on experimental trajectories of a small fish species to reproduce sudden and fast-turns [24]. However, the jump term in [24] randomly take either positive or negative values such that they do not favor a preferential heading direction as observed during *escape reaction*. This problem can be solved by selecting appropriate distribution to model the jumps.

2.2 Distribution of the Jumps to Recreate *escape Reaction*

Departing from the process model in (1), *escape reaction* is modeled by considering a biased distribution and notably the half-normal distribution defined as:

$$Z_j(t) = \begin{cases} z, & \text{if } Z = z \geq 0, \\ -z, & \text{if } Z = z < 0. \end{cases} \tag{2}$$

where Z is the standard normal distribution. The resulting skewed distribution induces turns in average in a preferential direction of motion as observed in the literature [13] after fish exposure to a strong stimulus such as a predator. Note that, other biased distributions including skewed distribution or symmetric distribution with non-zero mean can be considered.

Note that, fish initiating *escape reaction* and *follower* fish share identical interaction parameters such that, before the initiation of the *escape reaction*, *startled* fish are assimilated to *followers* as suggested in [3] where social subordinated fish are observed to be more favorable to exhibit such a behavior. In addition, the sign of the expected value of the distribution of the jump determines in which direction the *startled* fish will turn. In particular, using the distribution in (2), the expected mean of a fish i turn rate is evaluated as $\mathbf{E}\left[w_i, \Delta\tau\right] = \iota\delta\Delta\tau\sqrt{\frac{2}{\pi}} > 0$ while for a *follower* not subject to any additional disturbances, one has $\mathbf{E}\left[w_i, \Delta\tau\right] = 0$. Thus, at steady state, one expects that the coupled system in (1) will move either in a direction determined by the long-term mean of the abrupt turns introduced by fish initiating *escape reaction*, in a direction prescribed by the *leaders*, or will simply split-up with followers following either group *leaders* or *startled* fish. These transitional states tend to characterize the various responses observed during *escape reaction* in nature where the fast and large turns initiated by a few fish can significantly affect the collective response of a group [3,13,14,29].

2.3 Discrete Time Approximation

The stochastic system in (1) is solved using a discrete-time Euler-Maryuma approximation scheme [30]. In particular, assuming that $\Delta\tau$ is small enough such that at most a single jump is observed in a time step, the discrete time approximation of the turn rate process in (1) allows to estimate a fish i position at any time step $\tau_k, k \in \mathbb{N}$ by a forward Euler scheme [6,9]:

$$\begin{aligned} x_i(k+1) &= x_i(k) + \Delta\tau \cos\theta_i(k) \\ y_i(k+1) &= y_i(k) + \Delta\tau \sin\theta_i(k) \\ \theta_i(k+1) &= \theta_i(k) + \Delta\tau w_i(k), \\ w_i(k+1) &= w_i(k)\,(1-\Delta\tau) + w_i^\star(k)\Delta\tau + \varsigma_i\sqrt{\Delta\tau}\varepsilon(k) + \delta_i\Delta\nu_i(k)\zeta(k), \end{aligned} \tag{3}$$

where $\tau_{k+1} = \tau_k + k\Delta\tau$, $w_i^\star(k) = w_i^\star(\tau_k)$ is the social interaction function, $\varepsilon(k)$ and $\zeta(k)$ are i.i.d. Gaussian random variables, and for simplicity τ_k is replaced by k. The discrete time process $w_i(k)$, $k \in \mathbb{N}$ converges weakly to the continuous time process $w_i(\tau)$ [24]. Note that, using the above discrete-time scheme, except for the *startled* fish, when ι_i and δ_i are null, the expectation and variance of a *follower* fish turn rate are $\mu_i\left(w_i(k+1), \Delta\tau\right) = w_i(k)\,(1-\Delta\tau) + w_i^\star\Delta\tau$ and $\mathrm{V}_i\left(\omega_i(k+1), \Delta t\right) = \varsigma_i^2\Delta\tau$, respectively.

3 Analysis of Group Coordination

Group coordination is analyzed at the vicinity of a synchronized state when all fish tend to move in the same direction and shared a common heading angle denoted θ_0. The stability of the group is evaluated by introducing to the system in equations (1) small perturbations. To simplify the analysis, a single *leader* and a single *startled* fish are considered. Note that for such a system to achieve group coordination, all *leaders* should move in the same direction. Similarly, a single strong stimulus is considered and all *startled* fish swim in the same direction to maintain a consistent escaping route. With the above considerations, given that θ_0 represents the *leader* heading direction, the stability study of the local disagreement $\theta_i - \theta_0$ can be conducted for the rest of the $N-1$ fish.

3.1 Measure of Group Coordination

Group coordination is evaluated with a traditional order parameter introduced in [25,26] to measure group alignment in self-propelled particles and denoted as the polarization with expression determined as:

$$\mathrm{Pol} = \frac{1}{N}\Big\|\sum_{i=1}^{N}\mathbf{v}_i\Big\|, \tag{4}$$

where $\|(\cdot)\|$ defines the norm of the unit velocity vector $\mathbf{v}_i$. The values of the polarization order parameter [25,26] $P(\tau)$ range from 0 to 1 with values closer to 1 indicating that all fish moving in the same heading direction and values closer to 0 when they move in completely different heading directions.

3.2 Group Response to Small Perturbations

Using the polarization order parameter Pol, group response to small perturbations denoted as the Pol-susceptibility is equivalent up to a constant factor to the fluctuation of the order parameter Pol [31], that is:

$$N\left[\langle P^2\rangle - \langle P\rangle^2\right] = \frac{\partial P}{\partial x}_{|x=0}, \tag{5}$$

where $\langle \text{Pol}\rangle = \lim_{T\to\infty} 1/(T-T_r)\sum_{k=T_r}^{T} Pol(k)$, and x is related to the perturbation field. For a fixed value of ς, the perturbation field for the system in (1) is captured by the jumps parameters ι and δ as further elaborated below.

At closer proximity of the coordinated state, when all individuals share a similar heading direction, let say θ_0, $\theta_{ij} = \theta_i - \theta_j \simeq 0$ for all i, j. Denoting $\tilde{w}_i(k) = w_i(k) - w_i^\star(k)$, the discrete-time system in (3) reduced to:

$$\begin{aligned}\theta_i(k+1) &= \theta_i(k) + w_i^\star(k)\Delta\tau + \tilde{w}_i(k)\Delta\tau\\ \tilde{w}_i(k+1) &= \tilde{w}_i(k)e^{-\Delta\tau} + \varsigma\sqrt{\frac{1}{2}\left(1-e^{-2\Delta\tau}\right)}\varepsilon_i(k) + \delta\Delta\nu_i(k\Delta\tau)\zeta_i(k),\end{aligned} \tag{6}$$

where for simplicity, $w_i^\star(k) = w_i^\star(\tau_k)$. For smaller values of $\Delta\tau \ll 1$, the system can be further reduced to:

$$\begin{aligned}\theta_i(k+1) &= \theta_i(k) + w_i^\star(k)\Delta\tau + \tilde{w}_i(k+1)\Delta\tau\\ \tilde{w}_i(k+1) &= (1-\Delta\tau)\tilde{w}_i(k) + \varsigma\sqrt{\Delta\tau}\varepsilon_i(k) + \delta\Delta\nu_i(k\Delta\tau)\zeta_i(k),\end{aligned} \tag{7}$$

In addition, for small misalignment between pair of fish i and j, one can approximate $\sin(\theta_{ij}(k)) \simeq \theta_{ij}(k)$ and $\phi_{ij}(k) \simeq \frac{\pi}{2}$. These approximation allows in turn to get the following discrete time double integrator system:

$$\begin{aligned}\theta_i(k+1) &= \theta_i(k) + \kappa_v\Delta\tau \sum_{j\in\mathcal{N}_i(k)} \theta_{ij}(k) + \tilde{w}(k)\Delta\tau\\ \tilde{w}_i(k+1) &= (1-\Delta\tau)\,\tilde{w}_i(k) + \varsigma\sqrt{\Delta\tau}\varepsilon_i(k) + \delta\Delta\nu_i(k\Delta\tau)\zeta_i(k).\end{aligned} \tag{8}$$

The mean square stability analysis of a stochastic system similar to the one in (8) has been thoughtfully investigated in [32,33] for the vectorial network model (VNM) [34] which is considered as a simplification of the celebrated Vicsek model at the limit of large speed. In the VNM model, particles are thought to move fast enough such that they can interact at any time instant with any other randomly selected particles in the group. Such a consideration holds in particular for small particle sizes. Using the expression of the polarization in (4), writing the velocity vector in polar coordinates assuming a unit constant speed of 1, a linear approximation of the polarization is given for the remaining $N-1$ particles excluding the leader as [32,33]:

$$\text{Pol}(k) = \frac{1}{N-1}\sum_{i=1}^{N-1}\mathbf{v}_i = \frac{1}{N-1}\left|\sum_{i=1}^{N-1} e^{j\theta_i(k)}\right| \simeq 1 - \frac{1}{2(N-1)}\rho(k),$$

where $\mathbf{v}_i$ is the unit velocity vector; j is the imaginary index; and $\rho(k) = \mathbf{E}\left[\sum_{i=1}^{N-1}(\theta_i(k) - \theta_0)^2\right]$ defines the steady state deviation from the synchronized state θ_0.

3.3 Closed Form Expression

The steady state deviation ρ has been shown in [32,33] to converge towards a finite value at the vicinity of a synchronized state for $0 < (1 - \Delta\tau)^2 < 1$. For the dimensionless model considered here, such a condition holds for $0 < \Delta\tau < 2$. After identification from the expression obtained in [32], a closed form expression of the polarization order parameter can be obtained as:

$$\mathrm{Pol} \simeq 1 - \frac{(\varsigma^2 \Delta\tau + \lambda\gamma^2)}{2} \frac{N-2}{N-1}, \tag{9}$$

where the number of connected neighbors is set to $|\mathcal{N}_i| = N - 1$ for small group sizes. For group size of $N \gg 2$, the expression in (9) can be reduced to

$$\mathrm{Pol} \simeq 1 - \frac{(\varsigma^2 \Delta\tau + \iota\delta^2)}{2}$$

indicating that group response is mainly explained by the noise intensity ς and the jumps frequency ι and intensity δ. Doing a simple transformation to obtain $\iota\delta^2 \simeq 2(1 - \mathrm{Pol}) - \varsigma^2\Delta\tau$, the closed-form expression above suggests that both ι and δ tend to be related by a negative power law. In addition, the closed form expression obtained above suggests that the expected variance of the jumps computed as $\iota\delta^2$ can be leveraged to explain the transitions observed in the group in terms of a single parameter characterizing the level of noise in the system given a fixed value of the noise scaling coefficient ς.

Table 1. Model parameters retained in the simulations. The letter codes L, F, and S are used to indicate parameters required to model *leaders*, the *followers*, and *startled* fish behavior respectively.

Model	Behavior	Parameter	Description	Value
Individual	L-F-S	ς	Noise intensity	0.40
dynamics	S	ι	Jumps frequency	[0 0.28]
	S	δ	Jumps Intensity	[0 4.18]
	L-F-S	v	Average speed	3.546
Social	F-S	κ_p	Attraction gain	0.036
behavior	F-S	κ_v	Alignment gain	4.24
	F-S	R	Interaction distance	12.74

4 Numerical Analysis

Group coordination is analyzed for small group size of 10 fish which for simplicity included a single *leader* and a single *startled* fish initiating *escape reaction*. The reader is referred to [35] for results on larger group size. In all simulations, each fish type is set to share identical parameters as indicated in Table 1 with $\varsigma_i = \varsigma$, $\kappa_{vi} = \kappa_v$, $\kappa_{pi} = \kappa_p$, $\iota_i = \iota$, and $\delta_i = \delta$. The polarization order parameter Pol is evaluated at steady state by computing the average values over 100 sample trajectories by varying values of ι and δ sampled in a discrete 100×100 grid size. Only the last 180-time steps of the simulations were considered to allow the group to reach a steady state.

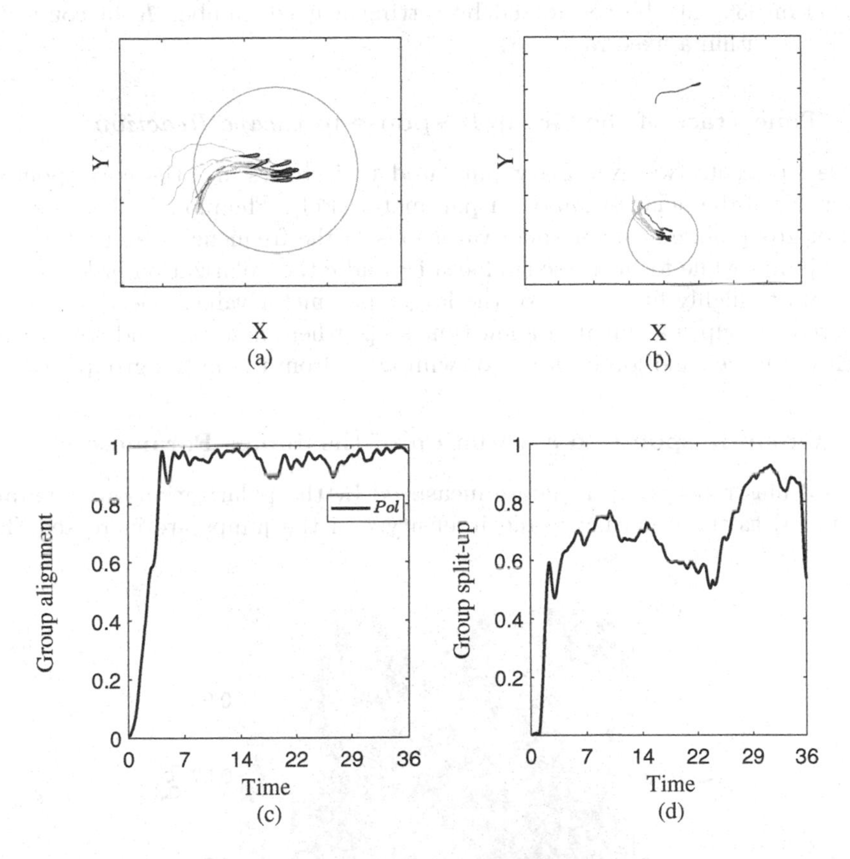

Fig. 2. Sample trajectories illustrating group response to (a) small jumps ($\iota = 0.04, \delta = 0.17$) and to (b) larger jumps ($\iota = 0.12, \delta = 0.50$) introduced by the *startled* fish and time trace of the corresponding group polarization Pol for small jumps (c) and for larger jumps (d). The circles in (a) and (b) indicates the interaction radius from the *followers* averaged position.

In the analysis, simulations are always started with individuals coordinating their motion before introducing the fish initiating *escape reaction* after a few time steps. In particular, departing from a random location within a circle of radius R

from the origin, fish initial positions are generated within the interaction radius of their neighbor. The *leader* position is set at a random location near the origin. The fish initiating *escape reaction* is introduced at a time step corresponding to 7.18 time units at a random position within a distance corresponding to half of the interaction radius of R from the group center. The initial heading direction of *followers* is randomly set between $[-\pi, \pi]$ rad. Simulations are conducted in an unbounded and non-periodic domain with the interaction radius set to 12.74 length units. The latter value was selected to avoid group splitting in the absence of jumps. Note that, the choice of an interaction radius rather than a vision cone allows group response to be independent of the position of individual fish within the group. In addition, for larger group sizes, a topological approach for the interaction [36] can be considered by setting a fixed number K of connected neighbors within a fixed radius R.

4.1 Time Trace of the Group Response to *escape Reaction*

Figure 2 presents two exemplary simulated trajectories and the corresponding time trace of the polarization order parameter (Pol). The plots indicates a high level of group alignment for small variations of the frequency ι and intensity δ of the jumps while for increased values of ι and δ the polarization order parameter tend to highly fluctuate. For the larger parameter values selected, one can observed a group split-up after a few time steps where *followers* and *startled* fish synchronize their motion in order to swim away from the initial group *leader*.

4.2 Group Response to a Variation of the Jumps Parameters

Figure 3 illustrates group response measured by the polarization order parameter in (4) as the frequency ι and intensity δ of the jumps are increased. The

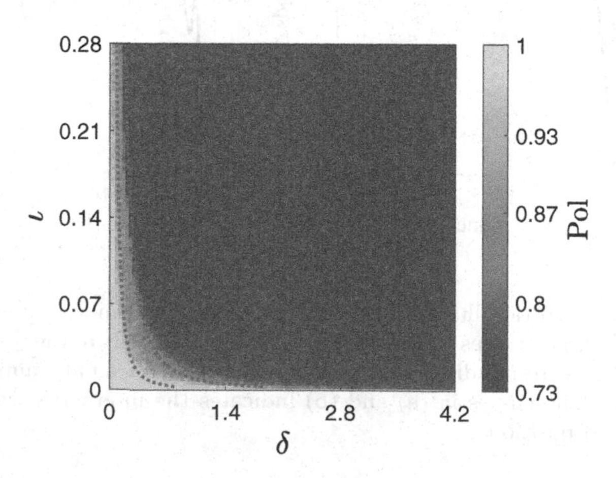

Fig. 3. Contour plot of group polarization (Pol) as the frequency ι and intensity δ of the jumps are varied. See Table 1 for model parameters.

contour plot of the polarization in Fig. 3 indicates a transition from a highly to a less coordinated state. As suggested by the closed form expression in (9), Fig. 3 indicates a negative power law relation between ι and δ for given values of the polarization order parameter Pol. In addition, the different coloration pattern in Fig. 3 tends to indicate that group coordination transitions through several states as ι and δ are increased from smaller to larger values. Two illustrative cases depicted by the red dashed lines are plotted in Fig. 3 to show how the closed form expression in (9) can be utilized to predict the level of order in the system as ι and δ varies. The first line corresponds to a value of Pol $= 0.965$ predicts the transition bound between the yellow area and the green area and the second line corresponding to a value of Pol $= 0.805$ predicts the transition bound between the green colored area and the blue colored area.

4.3 Transition States Observed as the Jumps Are Increased

Four distinct states illustrated in Fig. 4 are observed as the expected variance of the fast turns $\iota\delta^2$ is increased and characterized by the following behaviors:

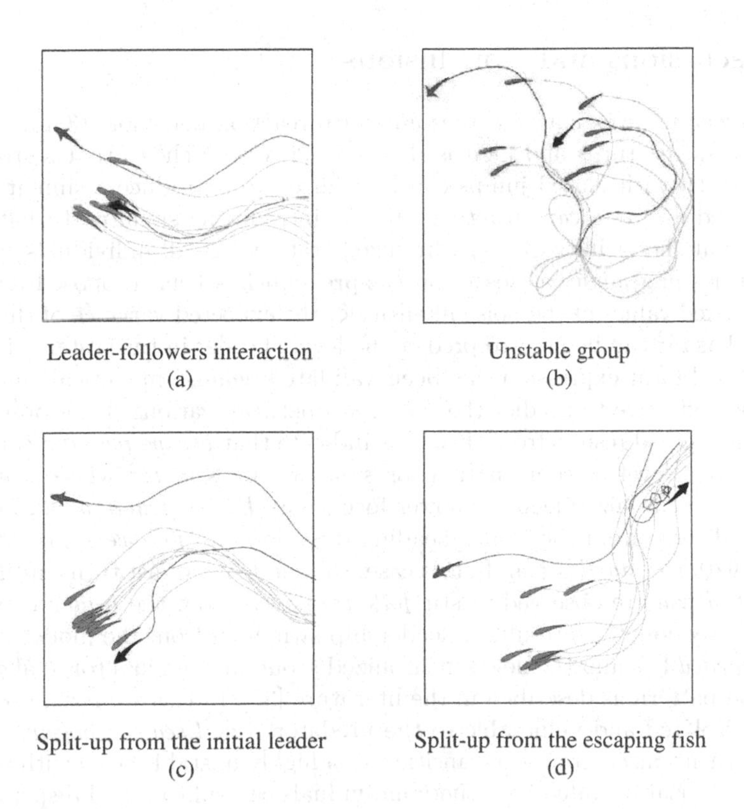

Fig. 4. Illustration of the transition states observed for selected value of the expected variance of the turn $\iota\delta^2$ in various colored areas of Fig. 3 including yellow for (a), between yellow to green and green to blue for (b), green for (c), and blue for (d).

- **(a) leader-followers relationship** illustrated in Fig. 4(a) depicted by a strongly coordinated state where all group members including *leader*, *followers*, and *startled* fish align their heading to move all-together in the same direction;
- **(b) unstable state** illustrated in Fig. 4(b) where the group collective response is highly unstable and fish group likely to split-up apart and disperse;
- **(c) group split-up and switching leadership** illustrated in Fig. 4(c) where the group split-up with the modelled *leader* and the *followers* tend to coordinate their motion in a direction prescribed by the *startled* fish;
- **(d) group split-up from the *startled* fish** illustrated in Fig. 4(d) where the fish initiating *escape reaction* cannot coordinate their motion with *followers* which tend to align their heading direction with the modelled *leader*.

Note that the states observed above result from simulations always departing from an ordered state where $\iota\delta^2 = 0$ prior to introducing *escape reaction* such that $\iota\delta^2 > 0$ after a few time steps. The values of $\iota\delta^2 > 0$ used in the illustrative cases in Fig. 4 are extracted from the colored area in the colormap in Fig. 3.

5 Discussions and Conclusions

Escape reaction investigated here is an alarm reaction behavior [13,14,29] characterized by fast turns and increased speed. This work shows that a stochastic process [9,24] with biased jumps can be utilized to reproduce a similar behavior observed during *escape reaction* [16]. By introducing small perturbations to the group at the vicinity of a synchronized state where all individuals shared a common heading angle, a closed-form expression has been proposed depicting that, for fixed values of the noise intensity ς, the expected variance of the jumps evaluated as $\iota\delta^2$ can be used to predict the level of order in the system. The proposed closed-form expression has been validated against numerical simulation and shown effective to predict the transition between various states observed.

The numerical results from this work indicate that *escape reaction* can maintain a strong level of coordination for smaller values of $\iota\delta^2$ where the group exhibits a synchronized leader-follower formation. *Escape reaction* can force the group to diverge from the *leader* heading direction and *followers* to align their heading with the *startled* fish. In this case, the sudden and fast turns initiated by the *startled* fish are observed to stir *followers* away from the *leader* prescribed heading direction. This results in leadership switching from the modeled *leader* to the *startled* fish and the new synchronized group moving in a toroidal circular formation pattern as described in the literature [37]. In this scenario, the leader becomes isolated and vulnerable to the predator [12]. *Escape reaction* can also cause a synchronized group to transition to a highly unstable state with a highly volatile polarization value and where individuals can split up and disperse. This situation can result in individuals being isolated and at risk in the presence of a predator. As the expected variance of the jumps become larger, *escape reaction* can force the *startled* fish to swim away from the group which then maintains its initial *leader-followers* synchronized state away from the disturbances introduced by the *startled* fish. This tendency to swim away from the group might put the *startled* fish into a vulnerable position.

The work proposed in this manuscript contrasts with other attempts to model *escape reaction* such as the one in [19] where the behavior is initialized by setting the *startled* fish a velocity significantly larger in amplitude and opposite direction to the averaged group heading. Such a modeling approach results in a propagation wave typical to cascading behaviors observed in nature on large herds [38]. The stochastic model proposed here results instead in circular or toroidal motion patterns typically observed in fish groups [37] in the presence of a predator [12]. The results from this work are thus in line with many prior works where *escape reaction* can initially destabilize a group of fish but can also favor strong group coordination in the long run [13] allowing groups to stay resilient when facing a predator.

The method proposed in this work to recreate *escape reaction* can be utilized in other model of collective behavior to recreate antagonistic interactions or switching leadership [39]. These behaviors are relevant when a group of fish is exposed to a strong stimulus or when the group must negotiate a drastic change of heading direction. The findings from this work also suggest that the expected variance of the turn can be utilized as a single parameter to characterize the level of disturbances introduced by the *startled* fish. This single parameter can be utilized to design appropriate controllers in order to either improve or disrupt group coordination in several unmanned multi-vehicle systems. In future works, the framework considered here can be extended to study group responses to various sources of disturbances resulting into several fish exhibiting sudden and fast turns with different probability distributions.

Acknowledgments. This work was supported by the United States Naval Academy. The author is grateful to Dr Franck Vernerey for useful discussions.

Data statement. All data used in this work are generated from the discrete time approximation (3) of the model in (1) using the parameters reported in Table 1.

References

1. Reebs, S.G., Leblond, C.: Individual leadership and boldness in shoals of golden shiners (Notemigonus crysoleucas). Behaviour **143**(10), 1263–80 (2006)
2. Miller, N., Garnier, S., Hartnett, A.T., Couzin, I.D.: Both information and social cohesion determine collective decisions in animal groups. Proc. Natl. Acad. Sci. **110**(13), 5263–8 (2013)
3. Miller, T.H., Clements, K., Ahn, S., Park, C., Ji, E.H., Issa, F.A.: Social status-dependent shift in neural circuit activation affects decision making. J. Neurosci. **37**(8), 2137–48 (2017)
4. Krause, J., Hoare, D., Krause, S., Hemelrijk, C., Rubenstein, D.: Leadership in fish shoals. Fish Fish. **1**(1), 82–9 (2000)

5. Krause, J., Godin, J.G.J.: Shoal choice in the banded killifish (Fundulus diaphanus, Teleostei, Cyprinodontidae): effects of predation risk, fish size, species composition and size of shoals. Ethology **98**(2), 128–36 (1994)
6. Mwaffo, V., Butail, S., Porfiri, M.: In-silico experiments of zebrafish behaviour: modeling swimming in three dimensions. Sci. Rep. **7**, 39877 (2017)
7. Berdahl, A.M., Kao, A.B., Flack, A., Westley, P.A., Codling, E.A., Couzin, I.D., et al.: Collective animal navigation and migratory culture: from theoretical models to empirical evidence. Philos. Trans. Royal Society B: Biol. Sci.. **373**(1746), 20170009 (2018)
8. Vernerey, F.J., Shen, T., Sridhar, S.L., Wagner, R.J.: How do fire ants control the rheology of their aggregations? A statistical mechanics approach. J. R. Soc. Interface **15**(147), 20180642 (2018)
9. Butail, S., Mwaffo, V., Porfiri, M.: Model-free information-theoretic approach to infer leadership in pairs of zebrafish. Phys. Rev. E **93**(4), 042411 (2016)
10. Nakayama, S., Harcourt, J.L., Johnstone, R.A., Manica, A.: Initiative, personality and leadership in pairs of foraging fish. PLoS ONE **7**(5), e36606 (2012)
11. Strandburg-Peshkin, A., Twomey, C.R., Bode, N.W., Kao, A.B., Katz, Y., Ioannou, C.C., et al.: Visual sensory networks and effective information transfer in animal groups. Curr. Biol. **23**(17), R709-11 (2013)
12. Domenici, P.: The visually mediated escape response in fish: predicting prey responsiveness and the locomotor behaviour of predators and prey. Mar. Freshw. Behav. Physiol. **35**(1–2), 87–110 (2002)
13. Kalueff, A.V., Cachat, J.M.: Zebrafish Models in Neurobehavioral Research. Springer, New York, USA (2011)
14. Speedie, N., Gerlai, R.: Alarm substance induced behavioral responses in zebrafish (Danio rerio). Behav. Brain Res. **188**(1), 168–77 (2008)
15. Egan, R.J., Bergner, C.L., Hart, P.C., Cachat, J.M., Canavello, P.R., Elegante, M.F., et al.: Understanding behavioral and physiological phenotypes of stress and anxiety in zebrafish. Behav. Brain Res. **205**(1), 38–44 (2009)
16. Frith, H., Blake, R.: Mechanics of the startle response in the northern pike. Esox lucius. Canadian J. Zool. **69**(11), 2831–9 (1991)
17. Song, J., Ampatzis, K., Ausborn, J., El Manira, A.: A hardwired circuit supplemented with endocannabinoids encodes behavioral choice in zebrafish. Curr. Biol. **25**(20), 2610–20 (2015)
18. Butail, S., Paley, D.A.: Three-dimensional reconstruction of the fast-start swimming kinematics of densely schooling fish. J. R. Soc. Interface **9**(66), 77–88 (2012)
19. Herbert-Read, J.E., Buhl, J., Hu, F., Ward, A.J., Sumpter, D.J.: Initiation and spread of escape waves within animal groups. Royal Society Open Sci. **2**(4), 140355 (2015)
20. Eaton, R.C.: Neural mechanisms of startle behavior. Springer Science & Business Media (1984)
21. Blumenthal, T.D.: Presidential address 2014: the more-or-less interrupting effects of the startle response. Psychophysiology **52**(11), 1417–31 (2015)
22. Couzin, I.D., Krause, J., Franks, N.R., Levin, S.A.: Effective leadership and decision-making in animal groups on the move. Nature **433**(7025), 513–6 (2005)
23. Couzin, I.D., Ioannou, C.C., Demirel, G., Gross, T., Torney, C.J., Hartnett, A., et al.: Uninformed individuals promote democratic consensus in animal groups. Science **334**(6062), 1578–80 (2011)
24. Mwaffo, V., Anderson, R.P., Butail, S., Porfiri, M.: A jump persistent turning walker to model zebrafish locomotion. J. R. Soc. Interface **12**(102), 20140884 (2015)

25. Vicsek, T., Czirók, A., Ben-Jacob, E., Cohen, I., Shochet, O.: Novel type of phase transition in a system of self-driven particles. Phys. Rev. Lett. **75**(6), 1226–9 (1995)
26. Vicsek, T., Zafeiris, A.: Collective motion. Phys. Rep. **517**(3–4), 71–140 (2012)
27. Mwaffo, V., Butail, S., Porfiri, M.: Analysis of pairwise interactions in a maximum likelihood sense to identify leaders in a group. Front. Robot. AI. **4**, 35 (2017)
28. Mwaffo, V., Vernerey, F.: Analysis of group of fish response to startle reaction. J. Nonlinear Sci. **32**(6), 1–26 (2022)
29. Jesuthasan, S.J., Mathuru, A.S.: The alarm response in zebrafish: innate fear in a vertebrate genetic model. J. Neurogenet. **22**(3), 211–28 (2008)
30. Kloeden, P.E., Platen, E.: A survey of numerical methods for stochastic differential equations. Stoch. Hydrol. Hydraul. **3**(3), 155–78 (1989)
31. Marconi, U.M.B., Puglisi, A., Rondoni, L., Vulpiani, A.: Fluctuation-dissipation: response theory in statistical physics. Phys. Rep. **461**(4–6), 111–95 (2008)
32. Mwaffo, V., Anderson, R.P., Porfiri, M.: Collective dynamics in the Vicsek and vectorial network models beyond uniform additive noise. J. Nonlinear Sci. **25**(5), 1053–76 (2015)
33. Mwaffo, V., Porfiri, M.: Group coordination in a biologically-inspired vectorial network model. In: Proceedings of the 9th EAI International Conference on Bio-inspired Information and Communications Technologies (formerly BIONETICS), pp. 303–310 (2016)
34. Aldana, M., Huepe, C.: Phase transitions in self-driven many-particle systems and related non-equilibrium models: a network approach. J. Stat. Phys. **112**(1–2), 135–53 (2003)
35. Mwaffo, V., Vernerey, F.: Analysis of group of fish response to startle reaction. J. Nonlinear Sci. **32**(6), 96 (2022)
36. Giardina, I.: Collective behavior in animal groups: theoretical models and empirical studies. HFSP J. **2**(4), 205–19 (2008)
37. Couzin, I.D., Krause, J., James, R., Ruxton, G.D., Franks, N.R.: Collective memory and spatial sorting in animal groups. J. Theor. Biol. **218**(1), 1–11 (2002)
38. Couzin, I.D., Krause, J.: Self-organization and collective behavior in vertebrates. Adv. Study Behav. **32**(1), (2003)
39. Mwaffo, V., Keshavan, J., Hedrick, T.L., Humbert, S.: Detecting intermittent switching leadership in coupled dynamical systems. Sci. Rep. **8**(1), 1–20 (2018)

Modeling and Simulation of a Bio-Inspired Nanorobotic Drug Delivery System

Qingying Zhao[1] and Lin Lin[2(✉)]

[1] Changshu Institute of Technology, Suzhou, China
qyzhao@cslg.edu.cn
[2] Tongji University, Shanghai, China
fxlinlin@tongji.edu.cn

Abstract. Current targeted drug delivery systems like passive targeting or active targeting are still inefficient because they mainly depend on blood circulation and extravasation. It is significant that drug delivery carriers are capable of autonomously swimming towards target site (e.g., diseased cells or tumors) and releasing drugs. In recent years, targeted drug delivery depending on autonomous swimmers such as nanorobot has been actively studied and a number of solutions have been proposed. In the paper, we propose a nanorobot-based system comprising of nanorobot behavior planning algorithm, drug reception model and adjusting method of release rate for a simulation of local targeted drug delivery. In this system, nanorobots can move and accumulate at target site by simulating bacterial chemotaxis, and determine the timing of drug release relying on quorum sensing. In addition, nanorobots can dynamically adjust the rate of drug release depending on the concentration of tumor biomarker. A simulation environment is established in order to evaluate the nanorobotic drug delivery system. The simulation results show that the nanorobotic drug delivery system can not only deliver drugs effectively at desired location but also enhance efficiency of drug utilization.

Keywords: Molecular communication · Nanorobotic · Targeted drug delivery

1 Introduction

During the past few decades, controlled drug delivery (CDD) has experienced an enormous upswing and brought a series of highly effective pharmaceutical preparations with low toxicity side effects and good compliance. The evolution of the controlled drug delivery mainly includes three stages from its origins to the present. The first stage was the "macro era" of zero-order "controlled" drug

Supported in part by the National Natural Science Foundation of China under Grant 61971314, in part by the Changshu Institute of Technology Scientific Research Fund under Grant KYZ2019003Q.

Y. Chen et al. (Eds.): BICT 2023, LNICST 512, pp. 64–76, 2023.
https://doi.org/10.1007/978-3-031-43135-7_7

delivery devices. Some CDD devices were designed in macroscopic scale in the early days. For example, the Ocusert designed by Alza Corp. was an ophthalmic insert that released the anti-glaucoma drug at a constant rate in the eye, and the Norplant developed by Population Council, Wyeth is a contraceptive subcutaneous implant comprised of six silicone rubber tubes filled with a contraceptive steroid, levo-Norgestrel which can be released at a constant rate [1]. The second stage is the "micro era" of sustained release, biodegradable microparticle delivery systems. Sustain-controlled release system could improve efficiency of drug utilization and plasma concentration by means of injectable, biodegradable drug-loaded microparticles. The third stage is the "nano era" of targeted drug delivery systems.

For the "nano era" of targeted drug delivery, PEGylation, the EPR effect and passive targeting, active targeting with antibodies, peptides and small molecule, cell-specific ligands are the three key technologies stimulating the development of drug nanocarriers as practical clinical realities. PEGylation is referred to as modifying the surface of nanoparticles (drug nanocarriers) with poly (ethylene glycol) [2]. The circulation time and stability of nanoparticles can be enhanced by the PEGylation in circulatory system [3]. The term "EPR" is abbreviation of enhanced permeation and retention first proposed by Professor Hiroshi Maeda [4,5]. Passive targeting is the process of extravasation drug accumulation in diseased tissues such tumors with leaky vasculature, which is commonly known as the enhanced permeation and retention (EPR) effect [6]. Although the longer systemic circulation time achieved by PEGylation [7] contribute to the EPR effect of drug nanocarriers, only a small percent ($< 10\%$) of administered drug nanocarriers actually reach diseased tissues or tumors [8,9]. If ligands are added to the surface of drug nanocarriers, the passive targeting can be improved through the interactions between drug nanocarriers and target receptors. Unfortunately, the specific interactions usually called active targeting [10,11] can occur only when drug nanocarriers and target receptors are in close proximity [2].

It is significant that drug delivery systems are capable of autonomously swimming towards disease site. Autonomous nanoswimmers or systems have been proposed to transport drug molecules in targeted drug delivery. The evolving nanoswimmers can be divided into biological such as modified bacteria, even sperms and synthetic such as nanorobots or nanomachines [12]. Bacteria have been used to carry therapeutic nanoparticles and diagnostic agent to disease site depending on the property of penetrating tissue, target tumors [13–15]. Nanorobots being able to perform tasks at nano-level could be used to travel through human blood vessels and microvasculature. For example, a chemical communication algorithm was proposed by Cavalcanti et al. to coordinate nanorobots to reach tumor site [16]. In addition, nanorobots carrying drugs can be directly injected into the target site of a patient body. After entering into the disease site, nanorobots would release drug molecules to treat disease cells. Since the drug molecules are usually expensive or where lost molecules may cause undesired side effects such as drug overuse or multi-drug resistance [17,18], the timing and rate of drug release become important issues.

Quorum sensing (QS) is a well-known cell-cell communication mechanism in bacteria [19–21]. Bacteria can use quorum sensing to coordinate timing of specific actions. Bacteria start quorum sensing behavior when the concentration of a specific molecules released by them in extracellular environment achieves a specific threshold. For example, when the bioluminescent marine bacterium Vibrio fischeri was alone, that is when they were in dilute suspension, they made no light [22]. When bacteria aggregated to grow to a certain cell number, all the bacteria turned on light simultaneously. Why they can do it is that they can talk to each other with a chemical language called autoinducer (AI). The autoinducers are acylated homoserine lactones (AHL) that regulate positively the lux operon in Vibrio fischeri. The lux can control bioluminescence and upregulate the expression of the AHL-synthase LuxI which produces the AI molecule. LuxR's dimerized complex with AHL can lead to transcriptional activation of LuxI. The concentration of the AHL is dependent not only on its production, but also on the permeability from outside the cell. Namely, the switch of QS is dependent on the AHL concentration just outside the cell. Since the AHL concentration outside is proportional to the number of bacteria, bacteria start quorum sensing when the bacteria density reaches a specific threshold.

This paper proposes a nanorobot-based drug delivery system comprising of Nanorobot Behavior Planning (NBP) algorithm, drug reception model and adjusting method of release rate in order to deliver drugs effectively at local of human body. In the system, nanorobots can accumulate at the target location and determine the timing of drug release relying on quorum sensing. In addition, the NBP algorithm can dynamically adjust the rate of drug release based on the concentration of tumor biomarker.

The rest of the paper is organized as follows. Section 2 presents the NBP algorithm based on bacterial chemotaxis and quorum sensing. The quorum sensing threshold is estimated in Sect. 3. Section 4 presents drug reception model and adjustable drug release based on the concentration of tumor biomarker. In Sect. 5, simulations are conducted and the obtained results are analyzed. Finally, in Sect. 6, the paper is conducted with a summary of results.

2 NBP Algorithm Based on Bacterial Chemotaxis and Quorum Sensing

In this section, the Nanorobot Behavior Planning (NBP) algorithm based on bacterial chemotaxis and quorum sensing is proposed. Firstly, nanorobots make chemotactic movement and determine whether to reach the tumor target site according to the concentration of tumor-related biomarker. Secondly, nanorobots release autoinducers (AI) when reaching tumor target site. When the concentration of AI reaches a specific threshold, nanorobots start release drug molecules. Finally, it is assumed that the concentration of tumor-related biomarker is proportional to the number of tumor cells. Nanorobots adjust drug release rate based on the biomarker concentration.

Table 1. Notations used in NBP algorithm.

Symbol	Description
i	Index of nanorobots
j	Index of motion steps (time)
C	Unit motion size
θ	Nanorobot location
φ	Nanorobot motion direction
T_t	Threshold of tumor biomarker concentration
J	Concentration of tumor biomarker
Q_A	Constant release rate of AI
τ	AI concentration
τ_{AIh}	High-threshold of AI concentration
τ_{AIl}	Low threshold of AI concentration (Quorum sensing threshold)
Q_D	Adjustable release rate of drug molecules

We formulate the NBP algorithm using the notations shown in Table 1. In the algorithm, each nanorobot is considered as a bacterium with specified chemotaxis and quorum sensing behavior. Let $i \in N$ represent each nanorobot. Let j be the index of time step of nanorobots. Let $\theta^i(j)$ be the location of the ith nanorobot at the jth motion step. Let $J(i, j)$ denote the concentration of tumor biomarker at the location $\theta^i(j)$ of the ith nanorobot. Let $C(i) > 0$ denote a basic motion size that we will use to define the step length during the ith nanorobot moves. Let $\varphi(j)$ denote a unit length random motion direction after a unit motion step. In particular, we let

$$\theta^i(j+1) = \theta^i(j) + C(i)\varphi(j) \tag{1}$$

represent a new position after a motion step of a nanorobot. Let T_t denote a specific threshold of tumor-related biomarker, and

$$J(i, j) \geq T_t \tag{2}$$

indicate that a nanorobot has reached the tumor target site. If a nanorobot detects tumor biomarker and its concentration is less than T_t, the nanorobot will move to a new position of higher concentration. That is, if $J(i, j)$ at position $\theta^i(j)$ is larger than that at $\theta^i(j-1)$, a motion step of size $C(i)$ in the same direction $\varphi(j-1)$ will be taken. Otherwise, the nanorobot moves a step with a random new direction $\varphi(j)$. If a nanorobot detects the concentration of tumor biomarker that meets to Eq. (2), it indicates that the nanorobot has reached the tumor target site. The nanorobot will release AI with a constant rate Q_A and detect the concentration of AI. Let $\tau(i, j)$ denote the concentration of AI at position $\theta^i(j)$ of the ith nanorobot. If the concentration of AI is larger than the high-threshold τ_{AIh}, it indicates that population of nanorobots in tumor target site is too many, and the nanorobot will move away from the tumor target site. That is, if $J(i, j)$ at position $\theta^i(j)$ is less than that at position $\theta^i(j-1)$, a motion

Algorithm 1: NBP Algorithm

```
   Input: i, j, J(i,j), τ(i,j), T_t, T_AIh, T_AIl
   Output: Nanorobot behavior, Drug release rate Q_D
 1 for i=1:N do
 2   if J(i,j) < T_t && J(i,j) > J(i,j-1) then
 3     θ^i(j+1) = θ^i(j) + C(i)φ(j-1) ;       // Move a step at the same
       direction
 4   end
 5   if J(i,j) < T_t && J(i,j) ≤ J(i,j-1) then
 6     θ^i(j+1) = θ^i(j) + C(i)φ(j) ; // Move a step at a random direction
 7   end
 8   if J(i,j) ≥ T_t && τ(i,j) > T_AIh then
 9     if J(i,j) ≤ J(i,j-1) then
10       θ^i(j+1) = θ^i(j) + C(i)φ(j-1);
11     end
12     if J(i,j) > J(i,j-1) then
13       θ^i(j+1) = θ^i(j) + C(i)φ(j);
14     end
15   end
16   if J(i,j) ≥ T_t && τ(i,j) ≤ T_AIh then
17     Δτ(i,j) = Q_A;
18   end
19   if J(i,j) ≥ T_t && T_AIl ≤ τ(i,j) ≤ T_AIh then
20     Drug release rate = Q_D;
21   end
22 end
```

step of size $C(i)$ in the same direction $\varphi(j-1)$ will be taken. Otherwise, the nanorobot will move a step following Eq. (1) with a random new direction. If the concentration of AI is less than the high-threshold T_{AIh}, the nanorobot at position $\theta^i(j)$ will release $\Delta\tau(i,j)$ doses of AI. $\Delta\tau(i,j)$ is an increment of AI, which can be expressed as Eq. (3).

$$\Delta\tau(i,j) = \begin{cases} Q_A, & \tau(i,j) \leq T_{AIh}, J(i,j) \geq T_t \\ 0, & \text{otherwise.} \end{cases} \tag{3}$$

The update of AI concentration is shown as Eq.(4).

$$\frac{\partial \tau}{\partial t} = D_A \nabla^2 \tau, \tag{4}$$

where D_A is diffusion coefficient of AI, ∇^2 Laplacian for the dimension considered. If the concentration of AI $\tau(i,j)$ is less than T_{AIh} and larger than the threshold T_{AIl}, the nanorobot will release drug molecules with rate Q_D based on the concentration of tumor-related biomarker. The pseudo code of the algorithm is described as Algorithm 1.

In the algorithm, the calculation of quorum sensing threshold T_{AIl} and the design of adjustable release rate Q_D are described at length in Sect. 3 and Sect. 4 respectively.

3 Calculation of Quorum Sensing Threshold

After AI molecules were released by nanorobots, the concentration of AI reaches a peak in an instant. AI molecules diffuse in the direction of low concentration as time goes on [23]. According to Fick's laws of diffusion, if a nanorobot releases Q AI molecules at time instant $t = 0$, the molecular concentration at any point in space is given by [24]:

$$c(r,t) = \frac{Q}{(4\pi D_A t)^{3/2}} \exp\left(\frac{-r^2}{4D_A t}\right), \tag{5}$$

where D_A is the diffusion coefficient of the AI molecules, t is time and r the distance from the nanorobot location. For AI continuous emission (i.e., releasing a train of bursts of size Q spaced by a period Δt) of single nanorobot, the molecular concentration at any point in space can be derived:

$$\begin{aligned} c_c(r,t) &\approx \frac{1}{\Delta t}\int_0^t c(r,\tau)d\tau \\ &= \int_0^t \frac{Q}{(4\pi D\tau)^{3/2}} \exp\left(-\frac{r^2}{4D\tau}\right)d\tau \\ &= \frac{Q}{\Delta t 4\pi D_A r} erfc\frac{r}{(4D_A t)^{\frac{1}{2}}}. \end{aligned} \tag{6}$$

When time t is large enough, we can obtain

$$c_c(r,t) \approx \frac{Q}{\Delta t 4\pi D_A r}. \tag{7}$$

When Δt is unit time (i.e., Δt =1), Eq.(7) becomes

$$c_c(r,t) \approx \frac{Q}{4\pi D_A r}. \tag{8}$$

For AI continuous emission of multi-nanorobot, we can apply the superposition principle (i.e., the addition of two received emissions will yield the same concentration than the reception of the addition of two emissions) to calculate AI molecular concentration at any point in space because the AI concentration is relatively low and in an scenario devoid of external forces (i.e., free diffusion) [24,25]. It is assumed that N nanorobots are distributed randomly in three-dimensional spherical space and form a cluster. Each nanorobot i locating at $\boldsymbol{p}_i$ releases AI molecules with a constant rate (Q molecules per unit time) and the concentration of AI molecules is low enough. It is assumed that each nanorobot i is at a distance $d_i = |\boldsymbol{p} - \boldsymbol{p}_i|$ of the evaluated point p. From Eq.(8), we can obtain AI concentration at $\boldsymbol{p}$

$$c_{multi}(\boldsymbol{p}) = \sum_{i=1}^{N} \frac{Q}{4\pi D d_i} = \frac{Q}{4\pi D} \sum_{i=1}^{N} \frac{1}{|\boldsymbol{p} - \boldsymbol{p}_i|}. \tag{9}$$

To simplify solving Eq.(9), we learn from the method in [26]. It is assumed that the nanorobots are arranged in a perfect tridimensional grid so that nanorobot positions are deterministic and symmetric with respect to the center origin $O(0, 0, 0)$. Therefore, the AI concentration at origin $O(0, 0, 0)$ can be calculated from Eq.(9):

$$c_{multi}(\boldsymbol{O}) = \frac{Q}{4\pi D} \sum_{i=1}^{N} \frac{1}{|\boldsymbol{p}_i|}, \tag{10}$$

and the AI concentration at a distance $r > 0$ from origin is

$$c_{multi}(\boldsymbol{p}_r) = \frac{Q}{4\pi D} \sum_{i=1}^{N} \frac{1}{|\boldsymbol{p}_r - \boldsymbol{p}_i|}. \tag{11}$$

Due to the distribution of perfect tridimensional grid, it can be concluded $c_{multi}(\boldsymbol{O}) > c_{multi}(\boldsymbol{p}_r)$, that is, the AI concentration decreases as the distance increases from origin $O(0, 0, 0)$. In addition, it has been demonstrated that the AI concentration of any evaluated point calculated from random distribution of nanorobots approaches the one of perfect tridimensional grid [26]. Hence, in order to ensure an activation of all nanorobots in quorum sensing, it is necessary that the quorum sensing threshold

$$T_{AI1} \geq c_{multi}(\boldsymbol{p}_R) = \frac{Q}{4\pi D} \sum_{i=1}^{N} \frac{1}{|\boldsymbol{p}_R - \boldsymbol{p}_i|}, \tag{12}$$

where the point $\boldsymbol{p}_R$ locates at the edge of the three-dimensional spherical space, i.e., R is the radius of the three-dimensional spherical space.

4 Drug Reception Model and Adjustable Release Rate

Many enzyme inhibitor drugs (EID), such as competitive inhibitor drugs, are noncovalent, reversible inhibitors [27]. Being similar to the substrate in structure, the kind of the inhibitor drugs can compete with the binding of substrates for the enzyme. The drug efficacy is produced as long as the drug is combined with the enzyme (producing E·EID). When the concentration of EID diminishes because of metabolism, the concentration of complex E·EID diminishes and the drug efficacy decreases. Administration of the drug several times a day is necessary to maintain the drug efficacy. In the section, the mechanism-based enzyme inactivator is adopted in drug reception process based on the Michaelis Menten enzymatic kinetics, which is formulated as Eq.(13).

$$E + EID \underset{k_{off}}{\overset{k_{on}}{\rightleftharpoons}} E \cdot EID \xrightarrow{k_2} E \cdot EID' \tag{13}$$

where k_{on}, k_{off} are the reaction rate constant, k_2 is turnover number in reaction, E·EID' an unreactive compound with the properties of specificity and low toxicity, EID the enzyme inhibitor drugs. The update of the concentration [EID] follows Fick's second law

$$\frac{\partial [EID]}{\partial t} = D_D \nabla^2 [EID] \tag{14}$$

where D_D is the diffusion coefficient of the enzyme inhibitor drug, ∇^2 Laplacian for the dimension considered. It is assumed that each tumor cell corresponds to an enzyme (e.g., tyrosine kinase) and it die when the enzyme is inactivated. If the total amount of enzyme E is constant, increasing drug release rate decreases the efficiency while results in higher reaction rate [28]. It is obvious that constant drug release rate decreases the efficiency while results in lower reaction rate when the total amount E decreases (i.e., the number of tumor cells decreases). It is necessary to adjust drug release rate in order to maintain efficiency. It is assumed that the molecular concentration of tumor biomarker is proportional to the number of tumor cells. The adjustable drug release rate based on the concentration of tumor biomarker is formulated as Eq.(15)

$$Q_D = \frac{L\,J(i,j)}{K + J(i,j)} \tag{15}$$

where K, L are constants associated with the release rate, $J(i,j)$ the concentration of tumor biomarker.

5 Simulations and Results

In the section, we in order to establish our simulations. First, we establish the simulation environment and configure the parameters in simulation. The next involves simulation results and analysis.

5.1 Simulation Setup

In this subsection, a three-dimensional space called simulation space is established with volume of $x \times y \times z$ = 100 patches × 100 patches × 100 patches, where patch is length unit in simulation. The origin of the simulation space is at $O(x, y, z) = O(0, 0, 0)$ and the coordinate range is $(-50 \leq x \leq 50, -50 \leq y \leq 50, -50 \leq z \leq -50)$. N nanorobots emerge randomly from $S(-40, -40, -40)$ in the simulation space with random motion directions. It is assumed that

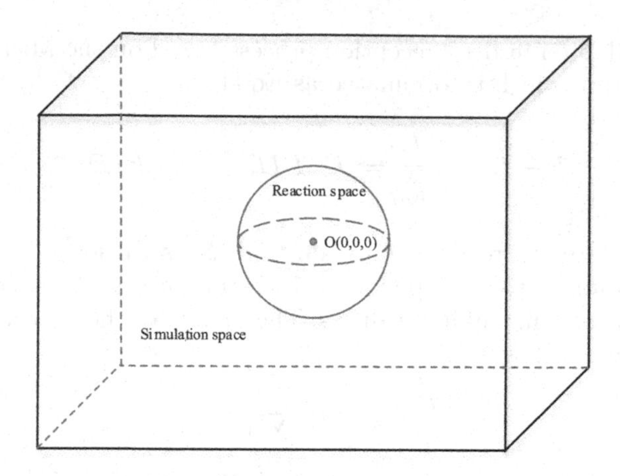

Fig. 1. Simulation and reaction space.

Table 2. Simulation parameters.

Symbol	Description	Value
N	Number of nanorobots	123
R	Radius of reaction space	3 patch
s	Nanorobot motion speed	0.2 patch/tick
Q_A	Release rate of AI	5 molecules/tick
D_A	Diffusion coefficient of AI	1 patch^2/tick
T_{AIl}	Quorum sensing threshold (Low threshold of AI)	15 molecules/patch^3
T_{AIh}	High threshold of AI	60 molecules/patch^3
K	Release rate constant	200
L	Release rate constant	500
D_D	Diffusion coefficient of drug	1 patch^2/tick
n_T	Number of tumor cells	1500
n_E	Number of enzymes	1500
k_2	Turnover number	0.02/tick
k_{on}	Reaction rate constant	0.2/tick
k_{off}	Reaction rate constant	0.03/tick

tumor cells are distributed randomly in a three-dimensional space named reaction space. The reaction space is a sphere of radius $R = 3$ patches and its center locates at $O(0, 0, 0)$ (see Fig. 1). The diffusion of tumor biomarker follows Fick's second law and Eq.(7), that is, the concentration of tumor biomarker is inversely proportional to the distance from the tumor target site.

Nanorobots simulate bacterial chemotaxis to approach reaction space with speed $s = 0.2$ patches/tick, where tick is time unit in simulation. Nanorobots release AI molecules as long as entering into reaction space. When AI molecules

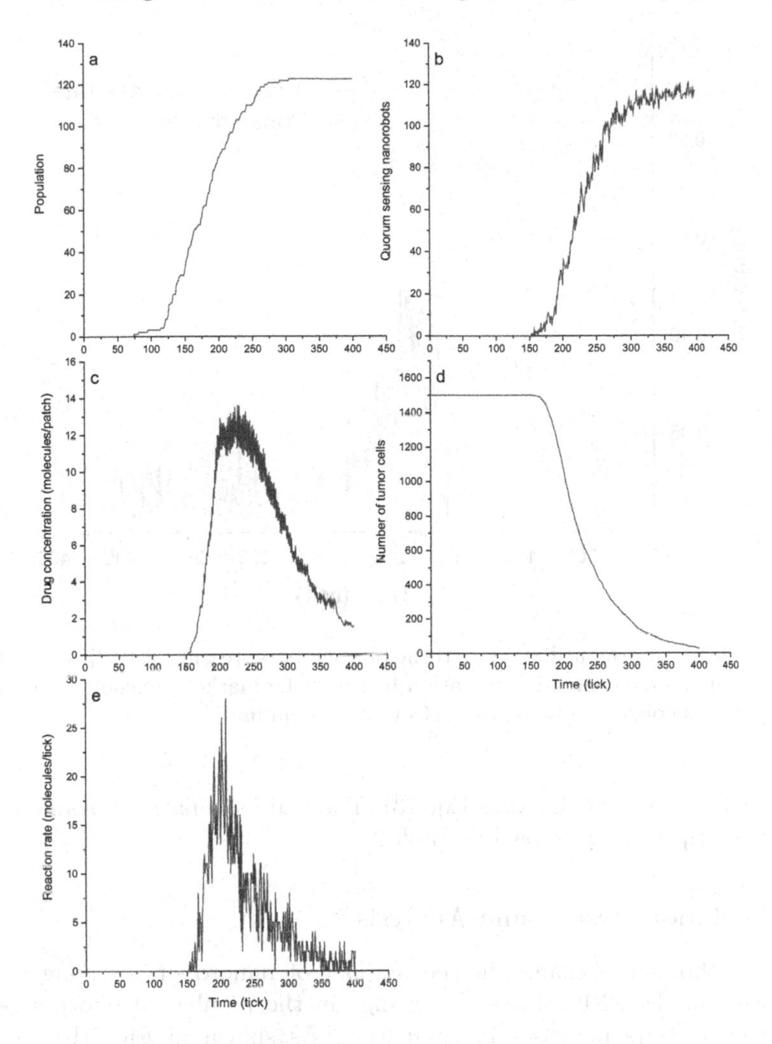

Fig. 2. Variation in simulation of drug delivery system.

exit simulation space, they are removed from the simulation, in order to limit the computational burden of the simulation. Nanorobots report the number of molecules found within the patch which the nanorobot locates at in each time step (tick). It is considered that molecular concentration is constant over such volume of V =1 patch^3 due to small dimensions, and nanorobots are arranged in a perfect tridimensional sphere grid with radius $R = 3$ patches and center coordinate $O(0, 0, 0)$, and the distance between adjacent nanorobots is $l = 1$ patch. Hence, the molecular concentration sensed by a nanorobot at a patch of volume V located at a distance r from the center of reaction space can be calculated by Eq.(11). The threshold of quorum sensing T_{AII} of nanorobots can be obtained from Eq.(12). The drug reception in reaction space follows Michaelis Menten

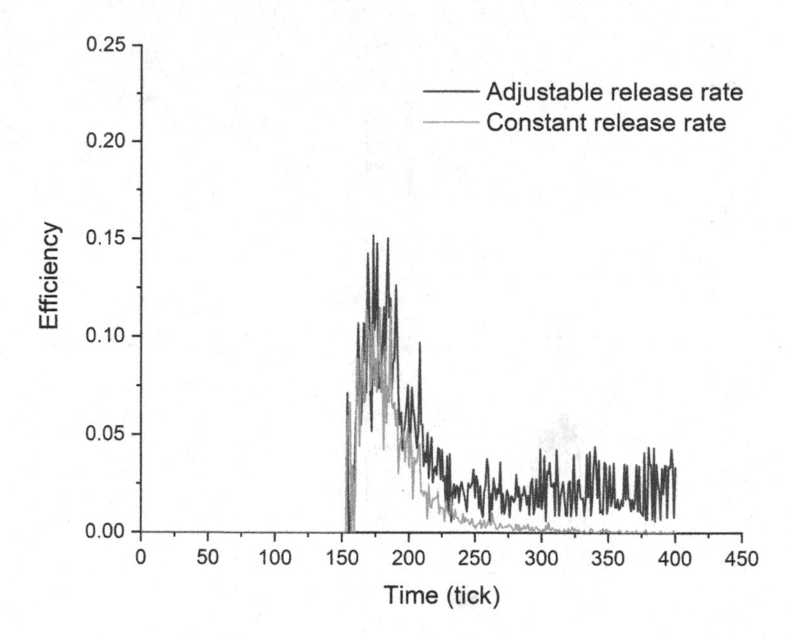

Fig. 3. Efficiency of drug utilization: (a) the blue curve represents the adjustable release rate based on the concentration variation of tumor biomarker molecules, (b) the red curve represents constant release rate. (Color figure online)

enzymatic kinetics formulated as Eq.(13). The main simulation parameters and physical descriptions are shown in Table 2.

5.2 Simulation Results and Analysis

Figure 2(a) shows the change in the number of nanorobots arriving in reaction space, and Fig. 2(b) shows the change in the number of quorum sensing nanorobots as time increases in simulation. As shown in Fig. 2(b), the first nanorobot reaches quorum at time $t = 153$ ticks, all nanorobots reach quorum at about $t = 400$ ticks, and the majority of nanorobots reach quorum in the time of $t = 200 \sim 275$ ticks. Obviously there is a delay from the first nanorobot to all nanorobots reaching quorum. This is because the concentration of AI molecules decreases as the distance increases from the center of reaction space, that is, when the nanorobots close to the center reach quorum, those close to the edge do not yet reach quorum due to the low AI concentration below the threshold T_{AI}. Figure 2(c) shows the variation of drug concentration as time increases. The drug concentration detected by a nanorobot is the number of drug molecules in a *patch*[2] where the nanorobot locates. From Figs. 2(a), (b) and (c), we observe that nanorobots rather release drugs until they reach quorum than as soon as they reach the reaction space. Figure 2(d) shows the change in the number of tumor cells as time increases. If release rate is constant, decreasing number of tumor cells decreases the reaction rate while results in lower efficiency according to enzymatic reaction kinetics formulated as Eq. (13). Figure 2(e) shows the variation of reaction rate as time increases in simulation. The reaction rate measured is the number of unreactive compound E·EID$'$ produced by reactions of drugs and enzymes per tick.

Figure 3 shows the variation of efficiency in simulation. The efficiency is the reaction rate divided by the sum of release rates of nanorobots reaching quorum. As shown in Fig. 3, the blue line represents the variation of efficiency under the adjustable release rate based on the concentration of tumor biomarker, while the red represents the variation under the constant release rate $Q_D = 30$ molecules/(2 ticks). Obviously the adjustable release rate enhances the efficiency compared with that constant release rate, thus saving drug and preventing drug overuse.

6 Conclusions

In this paper, we proposed a nanorobot-based system comprising of nanorobot behavior planning (NBP) algorithm, drug reception model and adjusting method of release rate. In NBP algorithm, each nanorobot is considered as a bacterium with specified chemotaxis and quorum sensing behavior. After reaching the tumor location by simulating bacterial chemotaxis, nanorobots determine the timing of drug release relying on quorum sensing to release drugs. The quorum threshold is calculated that ensure the quorum sensing of all of the nanorobots in reaction space. Under the drug reception based on enzymatic reaction kinetics, an adjustable rate of drug release is designed based on the concentration of tumor biomarker. Finally, we established the simulation environment reflecting the property of concentration, diffusion-rate of AI and drug molecles. The simulation results show that the nanorobot-based system not only enables nanorobots to reach target site and release drug molecules after they reach quorum, but also enhances the efficiency of drug utilization.

References

1. Hoffman, A.S.: The origins and evolution of "controlled" drug delivery systems. J. Control. Release **132**(3), 153–163 (2008)
2. Bae, Y.H., Park, K.: Targeted drug delivery to tumors: myths, reality and possibility. J. Control. Release **153**(3), 198–205 (2011)
3. Davis, F.F.: The origin of PEGnology. Adv. Drug Deliv. Rev. **54**(4), 457–458 (2002)
4. Maeda, H., Ueda, M., Morinaga, T., Matsumotog, T.: Conjugation of poly(styrene-co-maleic acid) derivatives to the antitumor protein neocarzinostatin: pronounced improvements in pharmacological properties. J. Med. Chem. **28**(4), 455–461 (1985)
5. Matsumura, Y., Maeda, H.: A new concept for macromolecular therapeutics in cancer chemotherapy:mechanism of tumoritropic accumulation of proteins and the antitumor agent smancs. Can. Res. **46**, 6387–6392 (1986)
6. Maeda, H., Matsumura, Y.: EPR effect based drug design and clinical outlook for enhanced cancer chemotherapy. Adv. Drug Delivery Rev. **63**(3), 129–130 (2011)
7. Harris, J.M., Martin, N.E., Modi, M.: Pegylation -a novel process for modifying pharmacokinetics. Clin. Pharmacokinet. **40**, 539–551 (2001)

8. Kizelsztein, P., Ovadia, H., Garbuzenko, O., Sigal, A.: Pegylated nanoliposomes remote-loaded with the antioxidant tempamine ameliorate experimental autoimmune encephalomyelitis. J. Neuroimmunol. **213**(1–2), 20–25 (2009)
9. Taurin, S., Nehoff, H., Greish, K.: Anticancer nanomedicine and tumor vascular permeability; where is the missing link? J. Contr. Release Off. J. Controll. Release Society **164**(3), 265–275 (2012)
10. Beduneau, A., Saulnier, P., Hindre, F., Clavreul, A., Leroux, J.C., Benoit, J.P.: Design of targeted lipid nanocapsules by conjugation of whole antibodies and antibody Fab' fragments. Biomaterials **28**(33), 4978–4990 (2007)
11. Deckert, P.M.: Current constructs and targets in clinical development for antibody-based cancer therapy. Curr. Drug Targets **10**(2), 158–175 (2009)
12. Zoabi, N., et al.: The evolution of tumor-targeted drug delivery: from the EPR effect to Nanoswimmers. Isr. J. Chem. **53**, 719–727 (2013)
13. Baban, C.K., Cronin, M., O'Hanlon, D., O'Sullivan, G.C., Tangney, M.: Bacteria as vectors for gene therapy of cancer. Bioengineered Bugs **1**(6), 385–394 (2010)
14. Akin, D., Sturgis, J., Ragheb, K., Sherman, D., Burkholder, K.: Bacteria-mediated delivery of nanoparticles and cargo into cells. Nat. Nanotechnol. **2**(7), 441–449 (2007)
15. Liu, Y., et al.: Bacteria-mediated in vivo delivery of quantum dots into solid tumor. Biochem. Biophys. Res. Commun. **425**(4), 769–774 (2012)
16. Cavalcanti, A., Hogg, T., Shirinzadeh, B., Liaw, H.C.: Nanorobot communication techniques: a comprehensive tutorial. In: 9th International Conference on Control, Automation, Robotics and Vision, pp. 1–6. Singapore (2006)
17. Tredan, O., Galmarini, C.M., Patel, K., Tannock, I.F.: Drug resistance and the solid tumor microenvironment. J. Natl Cancer Inst. **99**(19), 1441–1454 (2007)
18. Tannock, I., Lee, C., Tunggal, J.: Limited penetration of anticancer drugs through tumor tissue a potential cause of resistance of solid tumors to chemotherapy. Clin. Cancer **8**, 878–884 (2002)
19. Bassler, B.L., Losick, R.: Bacterially speaking. Cell **125**(2), 237–246 (2006)
20. Cobo, L.C., Akyildiz, I.F.: Bacteria-based communication in nanonetworks. Nano Commun. Netw. **1**(4), 244–256 (2010)
21. Akyildiz, I.F., Brunetti, F., Blazquez, C.: Nanonetworks: a new communication paradigm. Comput. Netw. **52**(12), 2260–2279 (2008)
22. Miller, M.B., Bassler, B.L.: Quorum sensing in bacteria. Annu. Rev. Microbiol. **55**(1), 165–199 (2001)
23. Philibert, J.: One and a half century of diffusion: Fick, Einstein, before and beyond. Diff. Fundamentals **4**(6), 1–19 (2006)
24. Bossert, W.H., Wilson, E.O.: The analysis of olfactory communication among animals. J. Theor. Biol. **5**(3), 443–469 (1963)
25. Shahmohammadian, H., Messier, G.G., Magierowski, S.: Optimum receiver for molecule shift keying modulation in diffusion-based molecular communication channels. Nano Commun. Netw. **3**(3), 183–195 (2012)
26. Abadal, S., Llatser, I., Alarcón, E., Cabellos-Aparicio, A.: Cooperative signal amplification for molecular communication in nanonetworks. Wirel. Netw. **20**(6), 1611–1626 (2014)
27. Silverman, R.B.: Mechanism-based enzyme inactivators. In: Purich DL Contemporary enzyme kinetics and mechanism. Academic Press, Burlington (2009)
28. Zhao, Q., Li, M., Luo, J.: Relationship among reaction rate, release rate and efficiency of nanomachine-based targeted drug delivery. Technol. Health Care **25**(6), 1119–1130 (2017)

Cooperative Relaying in Multi-hop Mobile Molecular Communication via Diffusion

Zhen Cheng(✉), Zhichao Zhang, and Jie Sun

Zhejiang University of Technology, Hangzhou 310023, China
chengzhen@zjut.edu.cn

Abstract. Molecular communication via diffusion (MCvD), in which molecules are used to transmit information by the movement of diffusion, is one of the most prominent systems in nanonetworks. In particular, the research on end-to-end mobile MCvD system is even more challenging. In this paper, we investigate the error probability of three dimensional (3D) multi-hop mobile MCvD system by proposing two relay schemes including multi-molecule-type (MMT) and single-molecule-type (SMT). Under MMT, the mathematical expression of optimal detection threshold can be derived. Especially under SMT, we propose the adaptive detection threshold method to alleviate the self-interference caused by the information molecules with the same type. Based on the two relay schemes, the mathematical expressions of error probability of this system are derived. Numerical results show the impacts of different parameters on the error probability performance.

Keywords: Molecular communication via diffusion · mobile · multi-hop · cooperative relaying

1 Introduction

The cooperative molecular communication via diffusion (MCvD) [1,2] attracts the attentions of many researchers because it has promising applications in biological environment by the cooperation of nodes, such as drug delivery [3]. In the cooperative MCvD, the researchers presented some relaying schemes. For example, Ahmadzadeh et al. [4] proposed a decode-and-forward (DF) relay strategy to improve the communication range of static multi-hop MCvD. Tiwari et al. [5] evaluated a static two-hop link by using an estimate-and-forward (EF) relaying scheme at relay nodes.

Recently, the scenarios of mobile MCvD [6,7], in which the transmitter and receiver nanomachines are mobile are more practical in biological environments compared with static MCvD. The mobile biological nanomachines in the fluid medium can be regarded as the components of nano-robots, which is expected to be applied in tracking specific targets [8]. In such a scenario, one source nano-robot can transmit the received signal to the destination nano-robot through the relay nano-robots. Thus, how to establish reliable and efficient communication between the nano-robots becomes an important research problem. In 2018, Ahmadzadeh et al. in [9] investigated the channel impulse response (CIR) of a single link mobile MCvD in the time-variant stochastic channels.

Y. Chen et al. (Eds.): BICT 2023, LNICST 512, pp. 77–90, 2023.
https://doi.org/10.1007/978-3-031-43135-7_8

In 2019, Varshney [10] analyzed the end-to-end probability of error and channel achievable rate by considering multiple cooperative nanomachine-assisted mobile MCvD in 1D flow channel. Cao et al. [11] performed a stochastic analysis of the CIR of a 3D diffusive mobile MC system with active absorbing nanomachines, and the obtained analysis results can be used for drug delivery problems with incomplete channel state information. In 2020, Huang et al. [12] studied the estimation of the initial distance in the point-to-point mobile MCvD system, and used the estimated initial distance to achieve signal detection. In 2021, Shrivastava et al. [13] analyzed the performance of one single link mobile MCvD system by implementing the neural networks to detect the received bits of time-varying parameters under different signal modes. Cheng et al. [14] adopted an amplify-and-forward (AF) relaying scheme to investigate the impacts of the amplification factor on the performance of mobile multi-hop MCvD. In 2022, the distance estimation and power control method for one-hop [15] and the Brownian motion of molecules in the multiuser mobile MCvD system [16] were studied.

However, the existed works [1–5] mainly focused on the static MCvD system with fixed transmitter nanomachine and fixed receiver nanomachine. But in mobile MCvD system, due to the random movements of the transmitter and receiver nano-machines, the statistics of CIR change over time. Therefore, the CIR in the mobile scenario is more complicated compared with the static scenarios. This motivates us to analyze the influence of the mobilities of nanomachines on the performance of mobile MCvD system. Second, the works [9–16] mainly focused on the research of the one-hop mobile MCvD system. In particular, the work [10] only studied 1D multi-hop mobile MCvD system with one relay scheme. The work [14] used AF relay scheme, Normal distribution and the maximum-a-posterior (MAP) detection method. Therefore, based on the above considerations, we investigate the mobilities of nodes and different relay schemes which have important impacts on the 3D multi-hop mobile MCvD system by using adaptive detection threshold. First, we propose two relay schemes including multi-molecule-type (MMT) and single-molecule-type (SMT). Considering each relay scheme, the adaptive detection threshold methods for multi-hop topology are presented. Then the mathematical expressions of the error probability of multi-hop networks under the two relay schemes are derived. Finally, the numerical results show that the parameters including the initial distance between two adjacent nodes, the number of molecules released in each time slot, the time slot duration, and the detection threshold schemes have impacts on the error probability performance of the 3D multi-hop mobile MCvD system.

The remainder of this paper is organized as follows. In Sect. 2, the multi-hop mobile MCvD system model is introduced. The error probability performances of this system using two relay schemes are evaluated in Sect. 3 and Sect. 4. Section 5 presents the numerical results. Section 6 concludes the paper.

2 System Model

The system model of mobile multi-hop MCvD network is consisted of a transmitter (node S), a receiver (node D), and Q relays (node $R_k, k = 1, 2, ..., Q$). Node S and node D are placed at locations (0, 0, 0) and (x_D, 0, 0) in a 3D space, respectively. And the Q relay nodes are equally spaced between node S and node D along the x-axis. It is assumed that nodes D and R_k are spherical with same radius ($r_D = r_{R_k}$) and volumes ($V_D = V_{R_k}$), and that they are passive observers. Also, it is assumed that the mobile nodes perform independent random walks. The walk path consists of a succession of random steps.

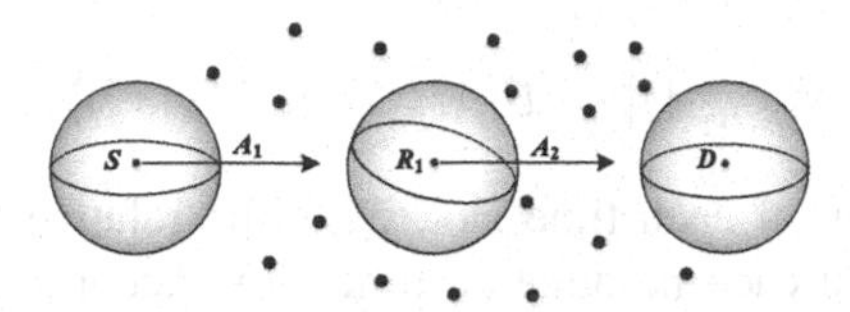

Fig. 1. The two-hop mobile MCvD network with the first relay scheme.

We adopt ON/OFF keying (OOK) modulation method which is commonly-used in the MCvD literature [17]. We assume that all nodes can be perfectly synchronized in terms of time [18]. Each relay node has full-duplex transmission and utilizes decode-and-forward strategy. Two relay schemes are used at the relay nodes including MMT and SMT. For the two-hop network $S \to R_1 \to D$ with the first relay scheme in Fig. 1, it is assumed that the movement of the nanomachines is not severe, meaning that the routing $S \to R_1 \to D$ does not change. Node S and relay node R_1 emit type A_1 molecules and type A_2 molecules, respectively. Each node has its diffusion coefficient $D_l (l = S, R_k, D)$. The nodes communicate as follows. At the beginning of the j-th time slot, node S transmits information bit which is 0 or 1, and node R_1 concurrently transmits the detected bit to node D. When the $(j+1)$-th time slot ends, node D determines the received information. For the second relay scheme, the molecules with same type are released by node S and node R_1.

In the propagation process, when both S and node D are mobile, the CIR $h(t, \tau_s)$ represents the probability that one information molecule emitted by node S is observed by node D at time τ_s [9] as follows:

$$h(t, \tau_s) = \frac{V_D}{(4\pi D_1 \tau_s)^{\frac{3}{2}}} \exp\left(-\frac{d^2(t)}{4D_1\tau_s}\right), \quad (1)$$

where $V_D = \frac{4}{3}\pi r_D^3$ is the volume of node D. r_D is the radius of node D. $D_1 = D_A + D_{R_1}$. D_A and D_{R_1} are the diffusion coefficients of the type A molecules and node R_1, respectively. $d(t)$ is the distance between node S and node D at time t. τ_s represents the relative time of t.

3 Analysis of Cooperative Relaying in MMT Multi-hop Network

In this section, we introduce the relay scheme which using MMT in each time slot to make analysis of error probability for multi-hop mobile MCvD network.

3.1 Two-Hop Network

Let M_S^j denote the number of information molecules released by node S in time slot j. β_S^i is the probability of transmission of bit 1 for node S at the beginning of time slot j. The number of molecules released by node S and received by node R_1 both in the current time slot j labelled as $N_{(S,R_1)}^C[j]$ follows a binomial distribution

$$N_{(S,R_1)}^C[j] \sim \mathcal{B}(M_S^j, \beta_S^i h(jT_S, \tau_s)), \tag{2}$$

where T_S is the duration of each time slot. τ_s is the relative time of the molecules released at the node S in each modulation time slot. According to the expression of CIR in formulas (2), $h(jT_S, \tau_s)$ corresponds to the receiving probability of a molecule for the link from node S to node R_1 in the current time slot. In addition, $\mathcal{B}$ represents the binomial distribution.

If M_S^j is large enough and the value of $h(jT_S, \tau_s)$ is around 0.1, the binomial distribution in Eq. (2) can be approximated by a Poisson distribution which is expressed as

$$N_{(S,R_1)}^C[j] \sim \mathcal{P}(M_S^j \beta_S^i h(jT_S, \tau_s)), \tag{3}$$

where $\mathcal{P}$ represents the Poisson distribution. Then the mean of $N_{(S,R_1)}^C[j]$ denoted by $\bar{N}_{(S,R_1)}^C[j]$ is computed by

$$\bar{N}_{(S,R_1)}^C[j] = M_S^j \beta_S^j h(jT_S, \tau_s). \tag{4}$$

For the transmission from node S to node R_1, node R_1 can receive the inter-symbol interference molecules from the previous $(j-1)$ time slots. Thus, in the current time slot j, the mean of the number of ISI molecules which is denoted by $\bar{N}_{(S,R_1)}^{ISI}[j]$ can be derived by

$$\bar{N}_{(S,R_1)}^{ISI}[j] = \sum_{i=1}^{j-1} M_S^i \beta_S^i h(iT_S, (j-i)T_S + \tau_s), \tag{5}$$

where $h(iT_S, (j-i)T_S + \tau_s)$ represents the probability that a molecule released by node S in the time slot i is received by node R_1 in the time slot j.

Furthermore, the mean of the number of received molecules at node R_1 in the j-th time slot is denoted by $\bar{N}_{(R_1)}[j]$. Therefore, according to (4) and (5), the mathematical expression of $\bar{N}_{(R_1)}[j]$ is written as

$$\bar{N}_{(R_1)}[j] = \bar{N}_{(S,R_1)}^C[j] + \bar{N}_{(S,R_1)}^{ISI}[j]. \tag{6}$$

Let H_0 and H_1 represent the events that the node S transmits bit 0 and 1 in the current time slot j, respectively. Under the two hypotheses, the number of received molecules in time slot j denoted by M_{H_0} and M_{H_1} obeys the following Poisson distributions, respectively.

$$M_{H_0} \sim \sum_{i=1}^{j-1} \mathcal{P}(M_S^i \beta_S^i h(iT_S, (j-i)T_S + \tau_s)), \tag{7}$$

$$M_{H_1} \sim \mathcal{P}(M_S^j h(jT_S, \tau_s)) + \sum_{i=1}^{j-1} \mathcal{P}(M_S^i \beta_S^i h(iT_S, (j-i)T_S + \tau_s)). \tag{8}$$

According to (7) and (8), the number of received molecules at node R_1 is written as the random variable Z_{R_1}. It follows the Poisson distributions based on the distribution of M_{H_0} and M_{H_1}, respectively.

$$\begin{aligned} H_0 : Z_{R_1} &\sim \mathcal{P}(\mu_{R_1}^0), \\ H_1 : Z_{R_1} &\sim \mathcal{P}(\mu_{R_1}^1), \end{aligned} \tag{9}$$

where the means of Poisson distribution $\mu_{R_1}^0$ and $\mu_{R_1}^1$ in (9) which can be obtained by (7) and (8) are

$$\mu_{R_1}^0 = \sum_{i=1}^{j-1} M_S^i \beta_S^i h\left(iT_S, (j-i)T_S + \tau_s\right), \tag{10}$$

$$\mu_{R_1}^1 = M_S^j h(jT_S, \tau_s) + \sum_{i=1}^{j-1} M_S^i \beta_S^i h(iT_S, (j-i)T_S + \tau_s). \tag{11}$$

The relay node R_1 decodes the received bit by comparing $\bar{N}_{(R_1)}[j]$ with the decision threshold at R_1 which is denoted by θ_{R_1}. With the hypothesis test model above, we can derive the optimal decision threshold that decreases the bit error probability using the maximum-a-posterior (MAP) decision method. Thus the received bit is determined by

$$\hat{y}_R[j] = \begin{cases} 1, & if\, \bar{N}_{(R_1)}[j] \geq \theta_{R_1}, \\ 0, & if\, \bar{N}_{(R_1)}[j] < \theta_{R_1}, \end{cases} \tag{12}$$

where $\hat{y}_{R_1}[j]$ is the information bit detected by node R_1 in the j-th time slot. The likelihood-ratio test is as follows:

$$\frac{P(z_{R_1}|H_1)}{P(z_{R_1}|H_0)} = \frac{f_{Z_{R_1}}^1(z_{R_1})}{f_{Z_{R_1}}^0(z_{R_1})} \underset{H_0}{\overset{H_1}{\gtrless}} \frac{P(H_0)}{P(H_1)}, \tag{13}$$

where $P(H_1) = \beta_S^j$ and $P(H_0) = 1 - \beta_S^j$ represent the probabilities of transmitting 1 and 0 in time slot j by node S, respectively. $f_{Z_{R_1}}^0(z_{R_1})$ and $f_{Z_{R_1}}^1(z_{R_1})$ are expressed as follows:

$$\begin{aligned} f_{Z_{R_1}}^0(z_{R_1}) &= \frac{(\mu_{R_1}^0)^{z_{R_1}}}{z_{R_1}!} e^{-\mu_{R_1}^0}, \\ f_{Z_{R_1}}^1(z_{R_1}) &= \frac{(\mu_{R_1}^1)^{z_{R_1}}}{z_{R_1}!} e^{-\mu_{R_1}^1}. \end{aligned} \tag{14}$$

Thus the solution of (13) is described as

$$z_{R_1} \underset{H_0}{\overset{H_1}{\gtrless}} \left\lceil \frac{\ln(P(H_0)/P(H_1)) + \mu_{R_1}^1 - \mu_{R_1}^0}{\ln(\mu_{R_1}^1/\mu_{R_1}^0)} \right\rceil \equiv \theta_{R_1}. \tag{15}$$

According to (15), from node S to node R_1 in the j-th time slot, the bit error probability of transmitting bit 1 is

$$\Pr(\hat{y}_{R_1}[j] = 1|x_S[j] = 0) = \Pr(N_{(S,R_1)}[j] \geq \theta_{R_1}|x_S[j] = 0) \\ = 1 - \sum_{w=0}^{\theta_{R_1}} e^{-\mu_{R_1}^0} \frac{(\mu_{R_1}^0)^w}{w!}, \tag{16}$$

where $x_S[j]$ is the j-th transmitted bit at node S. We also have

$$\Pr(\hat{y}_{R_1}[j] = 0|x_S[j] = 1) = \Pr(N_{(S,R_1)}[j] < \theta_{R_1}|x_S[j] = 1) \\ = \sum_{w=0}^{\theta_{R_1}} e^{-\mu_{R_1}^1} \frac{(\mu_{R_1}^1)^w}{w!}, \tag{17}$$

where $\mu_{R_1}^1$ are given in (11). According to (16) and (17), the error probability of the j-th bit for a single link can be expressed as

$$Pe_{R_1}[j] = \beta_S^j \Pr(\hat{y}_{R_1}[j] = 0|x_S[j] = 1) \\ + (1 - \beta_S^j) \Pr(\hat{y}_{R_1}[j] = 1|x_S[j] = 0). \tag{18}$$

Similarly, the error probability of the $(j+1)$-th transmitted bit from node R_1 to node D is

$$\Pr(\hat{y}_D[j+1] = 1|x_{R_1}[j+1] = 0) \\ = \Pr(N_{(D)}[j] \geq \theta_D|x_{R_1}[j+1] = 0) \\ = 1 - \sum_{w=0}^{\theta_D} e^{-\mu_D^0} \frac{(\mu_D^0)^w}{w!}, \tag{19}$$

$$\Pr(\hat{y}_D[j+1] = 0|x_{R_1}[j+1]) \\ = \Pr(N_{(D)}[j] < \theta_D|x_{R_1}[j+1] = 1) \\ = \sum_{w=0}^{\theta_D} e^{-\mu_D^1} \frac{(\mu_D^1)^w}{w!}, \tag{20}$$

where $\hat{y}_D[j+1]$ represents the detected bit at node D in the $(j+1)$-th time slot. θ_D is the optimal decision threshold at node D. μ_D^0 in (19) and μ_D^1 in (20) which are the means of the corresponding Poisson distributions are computed by

$$\mu_D^0 = \sum_{i=1}^{j-1} \beta_{R_1}^i M_{R_1}^i h\left(iT_S, (j-i)T_S + \tau_s\right), \tag{21}$$

$$\mu_D^1 = M_{R_1}^j h(jT_S, \tau_s) + \sum_{i=1}^{j-1} M_{R_1}^i \beta_{R_1}^i h(iT_S, (j-i)T_S + \tau_s), \tag{22}$$

The error probability of the j-th bit for the two-hop mobile MCvD system can be expressed as

$$Pe_D[j] = \beta_S^j \Pr(\hat{y}_D[j+1] = 0|x_S[j] = 1) \\ + (1 - \beta_S^j) \Pr(\hat{y}_D[j+1] = 1|x_S[j] = 0), \tag{23}$$

where $\Pr(\hat{y}_D[j+1]=0|x_S[j]=1)$ and $\Pr(\hat{y}_D[j+1]=1|x_S[j]=0)$ can be derived as follows:

$$\begin{aligned}&\Pr(\hat{y}_D[j+1]=0|x_S[j]=1)\\&\quad=\Pr(\hat{y}_{R_1}[j]=0|x_S[j]=1)\\&\qquad\times\Pr(\hat{y}_D[j+1]=0|x_{R_1}[j+1]=0)\\&\qquad+\Pr(\hat{y}_{R_1}[j]=1|x_S[j]=1)\\&\qquad\times\Pr(\hat{y}_D[j+1]=0|x_{R_1}[j+1]=1),\end{aligned}\tag{24}$$

$$\begin{aligned}&\Pr(\hat{y}_D[j+1]=1|x_S[j]=0)\\&\quad=\Pr(\hat{y}_{R_1}[j]=0|x_S[j]=0)\\&\qquad\times\Pr(\hat{y}_D[j+1]=1|x_{R_1}[j+1]=0)\\&\qquad+\Pr(\hat{y}_{R_1}[j]=1|x_S[j]=0)\\&\qquad\times\Pr(\hat{y}_D[j+1]=1|x_{R_1}[j+1]=1).\end{aligned}\tag{25}$$

3.2 Multi-hop Network

We use the recursive method to deduce the error probability of the $(k+1)$-hop mobile MCvD which is labelled as $Pe_{k+1}[j+1]$. Therefore we should compute the k-hop error probability $Pe_k[j+1]$ which is obtained by the error probability $Pe_k[j|x_S[j]=1]$ and $Pe_k[j|x_S[j]=0]$. Thus, assume that the error probabilities $Pe_k[j|x_S[j]=1]$ and $Pe_k[j|x_S[j]=0]$ are known. Under $x_S[j]=1$ and $x_S[j]=0$, the error probabilities of the first $(k+1)$ hops which are denoted by $Pe_{k+1}[j|x_S[j]=1]$ and $Pe_{k+1}[j|x_S[j]=0]$, respectively, are

$$\begin{aligned}&Pe_{k+1}[j|x_S[j]=1]\\&\quad=Pe_k[j|x_S[j]=1]\\&\qquad\times\Pr[N_{(R_{k+1})}[j+k]<\xi_{R_{k+1}}|x_{R_k}[j+k]=0]\\&\qquad+(1-Pe_k[j|x_S[j]=1])\\&\qquad\times\Pr[N_{(R_{k+1})}[j+k]<\xi_{R_{k+1}}|x_{R_k}[j+k]=1],\end{aligned}\tag{26}$$

$$\begin{aligned}&Pe_{k+1}[j|x_S[j]=0]\\&\quad=Pe_k[j|x_S[j]=0]\\&\qquad\times\Pr[N_{(R_{k+1})}[j+k]\geq\xi_{R_{k+1}}|x_{R_k}[j+k]=0]\\&\qquad+(1-Pe_k[j|x_S[j]=0])\\&\qquad\times\Pr[N_{(R_{k+1})}[j+k]\geq\xi_{R_{k+1}}|x_{R_k}[j+k]=0].\end{aligned}\tag{27}$$

Therefore, the error probability of the $(k+1)$-hop mobile MCvD network $Pe_{k+1}[j+1]$ is computed by

$$\begin{aligned}Pe_{k+1}[j+1]=\;&P_1Pe_{k+1}[j|x_S[j]=1]\\&+P_0Pe_{k+1}[j|x_S[j]=0].\end{aligned}\tag{28}$$

4 Analysis of Cooperative Relaying in SMT Multi-hop Network

In this section, the adaptive detection threshold is proposed to alleviate the self-interference caused by same type molecules from the source node and the relay nodes.

4.1 Mobile MCvD Network for a Single Link

The number of received molecules at the relay node R_1 in the j-th time slot, $N_{(R_1)}[j]$, is the sum of the number of molecules $N_{(S,R_1)}[j]$ and $N_{(R_1,R_1)}[j]$. $N_{(R_1,R_1)}[j]$. It can be computed by

$$N_{(R_1)}[j] = N_{(S,R_1)}[j] + N_{(R_1,R_1)}[j]. \tag{29}$$

We can get

$$\bar{N}_{(R_1)}[j] = \bar{N}_{(S,R_1)}[j] + \bar{N}_{(R_1,R_1)}[j], \tag{30}$$

where $\bar{N}_{(S,R_1)}[j]$ and $\bar{N}_{(R_1,R_1)}[j]$ denote the means of $N_{(S,R_1)}[j]$ and $N_{(R_1,R_1)}[j]$, respectively, which can be computed by (6).

Analogously, the number of received molecules at node D in the j-th time slot $N_{(D)}[j]$ is expressed as

$$N_{(D)}[j] = N_{(S,D)}[j] + N_{(R_1,D)}[j]. \tag{31}$$

And we also have

$$\bar{N}_{(D)}[j] = \bar{N}_{(S,D)}[j] + \bar{N}_{(R_1,D)}[j], \tag{32}$$

where $\bar{N}_{(S,D)}[j]$ and $\bar{N}_{(R_1,D)}[j]$ are the means of $N_{(S,D)}[j]$ and $N_{(R_1,D)}[j]$, respectively, which can be analogously computed by (6).

At the relay node R_1, we use the adaptive detection threshold method. Node R_1 can adjust its decision threshold in each time slot based on the detected bits. Thus, the adaptive decision threshold at node R_1 in the j-th time slot $\xi_{R_1}[j]$ can be written as [8]

$$\xi_{R_1}[j] = \xi + \xi_{R_1,Adaptive}[j], \tag{33}$$

where ξ is the fixed part which depends on the value of the number of molecules from the node S in previous time slots, and $\xi_{R_1,Adaptive}[j]$ is the adaptive part which changes adaptively based on the number of molecules from the node R_1 in the current time slot. It can be computed by

$$\xi_{R_1,Adaptive}[j] = \bar{N}_{(R_1,R_1)}[j]. \tag{34}$$

4.2 Multi-hop Network

In a SMT multi-hop network, the number of received molecules at node R_k in the j-th time slot $N_{(R_k)}[j]$ can be written as

$$N_{(R_k)}[j] = \sum_{q=1}^{Q} N_{(R_q,R_k)}[j]. \tag{35}$$

$N_{(R_k)}[j]$ is a Poisson random variable, then we have

$$\bar{N}_{(R_k)}[j] = \sum_{q=1}^{Q} \bar{N}_{(R_q,R_k)}[j], \tag{36}$$

where $\bar{N}_{R_k}[j]$ and $\bar{N}_{(R_q,R_k)}[j]$ are the corresponding means of $N_{(R_k)}[j]$ and $N_{(R_q,R_k)}[j]$, respectively.

According to the adaptive detection threshold method in two-molecule-type two-hop network introduced above, for full-duplex transmission, node R_k adjusts the value of $\xi_{R_k}[j]$ as

$$\xi_{R_k}[j] = \xi + \xi_{R_k,Adaptive}[j], \tag{37}$$

where ξ is the fixed part which depends on the value of the number of molecules from nodes $R_q(q = 1, 2, ..., k-1)$. $\xi_{R_k,Adaptive}[j]$ is the adaptive part which depends on the sum of the molecules from node R_k and the adjacent node R_{k+1} on the basis that the residual nodes $R_q(q = k+2, k+3, ..., Q)$ have less effects on the node R_k. It can be computed by

$$\xi_{R_k,Adaptive}[j] = \bar{N}_{(R_k,R_k)}[j] + \bar{N}_{(R_{k+1},R_k)}[j]. \tag{38}$$

5 Numerical Results

The parameters used for numerical results are set in Table 1. For the sake of simplicity, we assume the diffusion coefficient of each type molecules D_{A_i} and the number of molecules emitted by node S and relay nodes at the beginning of each time slot $M_l^j(l = S, R_k)$ be the same, respectively. D_A and $M_l(l = S, R_k)$ are used to represent D_{A_i} and $M_l^j(l = S, R_k)$, respectively.

Table 1. Simulation parameters.

Parameter	Value
(x_S^0, y_S^0, z_S^0)	(0,0,0)
(x_R^0, y_R^0, z_R^0)	(10 μm,0,0)
(x_D^0, y_D^0, z_D^0)	(20 μm,0,0)
$D_l(l = S, R_k, D)$	1×10^{-13} m^2/s
D_A	5×10^{-9} m^2/s
$M_l(l = S, R_k)$	$0 - 1 \times 10^5$
n	10
$\beta_l^j(l = S, R_k)$	0.5
r_{R_k}, r_{R_D}	1 μm
$\Delta x_l, \Delta y_l, \Delta z_l$	0.01 μm
T_S	0–400 ms

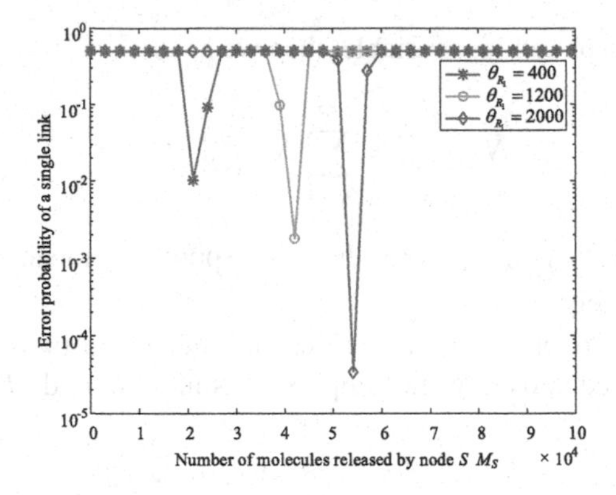

Fig. 2. The error probability vs M_S with different detection threshold at node R_1 for a single link.

In Fig. 2, the error probability is varying with M_S for different values of θ_{R_1} with $\theta_{R_1} = 400, 1200, 2000$. The parameters are set as $T_S = 300$ ms, $\tau_s = 5$ ms. We observe that for one single link, we set fixed detection threshold at node R_1 with different values. With the increasing value of M_S, the fixed detection threshold θ_{R_1} is also increasing to achieve minimum value of error probability. In particular, the larger value of θ_{R_1}, the smaller minimum value of error probability. It is because that more molecules are released by node S for one single link, the probability that one molecule released by node S is received by node R_1 is improved, then more molecules are received by node R_1, therefore smaller minimum value of error probability can be achieved.

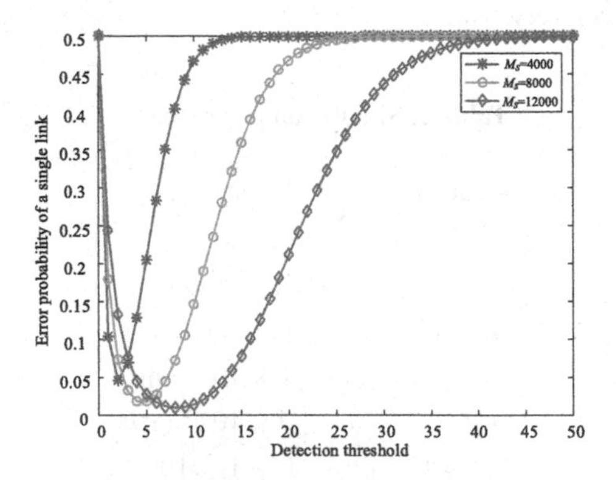

Fig. 3. The error probability vs the decision threshold at node R_1 with different values of M_S.

In Fig. 3, we analyze the performance of error probability from node S to R_1 as a function of θ_{R_1} at node R_1 with different values of $M_S = 4000, 8000, 12000$ by using the optimal detection threshold. The parameters are set as $T_S = 400$ ms, $\tau_s = 11$ ms. It is observed that when the value of M_S decreases, the minimum value of error probability is larger. In particular, when M_S is increasing, the optimal decision threshold θ_{R_1}

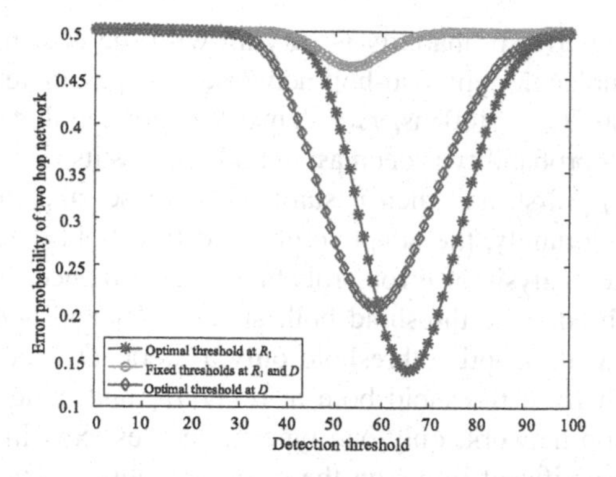

Fig. 4. The error probability of two-molecule-type two-hop network vs fixed detection threshold at node R_1.

which can minimize the error probability is also increasing. This is based on the fact that the more molecules are released in each time slot, the more molecules are received by node D. Then the optimal decision threshold θ_{R_1} should be increase to improve the error probability.

In Fig. 4, the error probability is varying with the decision threshold θ_{R_1} at node R_1 for two-molecule-type two-hop network $S \rightarrow R_1 \rightarrow D$. The parameters are set as $M_S = 10000$, $M_{R_1} = 10000$, $T_S = 400$ ms, $\tau_s = 2$ ms. We use the optimal and fixed threshold detection method at node R_1 and node D as the three cases: (1) optimal threshold at the relay R_1 and fixed threshold at D, (2) fixed threshold at R_1 and optimal threshold at the relay D, (3) fixed threshold both at R_1 and D. Then the error probability is decreasing from case (1) to case (3). For the two-molecule-type two-hop network, we should select the optimal threshold detection to reduce the error probability. Especially, the optimal threshold detection method for the first one link can be used to improve the error probability for this two-hop network.

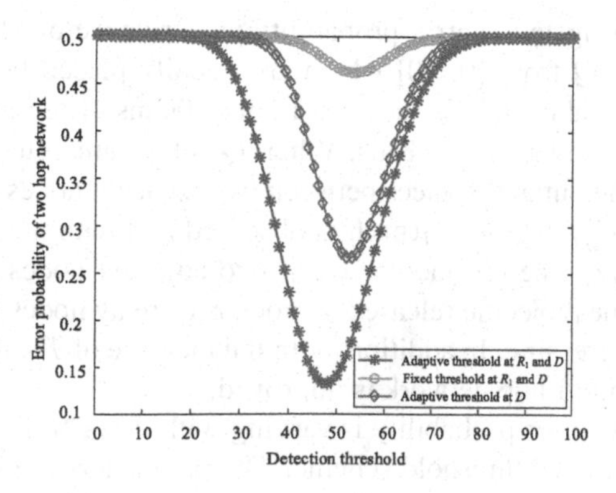

Fig. 5. The error probability performance of single-molecule-type two-hop network vs threshold detection at relay R_1 and node D.

Figure 5 shows the error probability is varying with the decision threshold θ_{R_1} at relay R_1 for single-molecule-type two-hop network. The parameters are set as $M_S = 10000$, $M_{R_1} = 10000$, $T_S = 400\,\text{ms}$, $\tau_s = 2\,\text{ms}$. It is noticed that with the increasing value of θ_{R_1}, the error probability is decreasing and reaches its minimum value at some particular value of θ_{R_1} first, and then it starts to increase, and finally arrives at the peak value. More importantly, the adaptive or fixed threshold detection method play important roles in the analysis of error probability performance. We can see that the error probability with adaptive threshold both at relay R_1 and node D is better than the error probability with adaptive threshold only at node D which is better than the error probability with fixed threshold both at relay R_1 and node D. For the single-molecule-type two-hop network, only one type molecules exist in the same medium, the ISI effects have significant limits on the error probability. Therefore the adaptive threshold detection method should be performed at the relay R_1 and node D to alleviate the ISI effects.

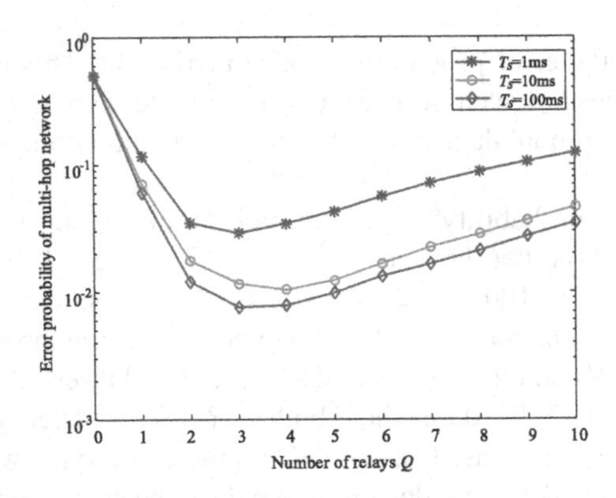

Fig. 6. The error probability performance of MMT multi-hop network as a function of the number of relays.

In Fig. 6, we investigate the error probability performance of MMT multi-hop network is varying with Q from [0, 10] which are equally placed between node S and node D for different values of $T_S = 1\,\text{ms}, 10\,\text{ms}, 100\,\text{ms}$. The parameters are set as $M_S = 1000$, $M_{R_1} = 5000$, $\tau_s = 3\,\text{ms}$. When Q relays are equally deployed in the multi-hop network, the initial distances between two adjacent nodes are the same which can be computed by $d_{SD}^0/(Q+1)\,\mu\text{m}$. It is observed that with the increasing value of the number of relays Q, the distance between two adjacent nodes is decreasing, then the probability that one molecule released by node S or relay nodes is received by relay nodes or node D is increasing. In addition, with the increase of T_S, the error probability performance of this multi-hop network is improved.

Figure 7 shows the error probability is varying with Q for SMT multi-hop network with the adaptive and fixed threshold scheme. The parameters are set as $M_S = 1000$, $M_{R_1} = 5000$, $\tau_s = 3\,\text{ms}$. It is observed that with the increasing value of Q, because the types of the molecules released by node S and relay nodes are the same, the self-interference increases. Then we use the adaptive threshold detection method to alleviate

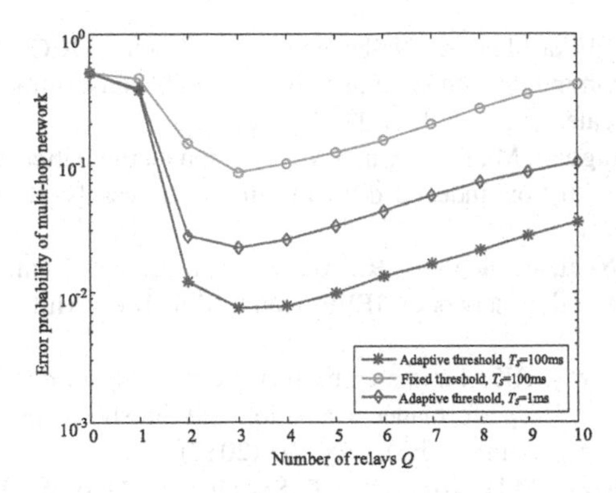

Fig. 7. The error probability of single-molecule-type multi-hop network vs the number of relays.

the self-interference. For the SMT multi-hop network, the adaptive detection threshold is composed of the fixed part and adaptive part. In Fig. 7, the fixed threshold is set as 5. We can see that the performance with the adaptive threshold scheme is improved comparing with the fixed threshold scheme. This is due to the fact that the self-interference effects caused by using the same type molecules can be mitigated by adaptive threshold scheme.

6 Conclusion

We consider multi-hop mobile MCvD system in 3D environment and investigate the error probability of this system. First, two relay schemes including MMT and SMT are proposed, respectively. Second, in the first relay scheme, the mathematical expression of optimal detection threshold can be derived. In the second scheme, the adaptive detection threshold is proposed to alleviate the self-interference caused by same type molecules existed in the same fluid medium. Third, we have derived the error probability of this system under MMT and SMT. Numerical results show that the parameters including the initial distance between two adjacent nodes, the number of molecules released in each time slot, the time slot duration, the detection threshold schemes have different effects on the analysis of the error probability performance of the multi-hop mobile MCvD system.

Acknowledgment. This work was supported in part by National Natural Science Foundation of China (Grant Nos. 62271446); in part by Zhejiang Provincial Natural Science Foundation of China (Grant Nos. LY23F020021).

References

1. Damrath, M., Hoeher, P.A.: Low-complexity adaptive threshold detection for molecular communication. IEEE Trans. Nanobiosci. **15**(3), 200–208 (2016)

2. Deng, Y., Noel, A., Elkashlan, M., Nallanathanet, A., Cheung, K.C.: Modeling and simulation of molecular communication systems with a reversible adsorption receiver. IEEE Trans. Mol. Biol. Multi-Scale Commun. **1**(4), 1–15 (2016)
3. Arjmandi, H., Zoofaghari, M., Rouzegar, S.V., et al.: On mathematical analysis of active drug transport coupled with flow-induced diffusion in blood vessels. IEEE Trans. Nanobiosci. **20**(1), 105–115 (2021)
4. Ahmadzadeh, A., Noel, A., Schober, R.: Analysis and design of multi-hop diffusion-based molecular communication networks. IEEE Trans. Mol. Biol. Multi-Scale Commun. **1**(2), 144–157 (2015)
5. Tiwari, S.K., Upadhyay, P.K.: Estimate-and-forward relaying in diffusion-based molecular communication networks: performance evaluation and threshold optimization. IEEE Trans. Mol. Biol. Multi-Scale Commun. **3**(3), 183–193 (2017)
6. Haselmayr, W., Aejaz, S.M.H., Asyhari, A.T., Springer, A., Guo, W.: Transposition errors in diffusion-based mobile molecular communication. IEEE Commun. Lett. **21**(9), 1973–1976 (2017)
7. Ahmadzadeh, A., Jamali, V., Noel, A., Schober, R.: Diffusive mobile molecular communications over time-variant channels. IEEE Commun. Lett. **21**(6), 1265–1268 (2017)
8. Douglas, S.M., Bachelet, I., Church, G.M.: A logic-gated nanorobot for targeted transport of molecular payloads. Science **335**(6070), 831–834 (2012)
9. Ahmadzadeh, A., Jamali, V., Schober, R.: Stochastic channel modeling for diffusive mobile molecular communication systems. IEEE Trans. Commun. **66**(12), 6205–6220 (2018)
10. Varshney, N., Patel, A., Haselmayr, W., et al.: Impact of intermediate nanomachines in multiple cooperative nanomachine-assisted diffusion advection mobile molecular communication. IEEE Trans. Commun. **67**(7), 4856–4871 (2019)
11. Cao, T.N., Ahmadzadeh, A., Jamali, V., et al.: Diffusive mobile MC with absorbing receivers: stochastic analysis and applications. IEEE Trans. Mol. Biol. Multi-Scale Commun. **5**(2), 84–99 (2019)
12. Huang, S., Lin, L., Guo, W., et al.: Initial distance estimation and signal detection for diffusive mobile molecular communication. IEEE Trans. Nanobiosci. **19**(3), 422–433 (2020)
13. Shrivastava, A.K., Das, D., Mahapatra, R.: Performance evaluation of mobile molecular communication system using neural network detector. IEEE Wireless Commun. Lett. **10**(8), 1776–1779 (2021)
14. Cheng, Z., Tu, Y., Yan, J., et al.: Amplify-and-forward relaying in mobile multi-hop molecular communication via diffusion. Nano Commun. Netw. **30**, 100375 (2021)
15. Jing, D., Li, Y., Eckford, A.W.: An extended Kalman filter for distance estimation and power control in mobile molecular communication. IEEE Trans. Commun. **70**(7), 4373–4385 (2022)
16. Chouhan, L., Alouini, M.: Rescaled Brownian motion of molecules and devices in three-dimensional multiuser mobile molecular communication systems. IEEE Trans. Commun. **21**(12), 10472–10485 (2022)
17. Noel, A., Cheung, K.C., Schober, R.: Optimal receiver design for diffusive molecular communication with flow and additive noise. IEEE Trans. Nanobiosci. **13**(3), 350–362 (2014)
18. Lin, L., Zhang, J., Ma, M., Yan, H.: Time synchronization for molecular communication with drift. IEEE Commun. Lett. **21**(3), 476–479 (2017)

Covid-19 Versus Monkeypox-2022: The Silent Struggle of Global Pandemics

Huber Nieto-Chaupis(✉)

Universidad Autónoma del Perú, Panamericana Sur Km. 16.3 Villa el Salvador, Lima, Peru
hubernietochaupis@gmail.com

Abstract. The theorem of Bayes is applied in a straightforward manner to investigate if Covid-19 and Monkeypox 2022 can coexist together. According to realistic scenarios and global data it was verified that Covid-19 is a kind of main pandemic whereas Monkeypox can be accepted a mini pandemic with a low lethality and a short period of existence. This would suggest that two global pandemics might not coexist at same time from the fact that people would acquire a disease belonging to all those pandemics with a strong capabilities of geographical translation and stability at long periods. From simulations, it is seen that Covid-19 would remain against Monkeypox that exhibits a noteworthy capability to produce infections but a weak lethality.

Keywords: Covid-19 · Monkeypox · Pandemic modeling

1 Introduction

At May 2022, it has been witnessed the apparition of the so-called Monkeypox virus (to be called along this paper Mkpx-22 in shorthand) [1–3] whose origin would be in central Africa [4,5]. According to global data, Mkpx-22 has bee aggressive at the very beginning showing high rates of infection basically in central Europe, mainly in England, Spain, Portugal, and reaching USA. The representative time evolution has been identified to be exponential:

$$N(t) = n_0 t^{\ell}, \tag{1}$$

as seen at Mkpx-22 in May 2022. However, as any pandemic, it will have to shown either capabilities or weakness. This can be seen in the second derivative of Eq. 1

$$\frac{d^3 N(t)}{dt^3} = n_0 \ell(\ell-1)(\ell-2)t^{\ell-3}. \tag{2}$$

Clearly, a interesting case is when $\ell = 2$ that convert to a polynomial that falls down in time. To guarantee this then Eq. 1 can be written as:

$$N(t) = \frac{n_0}{\ell-2} t^{\ell}, \tag{3}$$

Y. Chen et al. (Eds.): BICT 2023, LNICST 512, pp. 91–100, 2023.
https://doi.org/10.1007/978-3-031-43135-7_9

for $\ell > 2$. In this manner more than a polynomial evolution, the number of infections is also depending on the characteristics of virus to take humans as potential host as for example the case of Covid-19 [6,7]. Thus, this paper has opted to apply the well-know Bayes theorem to identify if there is coexistence of Covid-19 and Mkpx-22. The direct usage of Bayes equation might not be enough to find evidences between these two global pandemics. In this manner, it was explored closed-form solutions for the probabilities. Thus, the well-known diffusion equation could be the one that models the spatial propagation of virus in pandemic. The rest of this paper is as follows: In Sect. 2 are formulated all required probabilities. In Sect. 3, the probability of confirmation is derived. In Sect. 4, the direct confrontation between Covid-19 and Mkpx-22 is done. Finally, the conclusion of paper is presented (Fig. 1).

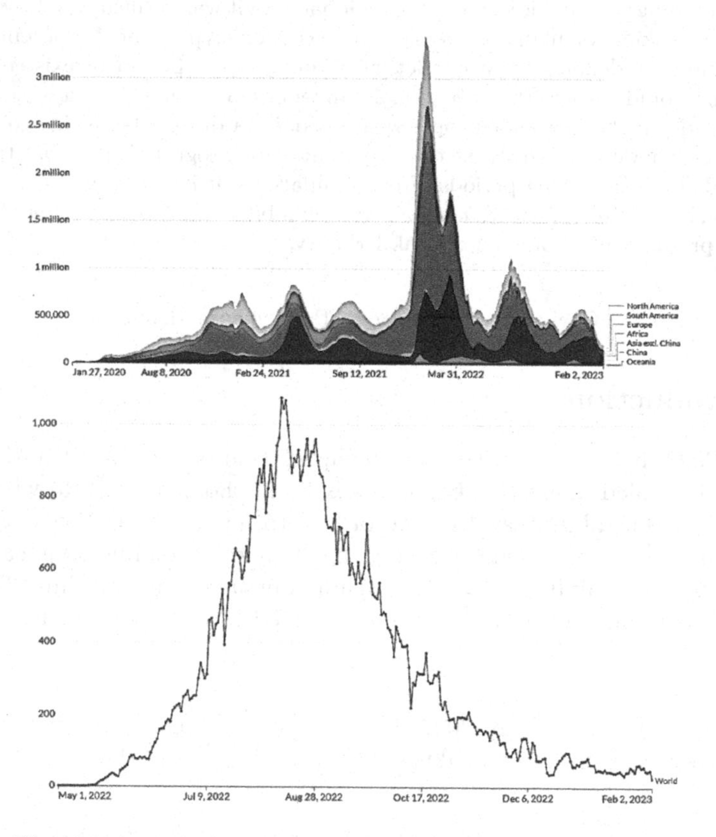

Fig. 1. Up: The global curve of infections for Covid-19 [8] and (Down) Mkpx-22 showing well-marked quantitative as well as qualitative differences. Data from [9]. Clearly Covid-19 imposes onto Mkpx-22 because the permanent intercontinental diffusion [9].

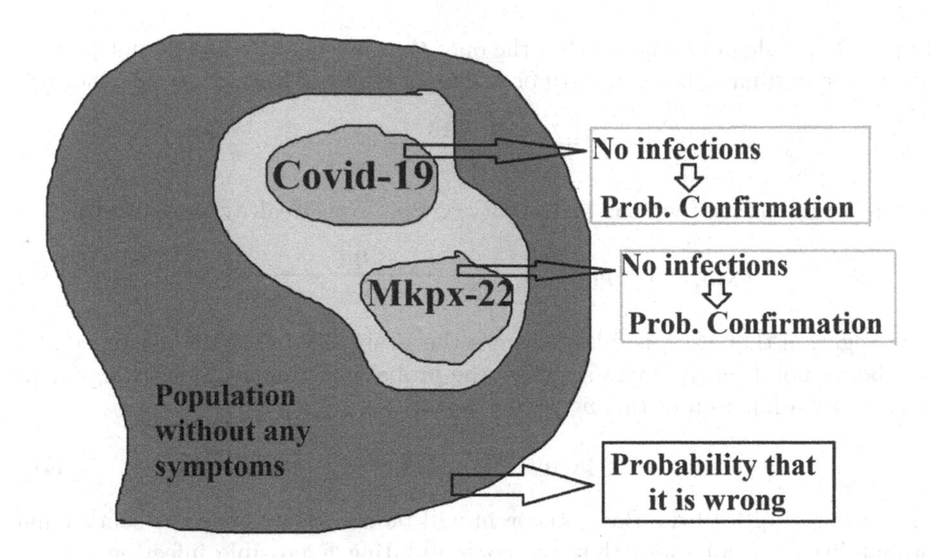

Fig. 2. Idea of paper: by knowing the rate of infection by Covid-19 and Mkpx-22, then emerges the question: What is the probability at a time t that a certain number of people acquires either Covid-19 or Mkpx-22 when both pandemics are superimposed each other? What is the implications about the confirmations for getting one of them if are true of wrong?.

2 Formulation of Probabilities

Consider a country with N_{TOT} the total number of habitants under the arrival of a first pandemic it is Covid-19 [10] and after a time (one or two years) there is evidence of arrival of a second pandemic: Mkpx-22 (see Fig. 2 above). Therefore there is time by which the pandemics are superimposed each other. In this time the fraction of people with symptoms like the ones of Covid-19 n_{SCV} and this fraction is seen as a probability p_{CV}:

$$p_{\text{CV}} = \frac{n_{\text{SCV}}}{N_{\text{TOT}}}. \tag{4}$$

Same idea is used to characterize the ones under the suspect of having Mkpx-22:

$$p_{\text{MK}} = \frac{n_{\text{SMK}}}{N_{\text{TOT}}}. \tag{5}$$

Now the it is noteworthy to add the probability that confirm that each suspect has absolutely the virus. Thus, for both Covid-19 and Mkpx-22 these confirmation probability are written as [11]:

$$p_{\text{CCV}} = C(p_{\text{CV}}) \tag{6}$$

$$p_{\text{CMK}} = C(p_{\text{MK}}). \tag{7}$$

In parallel, it should be noted that the ones that are healthy and do not present not any symptom either Covid-19 or Mkpx-22 is also a fraction and define by:

$$p_{\mathrm{HE}} = \frac{n_{\mathrm{HE}}}{N_{\mathrm{TOT}} - n_{\mathrm{SCV}} - n_{\mathrm{SMK}}}. \tag{8}$$

If it is a probability then the probability of the "Non-Healthy" is given by:

$$p_{\mathrm{NHE}} = 1 - p_{\mathrm{HE}} = 1 - \frac{n_{\mathrm{HE}}}{N_{\mathrm{TOT}} - n_{\mathrm{SCV}} - n_{\mathrm{SMK}}}. \tag{9}$$

The negation that Eq. 8 is false becomes the probability that the existence of a number of non-healthy is wrong. Thus, the probability that p_{NHE} is wrong p_{WHE} is given by a function of this p_{NHE}:

$$p_{\mathrm{WHE}} = W(p_{\mathrm{NHE}}). \tag{10}$$

In virtue to Eq. 4–10 the Bayes theorem will be used [12]. Thus the conditional probabilities are all those that are contemplating a possible infection due to either Covid-19 or Mkpx-22. In this manner the Bayes theorem is adjusted to the following question: What's the probability that at least a number of people have got the infection either Covid-19 or Mkpx-22 at the same period? To answer this the Bayes probability reads:

$$B = \frac{p_{\mathrm{CV}}C(p_{\mathrm{CV}}) + p_{\mathrm{MK}}C(p_{\mathrm{MK}})}{p_{\mathrm{CV}}C(p_{\mathrm{CV}}) + p_{\mathrm{MK}}C(p_{\mathrm{MK}}) + (1 - p_{\mathrm{HE}})W(p_{\mathrm{NHE}})} \tag{11}$$

3 Construction of Confirmation Probability from Trigonometrical Bayes Theorem

When one talk about According to data, cases of Mkpx-22 it has been propagated very fast particularly in central Europe in May 2022. In contrast to Covid-19, geographic propagation of Mkpx-22 [13,14] has exhibited to be continental in the sense that the first infections have been registered in central Europe, while Covid-19 has go ut from Wuhan, China at December 2019. Thus the spatial propagation becomes relevant.

3.1 Propagation of Confirmed Cases

It is again used the Bayes theorem as follows: For example if $\mathbf{S}^2$ the prior probability of having the virus at the normalized distance $\mathbf{S}$ as well as $(\frac{\mathbf{X}}{2})^2$ the posterior probability of getting the infection at $\mathbf{X}$ (with $\mathbf{X}$ also another normalized distance) with these definitions the Bayes's probability with obvious values ranging between 0 and 1, can be written as the square of a sinusoid function either Sin or Cos for instance:

$$\mathrm{Cos}^2\theta = \frac{\mathbf{S}^2}{\mathbf{S}^2 + \left(\frac{\mathbf{X}}{2}\right)^2}. \tag{12}$$

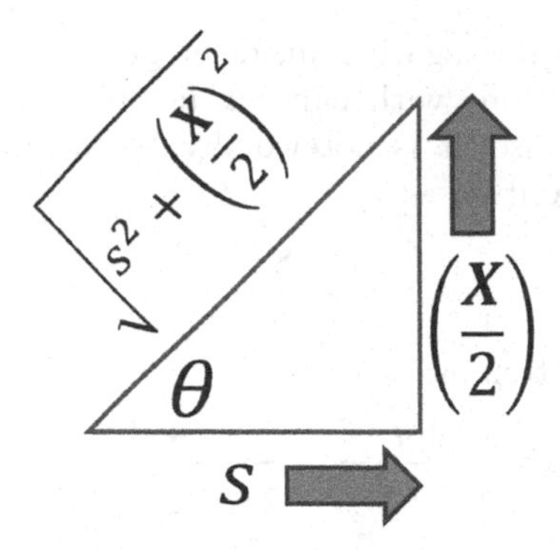

Fig. 3. Bayes's Path: For any traveler containing the virus at a time t and after of having moved S the subsequent step is random with a 50% of chance to go up or continue.

Under an inversion of roles so that one has in a similar manner as above:

$$\text{Sin}^2\theta = \frac{\left(\frac{\mathbf{X}}{2}\right)^2}{\left(\frac{\mathbf{X}}{2}\right)^2 + \mathbf{S}^2}. \tag{13}$$

It should be noted that both $\mathbf{X}$ and $\mathbf{S}$ are denoting spatial displacement of confirmed cases. It leads to define a rectangle triangle as the one sketched in Fig. 3 that explains the origin of distance for an infected traveler: Fig. 3 can be explained as follows: at a given time a traveler containing the infection has been displaced from rest to $\mathbf{S}$, just there the traveler has two options: (i) continue along the path or (ii) follow a perpendicular direction reaching to $\mathbf{X}$. Because the traveler has two options then the perpendicular path has a 50% of probability. Thus the segment at the triangle is written as $\frac{\mathbf{X}}{2}$. Of course it is a single event, in other words that fact that one has a rectangle triangle forming an angle θ it is a random event. In another event one can have ϕ with a value less than $\pi/2$.

3.2 Probability as Rate of Infection

Consider Fig. 3 in a generalized manner so that the probabilistic number of infections $\mathbf{n}$ that have been displaced of 0 to $\mathbf{X}$ can be written now as:

$$\mathbf{n} = \frac{\mathbf{S}^2}{\mathbf{S}^2 + \left(\frac{\mathbf{X}}{2}\right)^2}. \tag{14}$$

In order to introduce the time it is easy to verify that on can rewrite Eq. 14 when $\mathbf{S}^2 \Rightarrow 4\mathbf{S}^2 t/t$:

$$\mathbf{n} = \frac{4\mathbf{S}^2 \frac{t}{t}}{4\mathbf{S}^2 \frac{t}{t} + \left(\frac{\mathbf{X}}{2}\right)^2}. \tag{15}$$

Once the infections are propagating inside a geographical area then one can invoke to diffusive equations with capabilities to describe diffusion of infections [15,16]. In such sense, it is noteworthy to introduce a kind of diffusive parameter that can be written as:

$$\gamma = \frac{\mathbf{S}^2}{t}. \tag{16}$$

With this Eq. 15 is rewritten as:

$$\mathbf{n} = \frac{4\gamma t}{4\gamma t + \mathbf{X}^2} = \frac{4\gamma t}{4\gamma t\left(1 + \frac{\mathbf{X}^2}{4\gamma t}\right)}. \tag{17}$$

It should be noted that $4\gamma t$ has equal units than $\mathbf{X}^2$ so that at the approximation dictated by $(\frac{\mathbf{X}^2}{4\gamma t})^{\ell} \approx 0$ if only if $\ell \geq 2$. Then in Eq. 17:

$$\mathbf{n} = \frac{4\gamma t}{4\gamma t\left(1 + \frac{\mathbf{X}^2}{4\gamma t} + \frac{\mathbf{X}^4}{16\gamma^2 t^2}\right) + ...} = \mathrm{Exp}\left(-\frac{\mathbf{X}^2}{4\gamma t}\right). \tag{18}$$

Actually one can multiply to $\mathbf{n}$ by $\sqrt{4\pi\gamma t}$ so that solving for $\mathbf{n}$ one arrives to:

$$\mathbf{n}(\mathbf{X}, t) = \frac{1}{\sqrt{4\pi\gamma t}}\mathrm{Exp}\left(-\frac{\mathbf{X}^2}{4\gamma t}\right), \tag{19}$$

that is actually direct solution of the well-known diffusion equation.

$$\frac{d\mathbf{n}}{dt} = \gamma\frac{d^2\mathbf{n}}{d^2\mathbf{X}}, \tag{20}$$

with γ the diffusion coefficient. The standard solution is given by:

$$\mathbf{n}(x, t) = \frac{1}{\sqrt{4\pi\gamma t}}\mathrm{Exp}\left[-\frac{\mathbf{X}^2}{4\gamma t}\right]. \tag{21}$$

Thus, Eq. 21 is interpreted as the probability of having $\mathbf{n}$ infections at the time t from an distance $\mathbf{X}$. Turning now to Fig. 3 one has that $\mathrm{Tan}\theta = \frac{\mathbf{X}}{2s} \Rightarrow \mathbf{X} = 2s\mathrm{Tan}\theta$, so that Eq. 21 can be generalized to one depending on the angle θ as:

$$\mathbf{n}(\theta, t) = \frac{1}{\sqrt{4\pi\gamma t}}\mathrm{Exp}\left[-\frac{s^2\mathrm{Tan}^2\theta}{\gamma t}\right], \tag{22}$$

with θ a random angle. It is emphasized the randomness of angle θ from the fact that there is not any knowledge about the route of infections once the pandemic has started. The case for calculating the probability of fatality can be modeled to a good level of approximation through the well-known theorem of Bayes [17] that states the estimation of a posterior probability $P(h|d)$ depends on the prior one $p(h)$ and $P(D) = \sum_Q P_Q(D)$ with $P_Q(D)$ the probability of having prior data that will be observed for Q cases or scenarios. In this manner the conditional probability of Bayes is given as:

$$P = \frac{P(h|d)}{\sum_Q P_Q(D)}. \tag{23}$$

4 Covid-19 Versus Mkpx-22

To normalize the probabilities, the constant $1/\sqrt{4\pi}$ is applied. The probability that $\mathbf{n}(t)$ is confirmed at time t reads for Covid-19 and Mkpx-22 respectively as:

$$\mathbf{C}_{\text{CV}} = \frac{N}{\sqrt{\gamma t}} \text{Exp}\left[-\frac{x_0^2}{4\gamma t}\right] \tag{24}$$

$$\mathbf{C}_{\text{MK}} = \frac{N}{\sqrt{\gamma t}} \text{Exp}\left[\frac{x_0^2}{4\gamma t}\right] \tag{25}$$

$$p_{\text{NHE}} = 1 - p_{\text{HE}} = 1 - \frac{N}{\sqrt{\gamma t}} e^{\left[-\frac{x_0}{4\gamma t}\right]} \tag{26}$$

It should be noted that apart that $\mathbf{n}(t)$ is expressing the instantaneous number of infections, the normalization through the constant N makes it acquire values between 0 and 1, so that $\mathbf{n}(t)$ now can describe probabilities. This is inserted for each case of Master Equation Eq. 11 that can be explicitly written by:

$$P_{\mathbf{B}} = \frac{\left(p_{\text{CV}} \frac{N}{\sqrt{\gamma t}} \text{Exp}\left[-\frac{x_0^2}{4\gamma t}\right] + p_{\text{MK}} \frac{N}{\sqrt{\gamma t}} \text{Exp}\left[\frac{x_0^2}{4\gamma t}\right]\right)}{\left(p_{\text{CV}} \frac{N}{\sqrt{\gamma t}} \text{Exp}\left[-\frac{x_0^2}{4\gamma t}\right] + p_{\text{MK}} \frac{N}{\sqrt{\gamma t}} \text{Exp}\left[\frac{x_0^2}{4\gamma t}\right]\right) + \left(1 - \frac{N}{\sqrt{\gamma t}} \text{Exp}\left[-\frac{x_0}{4\gamma t}\right]\right) W(p_{\text{NHE}})} \tag{27}$$

One can see that all variables can be known. Particular attention is paid onto the ones given by p_{CV} and p_{MK} that are denoting the fractions of all those that are exhibiting symptoms. The spatial variable x_0 is nominal, γ is an input that can be roughly calculated to priori from global data. N the normalization constant, and $W(p_{\text{NHE}})$ the probability by which discards the scenario where the healthy people is apparent and there is a latent risk that all of them are carrying the virus. In order to illustrate Eq. 27 in Fig. 4 a bivariate plot has been done. In one hand Eq. 27 with solely Covid-19 and on the other side the possible behavior of Mkpx-22. Thus, by using Wolfram [19] the smooth density histograms are depicted. Because it exhibited a fast up at the number of infections, the associate function follows t^{ℓ} with ℓ an integer number $\geq$2. In Fig. 4 are seen the ellipses corresponding to the Covid-19 risk and in a minor extent the ones belonging to Mkpx-22. Thus, as specified above, these distributions of probabilities are denoting in an explicit manner the number of infections. Thus, in left one can see that the small ellipse is losing a certain volume of infections, fact that is interpreted as the absorption of infections by the Covid-19 pandemic, and leaving a reduce ellipse. A similar case happens in middle plot where the Mkpx-22 transfers some infections to big ellipse governed by Covid-19. In right-side panel, the Mkpx-22 cases becomes to be separated and their infections although might back again as seen at May-2022, the plot is telling to us the although Mkpx-22 has a small fatality ratio, the people with Mkpx-22 symptoms although are released from Mkpx-22, some or all of them might to pass from a recovery state from having Mkpx-symptoms.

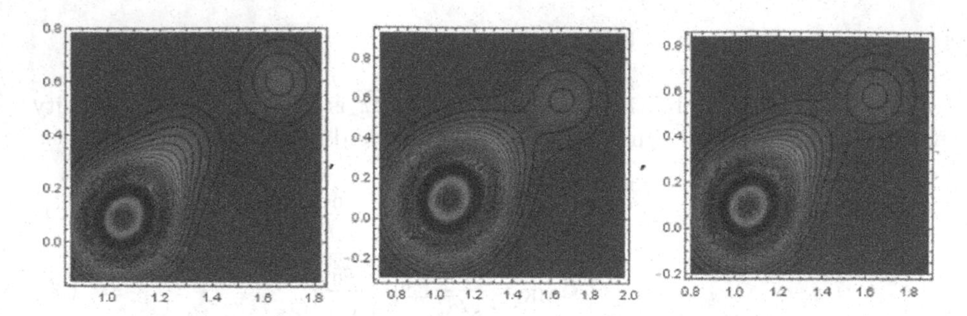

Fig. 4. The smooth density histogram [19] for a probability distribution function following Eq. 27 and one dictated by the polynomial t^{ℓ} in according to the very beginning of Mkpx-22. The ellipses begins to be closer each other.

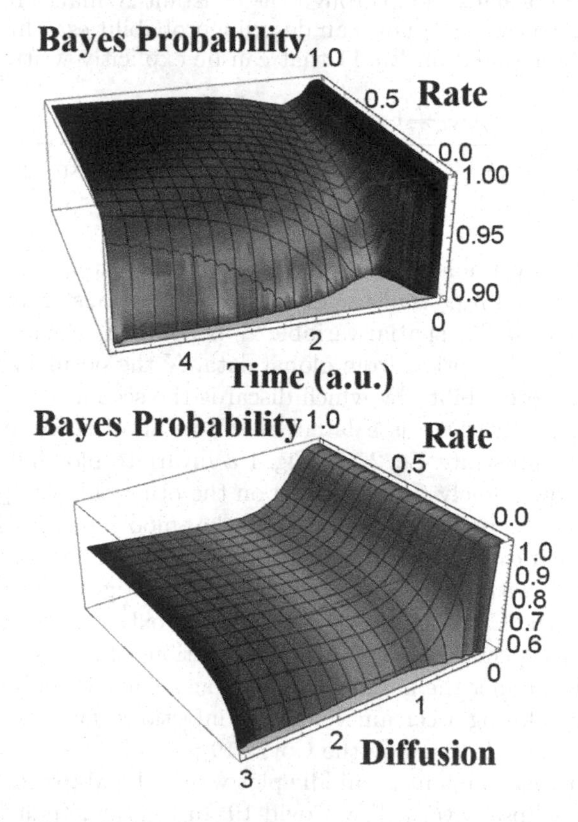

Fig. 5. The probability of Bayes as function of "Rate" and time (up) as well as "Rate" and the diffusion constant. In both cases the arrow is indicating the characteristic of Mkpx-22 and Covid-19. Here it is noted the infections and probabilities for the case of Covid-19 remain in its 50% of getting the virus [19].

In Fig. 5 up panel, the case of Eq. 27 when $p_{CV} = p_{MK} = u$ thus Eq. 27 is actually a 3D surface $P_{\mathbf{B}}(u,t)$. Thus the variable u defined as "Rate" is varied between 0 and 1. The highest value means that Covid-19 establishes a scenario of pandemic (first and second wave for example) and for the lowest values, it is clearly indicating the apparition of Mkpx-22. Thus for example, the reader can see the arrow is indicating the lowest values of "Rate" and for short periods, the Bayes probability Eq. 27 falls to the smallest values of that. It is actually of Mkpx-22 that only would emerge for short periods as seen at May 2022 at the very beginning at UK, Spain and Portugal. On the other side, one can see that Covid-19 keeps up to a 50% of lethality as seen in 3D plot. In down panel the case when Eq. 27 is plotted as function of "Rate" and diffusion constant. The arrow is indicating that for small rates and small diffusion constant the Bayes probability is also small. Although it is associated to Mkpx-22 one can also assume that this scenario encompasses well to Covid-19 in the sense that the Bayes probability is opting for small values as a response to the end of wave of pandemic but always by keeping at least a 50% of probability for getting the infection even in the case of having received a certain number of shots (vaccinations).

5 Conclusion

In this paper, the usage of Bayes theorem has been directly for testing in the case that two pandemics can coexist simultaneously. According to simulations, it has been verified in terms of probabilities, the Covid-19 is stable against Mkpx-22 fact that makes us to suppose that for capabilities to be scattered geographically, Covid-19 can remains in time, while Mkpx-22 although it is more diffusive, its lifetime as global pandemic is short as seen in global data. Simulations, have demonstrated that while Mkpx-22 ends in a full cycle, Covid-19 is indicating the end of a wave as experienced along 2020–2022. This would constitute the main argument to claim that Covid-19 still will manifest more pandemics, whereas Mkpx-22 or another pandemic would be active for short periods (months but not years), confirming that Covid-19 is the main pandemic whereas the other ones, are solely mini pandemics with seasonal apparition.

References

1. Petersen, E., et al.: Human Monkeypox: epidemiologic and clinical characteristics, diagnosis, and prevention. Infect. Dis. Clin. North Am. **33**(4), 1027–1043 (2019)
2. Kluge, H., Ammon, A.: Monkeypox in Europe and beyond - tackling a neglected disease together. Euro. Surveill. **27**(24), 2200482 (2022)
3. Kumar, S., et al.: Human Monkeypox outbreak in 2022. J. Med. Virol. **95**, e27894 (2022)
4. Kozlov, M.: Monkeypox outbreaks: 4 key questions researchers have. Nature **606**, 238–239 (2022)
5. Vandenbogaert, M., Kwasiborski, A., Gonofio, E., et al.: Nanopore sequencing of a Monkeypox virus strain isolated from a pustular lesion in the Central African Republic. Sci. Rep. **12**, 10768 (2022)

6. Tian, H., et al.: An investigation of transmission control measures during the first 50 days of the COVID-19 epidemic in China. Science **368**, 638–642 (2020)
7. Chakraborty, I., Maity, P.: Covid-19 outbreak: migration, effects on society, global environment and prevention. Sci. Total Environ. **728**, 138882 (2020)
8. Mathieu, E., et al.: Coronavirus Pandemic (COVID-19) (2020). Published online at OurWorldInData.org. https://ourworldindata.org/coronavirus
9. Roser, M., Mathieu, E., Spooner, F., Dattani, S., Ritchie, H., Roser, M.: Mpox (Monkeypox) (2022). Published online at OurWorldInData.org. https://ourworldindata.org/monkeypox
10. Nieto-Chaupis, H.: Pandemic of Covid-19 as global entropy: when Shannon theory fits to-date data of new infections. In: 2021 International Conference on Electronic Communications, Internet of Things and Big Data (ICEIB), pp. 298–301 (2021). https://doi.org/10.1109/ICEIB53692.2021.9686434
11. Nieto-Chaupis, H.: Identifying second wave and new variants of Covid-19 from Shannon entropy in global pandemic data. In: 2021 Fifth World Conference on Smart Trends in Systems Security and Sustainability (WorldS4), pp. 289–293 (2021). https://doi.org/10.1109/WorldS451998.2021.9514017
12. Gaglione, D., et al.: Adaptive Bayesian learning and forecasting of epidemic evolution-data analysis of the COVID-19 outbreak. IEEE Access **8**, 175244–175264 (2020). https://doi.org/10.1109/ACCESS.2020.3019922
13. Alakunle, E., Moens, U., Nchinda, G., Okeke, M.I.: Monkeypox virus in Nigeria: infection biology, epidemiology, and evolution. Viruses **12**(11), 1257 (2020)
14. Kumar, N., Acharya, A., Gendelman, H.E., Byrareddy, S.N.: The 2022 outbreak and the pathobiology of the Monkeypox virus. J. Autoimmun. **131**, 102855 (2022)
15. Jin, T., Xiong, J.: Singular extinction profiles of solutions to some fast diffusion equations. J. Funct. Anal. **283**(7), 109595 (2022)
16. Anacleto, M.A., Brito, F.A., de Queiroz, A.R., Passos, E., Santos, J.R.L.: Diffusive process under Lifshitz scaling and pandemic scenarios. Phys. A **559**, 125092 (2020)
17. Chan, G.M., et al.: Bayes' theorem, COVID19, and screening tests. Am. J. Emerg. Med. **38**(10), 2011–2013 (2020)
18. Liu, Y., Yangyang, Yu., Zhao, Y., He, D.: Reduction in the infection fatality rate of Omicron variant compared with previous variants in South Africa. Int. J. Infect. Dis. **120**, 146–149 (2022)
19. https://www.wolframalpha.com/

Monte Carlo Simulation of Arbitrium and the Probabilistic Behavior of Bacteriophages

Huber Nieto-Chaupis(✉)

Universidad Autónoma del Perú, Panamericana Sur Km. 16.3 Villa el Salvador, Lima, Peru
hubernietochaupis@gmail.com

Abstract. From the fact that phage takes a random decision to opt either by lysis or lysogeny, this paper has carried out a Montecarlo simulation of the attack of phage against bacteria. When it is assumed random variables along the attack, then it would produce a indecision for having lysis or lysogeny that is quantitatively seen as a small probability for both scenarios. Thus a delay emerges that is favorable to bacteria in the sense that phages might to be disorganized and destabilize because their ambition. This can also be understood as a scenario of interference by which the molecular messengers emitted by phages is abundant, so that more subprocesses together to lysis and lisogeny might be manifested. It has been assumed that messengers become negatively or positively charged, yielding to attraction or repulsion among genes. Thus, the decision for lysis and lysogeny would have a random origin in conjunction to Coulomb's forces.

Keywords: Arbitrium · Bacteriophage · Bayes theorem

1 Introduction

With the discovery of system baptized as Arbitrium by Rotem Sorek in 2017 [1], it was seen so far a progressive increasing of literature corresponding to the ways as phages are communicate each other in order to carry out a decision as to go through lsysis or lysogeny. In fact, it was discovery that bacteriophage makes a kind of coordination previous to the "final assault" in hostage bacteria. Although not clear in all if these coordinations have as central purpose the saving of time or energy, the phage might be behaving as a perfect criminal microbial targeting the extermination of their prey guided by randomness to some extent, so that deterministic aspects that are unknown emerge as a interesting window whose arguments might not be falling in a biological territory but also, in aspects that are totally determined by laws of physics and probabilities. As seen in [2,3] such decision might be as a group more than individual. Nevertheless, this apparent democracy exhibits a lack of consistency about the origin and true purpose of system Arbitrium. In other words, it is not clear if it is totally random or

Y. Chen et al. (Eds.): BICT 2023, LNICST 512, pp. 101–109, 2023.
https://doi.org/10.1007/978-3-031-43135-7_10

partially deterministic. In one phase of system, gen aimP is emitted so that, the accumulation of these genes in an extracellular manner to initialize the actions towards the decision might be in according to a concrete statistics. Thus, while not any manifestation of segregated peptides, then one can assign a probabilistic role at the sense that not all peptides would be received by aimP receptors along the DNA integration [4]. Despite the fact that the quorum sensing might be a very sophisticated regulatory systems, the presence of electrical factors either in phages or peptides makes the system arbitrium in one sign-dependent so that one would expect either attraction or repulsion. The case of repulsion might be a drastic scenario because peptides are displacing each other without to reach its aimR in the hijacked bacteria [5,6].

The rest of this paper is a follows: In second section, once the event of substrate depletion by a population of bacteria is defined, each subprocess is mathematically projected onto a scenario of Bayes's theorem. From this probability, some illustrative examples based in simulations are presented. Interestingly is found that the Bayesian probability is periodic in some ranges of space, fact that would trigger the idea that diffusivity and Bayes's theorem would have a short matching in space and time.

2 The Bayesian Interpretation

Before of going to a fully Bayesian interpretation of Arbitrium, it would be pedagogical to write in a nutshell about it [7]. In this manner the sequence can be listed as follows: (i) Phage adheres to a bacteria in order to take it as a pure hostage without any chance to be released in an independent manner. One can note that this hijacking has a only purpose: Destruction of bacteria or keep them alive to improve the phage population [8]. (ii) This adherence targets to deposit the AimR and AimP genes. It is noteworthy that AimR express the key gen AimX. (iii) When AimX is activated, the action of lysogeny is blocked out, so that lysis begins. (iv) AimP releases peptides (six amino acids long) as a form of secret communication. (v) Peptides can be taken by oligopeptide permease (OPP) transporter. Once inside hostage bacteria, AimR binds peptides so that the gene cannot activate AimX anymore. (vi) Lysogeny can be carry out. With this sequence one can define a list of probabilities: Probability of activation AimX because AimR P_{RX}. Probability of releasing peptides by AimP P_{PP}. Probability of uptake peptides by AimR P_{RPP}. In this manner, one can apply the conditional concepts of Bayes's theorem in fully relation to Arbitrium as follows: Probability for a decision of lysis: $P_{\mathrm{RX}} \times P_{\mathrm{PP}}$. Probability for lysogeny: $1 - P_{\mathrm{RX}} \times P_{\mathrm{PP}}$ and the probability to confirm lysogeny: P_{RPP}. Here one can see that the sum of all probabilities both lysis and lysogeny is equal to unity. The mathematical transcription of Bayes's theorem [9] with all these probabilities can be written as (Fig. 1):

$$P_{\mathrm{B}} = \frac{P_{\mathrm{RX}} \times P_{\mathrm{PP}}}{P_{\mathrm{RX}} \times P_{\mathrm{PP}} + (1 - P_{\mathrm{RX}} \times P_{\mathrm{PP}}) \times P_{\mathrm{RPP}}}. \tag{1}$$

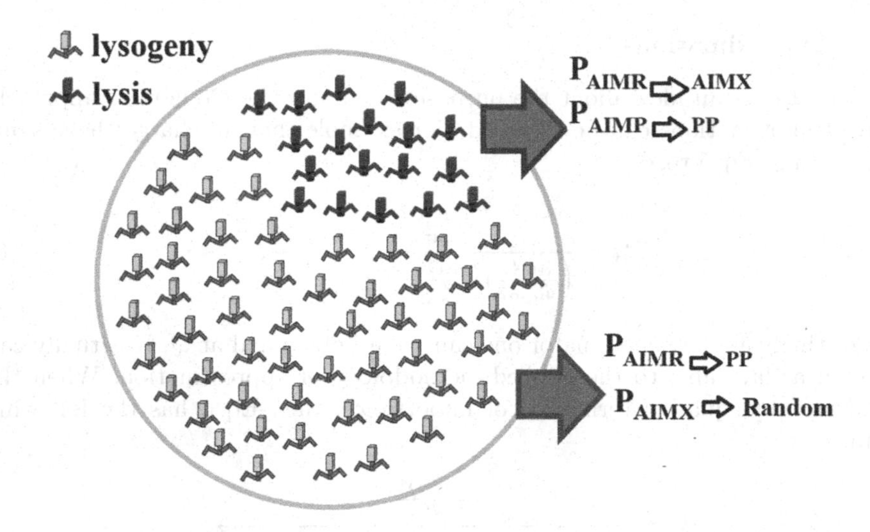

Fig. 1. Sketch of a scenario for Arbitrium where the option lysis is favorable. Thus the probabilities are generated in according to the conditional probability of Bayes's theorem. The main question associated to this sketch is: What is probability that a fraction of viruses take the option of lysis? The meaning of the acronyms is provided in text.

3 The Physics of Arbitrium

Equation 1 can be considered as a master equation in the sense that from it, various interpretations can be done. A possible is inside the territory of physics. In fact, from the fact that genes and peptides have electrical charges in their surface, then them can behave in according to well-known Coulomb's law, it is only can exist either attraction or rejection [10,11]. In this manner and by following the observations of Sorek one can define a kind of electrical origin to the actions for each one of probabilities defined above. In this manner one can write down:

$$P_{\mathrm{AimR}\rightarrow\mathrm{AimX}} = P_{\mathrm{RX}} = \frac{\beta Q_{\mathrm{R}} Q_{\mathrm{X}}}{d^2_{\mathrm{RX}}} \tag{2}$$

$$P_{\mathrm{AimP}\rightarrow\mathrm{Pep}} = P_{\mathrm{PP}} = \frac{\gamma Q_{\mathrm{P}} Q_{\mathrm{PP}}}{d^2_{\mathrm{PP}}} \tag{3}$$

$$P_{\mathrm{AimR}\rightarrow\mathrm{Pep}} = P_{\mathrm{RPP}} = \frac{\alpha Q_{\mathrm{R}} Q_{\mathrm{PP}}}{d^2_{\mathrm{RPP}}}, \tag{4}$$

with Q_{R} and Q_{P} the charges of AimR and AimP genes, respectively. Indeed, Q_{X} and Q_{PP} charges of AimX gen and peptides, respectively. In this manner Eq. 1 denotes the probability for a decision for lysis.

$$P_{\mathrm{B}} = \frac{\frac{\beta Q_{\mathrm{R}} Q_{\mathrm{X}}}{d^2_{\mathrm{RX}}} \otimes \frac{\gamma Q_{\mathrm{P}} Q_{\mathrm{PP}}}{d^2_{\mathrm{PP}}}}{\frac{\beta Q_{\mathrm{R}} Q_{\mathrm{X}}}{d^2_{\mathrm{RX}}} \otimes \frac{\gamma Q_{\mathrm{P}} Q_{\mathrm{PP}}}{d^2_{\mathrm{PP}}} + \left(1 - \frac{\beta Q_{\mathrm{R}} Q_{\mathrm{X}}}{d^2_{\mathrm{RX}}} \otimes \frac{\gamma Q_{\mathrm{P}} Q_{\mathrm{PP}}}{d^2_{\mathrm{PP}}}\right) \otimes \frac{\alpha Q_{\mathrm{R}} Q_{\mathrm{PP}}}{d^2_{\mathrm{RPP}}}} \tag{5}$$

3.1 Approximations

In order to get an idea about the purpose to employ the Coulomb's approach, some approximations can be assumed. For example that all charges have same value. Thus, Eq. 5 reads:

$$P_{\mathrm{B}} = \frac{\frac{\beta\gamma Q^4}{d_{\mathrm{RX}}^2}}{\left(\frac{\alpha Q^2}{d_{\mathrm{RPP}}^2} + \frac{\beta\gamma Q^4}{d_{\mathrm{RX}}^2} - \frac{\beta\gamma\alpha Q^6}{d_{\mathrm{RX}}^2 d_{\mathrm{RPP}}^2}\right)}. \tag{6}$$

Interestingly at the denominator one can see a polynomial at Q. It actually can be seen as favorable to the applied methodology of approximation. When the quadratic and quarter terms are of importance, then Eq. 6 has the following form:

$$P_{\mathrm{B}} = \frac{\frac{\beta\gamma Q^4}{d_{\mathrm{RX}}^2}}{\left[\frac{\alpha}{d_{\mathrm{RPP}}^2}\left(Q^2 + \frac{\beta\gamma d_{\mathrm{RPP}}^2 Q^4}{\alpha d_{\mathrm{RX}}^2}\right) - \frac{\beta\gamma\alpha Q^6}{d_{\mathrm{RX}}^2 d_{\mathrm{RPP}}^2}\right]}. \tag{7}$$

Thus with the scale:

$$\frac{\beta\gamma d_{\mathrm{RPP}}^2}{\alpha d_{\mathrm{RX}}^2} = 1/2! \tag{8}$$

then one can see that an exponential function exists there in the form as: $Q^2 + 1/2!Q^4 + 1/3!Q^6 + ... \approx \mathrm{Exp}Q^2$. An additional assumption is that of $Q^{2\ell} \approx 0$ if $\ell \geq 3$. With this one can arrive to:

$$P_{\mathrm{B}} = \frac{\frac{\beta\gamma Q^4}{d_{\mathrm{RX}}^2}}{\left[\frac{\alpha}{d_{\mathrm{RPP}}^2}(\mathrm{Exp}Q^2)\right]} = \frac{\beta\gamma Q^4 d_{\mathrm{RPP}}^2}{\alpha d_{\mathrm{RX}}^2}\mathrm{Exp}(-Q^2). \tag{9}$$

With Eq. 8 the scale allows to write a Weibull-like function for the Bayes's probability that reads:

$$P_{\mathrm{B}} = \frac{Q^4}{2!}\mathrm{Exp}(-Q^2). \tag{10}$$

In this way, Eq. 10 represents the probability to decide on lysis. A point of interest turns out to be the derivatives of P_{B}. In Fig. 2 (left panel) the case of Eq. 10 (labeled by 0), and its first (labeled by 1) as well as second derivative (labeled by 2) are shown. The first and second have plotted as their square to put away the negative values. A tentative generalization of Eq. 10 can be written as: $P_{\mathrm{B}} = \frac{Q^{2m}}{2!}\mathrm{Exp}(-Q^m)$ with $m \geq 2$. In virtue of distributions of Fig. 2 and the factor 1/2!, the one can define the probability for decision on lysis is a n-th derivative of Bayes's probability is given by $P_{\mathrm{B,n}}^{\mathrm{SY}} = 0.5\frac{d^n}{dQ^n}\left(Q^{2m}\mathrm{Exp}(-Q^m)\right)$.

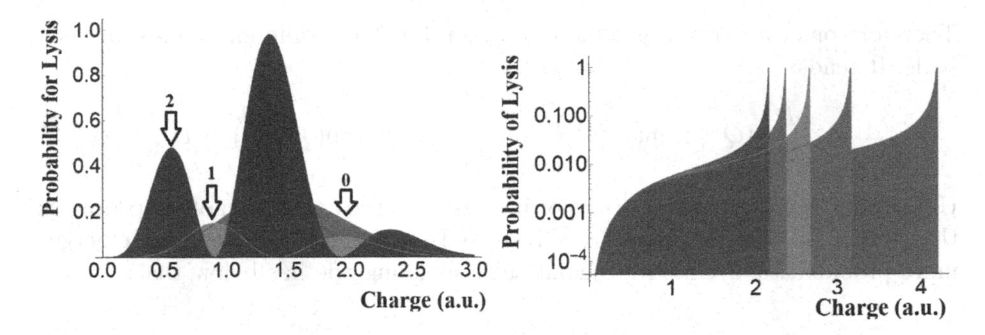

Fig. 2. Left-side: Plotting of Eq. 10 (green color) with first (magenta color) and second derivative (blue color). Right-side: Probability to execute lysis as function of electric charge. A polynomials $10^{-2}Q^4/(Q^2 + Q^4 - (0.01 + j * 0.05)Q^6)$ was used for this illustration (details see text below). (Color figure online)

On the other hand one can also determine in a logic way the probability for lysogeny is given for a similar expression= $P^{\mathrm{SG}}_{\mathrm{B,k}} = 0.5\frac{d^k}{dQ^k}\left(Q^{2m}\mathrm{Exp}(-Q^m)\right)$ for the k-th derivative of Bayes's probability. Therefore the total probability of "indecision" is written in a straightforward manner as:

$$0.5\frac{d^n}{dQ^n}\left[Q^{2m}\mathrm{Exp}(-Q^m)\right] + 0.5\frac{d^k}{dQ^k}\left[Q^{2m}\mathrm{Exp}(-Q^m)\right] = 1. \tag{11}$$

A compact form for writing Eq. 11 is then given by:

$$0.5\frac{d^h}{dQ^h}\left[\delta_{n,h}Q^{2m}\mathrm{Exp}(-Q^m) + \delta_{k,h}Q^{2m}\mathrm{Exp}(-Q^m)\right] = 1, \tag{12}$$

where $\delta_{n,h}$ and $\delta_{k,h}$ the Dirac-delta functions whose purpose here is the factorization of the generalized derivative $\frac{d^h}{dQ^h}$. In Fig. 2 (right panel) Eq. 6 was illustrated as: $10^{-2}Q^4/(Q^2 + Q^4 - (0.01 + j * 0.05)Q^6)$ up to for 5 values of j (red j = 1, green j = 2, yellow j = 3, red j = 4 and purple j = 5). It is noted that two well-defined phases are seen. When charges acquire their high values, the probabilities for all cases are reaching to 1. It might be the case that AimX is blocked out and to guarantee lysys, AimP releases large volumes of peptides.

4 The Cost for Indecisions

The factor 1/2 in Eq. 12 defined as a kind of scale (see Eq. 8) to balance the physical units because the Coulomb's force has been introduced as part of probabilistic model can also be generalized in conjunction of derivatives. It is actually seen in Fig. 2 by which the peaks of distributions are changing with the order of derivative. When not any decision is executed then one expects that the factors 1/2 might be falling down without any chance to back to the initial values.

Therefore one can write a general expression involving different values for such scale. It reads:

$$\xi_1 \frac{d^n}{dQ^n} \left(Q^{2m}\mathrm{Exp}(-Q^m)\right) + \xi_2 \frac{d^k}{dQ^k} \left(Q^{2m}\mathrm{Exp}(-Q^m)\right) = 1. \tag{13}$$

Unfortunately for phages the cost for indecision is the apparition of subprocesses that might be different away from lysis and lysogeny. It implies that emerges more probabilistic events. Mathematically speaking one has below that:

$$\sum_{\ell}^{L} \xi_\ell \frac{d^\ell}{dQ^\ell} \left(\omega^2 Q^{2m}\mathrm{Exp}(-Q^m)\right) = 1. \tag{14}$$

with ξ_ℓ a probability of event including lysis and lysogeny, ranging between 0 and 1, where ω is a constant related to charge Q. Again one can generalize this in the sense that $\omega^2 Q^{2m}$ is part of a infinite sum: $\omega Q^m + \omega^2 Q^{2m} + \omega^3 Q^{3m} + \ldots$. Thus one gets:

$$\sum_{\ell}^{L} \xi_\ell \frac{d^\ell}{dQ^\ell} \left([\mathrm{Exp}(\omega Q^m)]\mathrm{Exp}(-Q^m)\right) = \sum_{\ell}^{L} \xi_\ell \frac{d^\ell}{dQ^\ell} \mathrm{Exp}[Q^m(\omega - 1)] = 1. \tag{15}$$

One can see that when $L >> 1$ then both lysis as lysogeny might be strongly affected and not any action is going to be taken. This can happen in events where AimP cannot segregate peptides neither the extracellular conditions appear to be favorable for the propagation of them. Another possibility is the clogging of peptides to go out from the hijacked bacteria. Another fact is the possible interference among the events inside bacteria. It is remarked this since virus might not be so efficient to manage various actions, and instead opt by the ones that can guarantee the fast replication without the necessity to destroy its hostage. For example when AimR activates AimX to inhibit lysogeny, the action that AimP releases peptides through the extracellular zone might origin some interference towards the activation of AimX. Thus emerges the question: Can peptides also to be able to activate AimX. This might to confirm that if ξ_ℓ does not belong to either lysis or lysogeny then one can speak about the possible existence of latency state by which not any decision due to interference has been taken.

5 Monte Carlo Simulation

So far, it is clear that the role of released peptides is twofold, in one hand them are ejected from AimP in a lysis event, and on the other hand them are binded to AimR to deactivate AimX so that lysogeny is executed. These events can be part of a computational code as presented at the end of this paper. In fact, a main loop runs over a sample of phages that are adhered to bacteria. Thus the

Coulomb's force is used. The probability that AimR activates the inhibitor AimX is estimated and actually it is dependent on the normalization constant β. Actually here it should be noted that the total charge would come from an operation of integration on the charge density. In other words: $Q = \int \rho(s)dV$. Aside one can realize that the solving of density would be dictated by the diffusion equation in the sense that peptides are under extracellular propagation in both when (i) are released by AimP and (ii) shall be bonded by AimR. In fact $d\rho/dt = D\nabla^2\rho$ can be solve in a model 1D by assuming that the virus and bacteria can move in a 2D space. In radial symmetry the Bessel functions constitute an interesting solution. Also, from an initial amount of bacteria, the decisions for lysis or lysogeny are in function of charge density of peptides. In Fig. 3 up to 4 cases de charge density have been plotted. Here the percent of depleted bacteria is done as function the density of peptides in the extracellular case. The Monte Carlo [12] has been built with the Metropolis algorithm: accept or rejection. Thus once the peptide density has been solved then it is normalized. In conjunction to this the volume of bacteria has been estimated in according to peptide density [13]. The questions: (i) Is the normalized volume of peptides greater than a random number, then accept the depletion for fractions. The code that has been built takes a percent per each time that the acceptation or rejection is formulated. The results in Fig. 3 reveals that randomness might be behind the releasing and uptake of peptides. This stochastic view of Arbitrium is dependent on electric charge of genes.

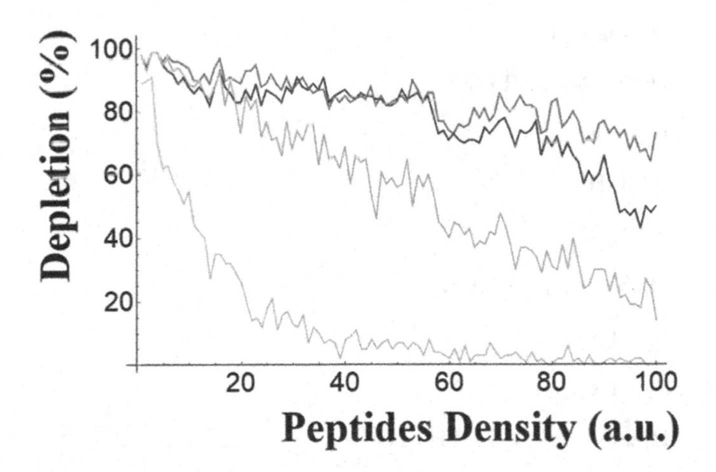

Fig. 3. Results of Monte Carlo simulation showing the percent of depletion as function of peptides density. At abundance of peptides the depleted population of bacteria does not follow a linear relation. This simulation that randomness is a key element in Arbitrium.

PSEUDO CODE TO ESTIMATE PHAGE DECISION

1	**DO N** $=1, N_{TOT}$
2	$P_{RX} = \frac{\beta q_R q_X}{d^2}$ (activates inhibidor)
3	CALL random(rnd1)
4	**IF** $P_{RX} >$ rnd1 (decision on lysis)
5	lx = lx + 1
6	**IF** $N = N_{MIN}$ **THEN**
7	lyx = lx
8	**ENDIF**
9	**ENDIF**
10	$P_{RPP} = \frac{\gamma q_P q_R}{d_{PP}^2}$ (untake peptides)
11	CALL random(rnd2)
12	**IF** $P_{RPP} >$ rnd2 (decision on lysogeny)
13	lg = lg + 1
14	**IF** $N = N_{MAX}$ **THEN**
15	lyg = lg
16	**ENDIF**
17	**ENDIF**
18	**IF** $N = N_{TOT}$ **THEN**
19	$x_1 =$ lyx/TOT
20	$x_2 =$ lyg/TOT
21	**IF** $x_1 + x_2 < 0.90$ **THEN**
22	Print $\frac{\text{Log}(\beta \otimes \gamma)}{\text{rnd1} \otimes \text{rnd2}}$
23	**ENDIF**
24	**ENDIF**
25	**END**

6 Conclusion

The phenomenon of Arbitrium discovered by Rotem Sorek at 2017 has been simulated with the Monte Carlo technology. To accomplish this a theory based at the Bayes theorem has been developed. The construction of this theory has demanded to assign some physical roles to the genes interaction as well as to the releasing and uptake of peptides. Thus, the states of lysis and lysogeny might not be the only ones, but also more gene-based events also exist and have a probability, fact the minimizes the priority for phages to carry out Arbitrium. The physics has been modeled by the Coulomb-like forces that allows attraction and repulsion in fully concordance to the discovery of Rotem Sorek. Finally the algorithm Monte Carlo has been designed under the view that bacteria population exhibits depletion in response to phages through a decision as to lysis or lysogeny. Thus, the probability of Bayes has been directly compared to random numbers. Simulated distributions are presenting a decreasing at the percent of bacteria population when the peptides density increases, thus phage requires of a large density of these amino acids to accomplish the annihilation of bacteria after their hijacked without negotiation. In future work, a 3D simulation shall be done.

References

1. Erez, Z., et al.: Communication between viruses guides lysis-lysogeny decisions. Nature **541**, 488–493 (2017)
2. Bruce, J.B., et al.: Regulation of prophage induction and lysogenization by phage communication systems. Curr. Biol. **31**, 5046–5051.e7 (2021)
3. Dou, C., et al.: Structural and functional insights into the regulation of the lysis-lysogeny decision in viral communities. Nat. Microbiol. **3**, 1285–1294 (2018)
4. Guan, Z., Pei, K., Wang, J., et al.: Structural insights into DNA recognition by AimR of the arbitrium communication system in the SPbeta phage. Cell Discov. **5**, 29 (2019)
5. Gallego del Sol, F., Quiles-Puchalt, N., Brady, A., et al.: Insights into the mechanism of action of the arbitrium communication system in SPbeta phages. Nat. Commun. **13**, 3627 (2022)
6. Neiditch, M.B., et al.: Genetic and structural analyses of RRNPP intercellular peptide signaling of gram-positive bacteria. Annu. Rev. Genet. **51**, 311–333 (2017)
7. Cui, Y., et al.: Bacterial MazF/MazE toxin-antitoxin suppresses lytic propagation of arbitrium-containing phages. Cell Rep. **41**(10), 111752 (2022)
8. Silpe, J.E., Bassler, B.L.: A host-produced quorum-sensing autoinducer controls a phage lysis-lysogeny decision. Cell **176**(1–2), 268–280.e13 (2019)
9. Costa, A.C.S., Angelo, R.M.: Bayes' rule, generalized discord, and nonextensive thermodynamics. Phys. Rev. A **87**, 032109 (2013)
10. Hu, T., Shklovskii, B.I.: Kinetics of viral self-assembly: role of the single-stranded RNA antenna. Phys. Rev. E **75**, 051901 (2007)
11. Safdari, M., et al.: Effect of electric fields on the director field and shape of nematic tactoids. Phys. Rev. E **103**, 062703 (2021)
12. Lyubartsev, A.P., et al.: electrostatically induced polyelectrolyte association of rodlike virus particles. Phys. Rev. Lett. **81**, 5465 (1998)
13. Chen, T., Glotzer, S.C.: Simulation studies of a phenomenological model for elongated virus capsid formation. Phys. Rev. E **75**, 051504 (2007)

Instant Messaging Application for 5G Core Network

Fan-Hsun Tseng(✉), Tung-Yi Wu, I-Lung Chang, Chia-Chen Hsu, and Yi-Cen Chen

Department of Computer Science and Information Engineering, National Cheng Kung University, Tainan 70101, Taiwan
skittles2567@gmail.com

Abstract. Instant messaging (IM) has been widely used for many years since Internet technology developed and all-IP network architecture proposed. 5G mobile network launched worldwide a few years ago for pursuing ultra-low latency and high speed data transmission. In this paper, an IM application is developed in an open source 5G core (5GC) network framework, i.e. the free5GC. To validate the developed IM application operates properly in free5GC, we utilized the Tshark to examine and capture packets. Experimental results showed that the messages of the developed IM application passed through the free5GC network.

Keywords: 5G · Core Network · Instant Messaging

1 Introduction

The fifth-generation mobile networks (5G) is the latest generation of mobile communications technology with the purposes of high data rates, reduced latency, energy savings, reduced costs, increased system capacity, and large-scale device connectivity [1]. Due to the overwhelming network demands along with rapid development of network, Internet service providers worldwide upgrade the access network and core network to 5G. The cost of purchasing and upgrading network hardware has increased dramatically to meet the different functions of information system, such as computing, storage, networking, security. Therefore, some technologies such as Network Function Virtualization (NFV) [2, 3] and Software-Defined Networking (SDN) [4] are flourishing. With these technologies, the cost of investment become lower compared to the dedicated hardware provided by the supplier. In addition, NFV can automate a variety of management tasks [5] to reduce human errors, shorten deployment time, and provide better network productivity and efficiency. The above conditions can noticeably reduce the procurement and operating costs.

International Mobile Communications (IMT) standardized the IMT-2020 for pursuing 5G standard. In IMT-2020, the three ultimate goals of 5G mobile network are Enhanced Mobile Broadband (eMBB), Ultra-Reliable Low Latency Communications (uRLLC) and Massive Machine-Type Communications (mMTC). However, these three characteristics belong a trade-off problem and are partially contradictory. For instance,

Y. Chen et al. (Eds.): BICT 2023, LNICST 512, pp. 110–121, 2023.
https://doi.org/10.1007/978-3-031-43135-7_11

mMTC services require low energy consumption and ultra-low latency [6] but may not need high transmission data rate.

A mobile communication network system is composed of core network (CN), backbone network, Radio Access Network (RAN), and user equipment (UE), and so on. The core network is a network system consisting of various communication systems, shared by all users, and is responsible for transmitting backbone data to achieve large data transmission. The core network for 5G is known as the Next Generation Core (NGC). The 5G core network framework in this paper is based on the above-mentioned virtualized network technology. In the 21st century, instant messaging (IM) [7] has long become an indispensable tool in our life. Instant messaging is a system, application, service for communicating over the Internet, and the content transmitted can be text, files, and audio/video. Unlike the early days of email service, instant messaging emphasizes that communication between two parties is real-time. In recent years, IM service is no longer limited to text service, IM becomes more popular due to the rapid development of Voice over IP (VoIP) and video conferencing services [8].

In this paper, we propose and design an IM application with Python programming language. The designed IM application is implemented in an open source 5G core network project, i.e. the open-source project free 5G core network (free5GC throughout the paper). The designed IM application is implemented in the free5GC platform. Then, the designed IM application is validated its functionality through capturing packets by using Tshark. Experimental results showed and proved that the designed IM application works correctly in the free5GC platform. Further IM applications such as VoIP and video conferencing services can be founded on the proposed IM application in this paper.

The rest of the paper is organized as follows. Section 2 discusses core network applications and related works of instant messaging in 5G. In Sect. 3, the system model of free5GC and its installation are introduced. The design of the proposed instant messaging application is presented in Sect. 4. Finally, Sect. 5 concludes this work and provides its future directions.

2 Background and Related Work

There are existing three open source 5G core network projects, i.e., the free5GC [9] hosted by the Communication Service/Software Laboratory in National Yang Ming Chiao Tung University in Taiwan, the Open5GS [10] developed in Korea, and the Open Mobile Evolved Core (OMEC) [11] developed by the Open Networking Foundation (ONF). The three open source 5G core networks are compared as follows. The free5GC project is stated that it was the first open-source 5G core network in the world. The free5GC supports for setting up physical networks and low hardware requirements. The characteristic of Open5GS is that it can choose either 4G core-based configuration or 5G core-based configuration in the core network architecture, but the Open5GS needs higher hardware requirements than the free5GC. The particularity of OMEC project is that it can be integrated with other ONF projects, but suffers from the highest hardware requirements with compared to the free5GC and Open5GS. After evaluating the hardware requirement and system capability of the three open 5G core network projects, the proposed IM application is implemented in the free5GC platform.

The technology of network virtualization brings many advantages to the research field of 5G core networks, but may also results in security concerns [12]. In [13], the authors analyzed the 5G proprietary mobile network architecture based on the 3rd Generation Partnership Project (3GPP) technical specifications. They explored three different scenarios, and the establishment of a core network focus on private can effectively increase information security. In [14], the authors described the Open Network Automation Platform (ONAP) architecture supporting network slicing in 5G communication systems and its new capabilities for creating network slices. The ONAP not only manages and controls complex infrastructures based on network virtualization and software-defined networks but also slices the core network into multiple virtual core networks for supporting different functions. Each network slice is able to operate independently. In [15], the authors considered that the 5G system operators cannot know whether the methods of network traffic delivery such as switching and branching are performed correctly. Therefore, a 5G core network traffic monitoring system was proposed that allows operators to easily identify switching and branching traffic using traffic monitoring messages delivered by 5G core network functions. In addition to establish a secure private 5G core network environment, this paper also concentrates on developing a real-time communication application that can operate in this environment.

In [16], the authors believed that client-based IM applications have some drawbacks, such as scalability, single point of failure (SPOF), and vulnerability to Distributed Denial-of-Service (DDoS) attacks. As a result, they proposed a semantic point-to-point (P2P) approach to build an IM application. The designed P2P-based IM is a simpler design approach with the shortcoming of insufficient security. However, using the designed approach to implement on private networks can solved this problem with some restrictions. Due to the different platforms between users, private networks may suffer from information barriers after successful setup. Therefore, the authors in [17] proposed a technique to implement a cross-platform real-time communication system through HTTP, XMPP and TCP protocols, and through frameworks such as SSH, DWR and ExtJS. In addition, the authors studied the quality of service (QoS) mechanism when multiple terminals are accessed by multiple users at the same time.

In [18], the authors stated that IM communication is widely used but users' topics are diverse. In order to assist users in capturing the topics and contents of IM communication without reading all messages, they proposed a new method for detecting topics of instant messaging. The concept is suitable for certain situations, such as useless words keep appearing, the instant messages are very short, or multiple languages are used. In [19], the authors discussed the expansion and extension of IM technologies in Internet of Things (IoT). IoT IM aims to allow users communicate with each other through mobile phones or terminal equipment no matter where they are. Then, the authors proposed some key technologies to build a powerful real-time communication framework that can handle a large scale IoT platform. Furthermore, they have developed some applications based on above-mentioned technologies, e.g. the smart home system.

3 Experiment Environment Setup

In this section, the experimental environment of core network is introduced. The core network is implemented on a virtual machine and the used OS version is Ubuntu 20.04.3 LTS. In the beginning, we create the first virtual machine as a carrier for the core network (CN) and make sure it can connect to the external network. The ping command is applied to contact the Google server, and waits for the responded packets sent from Google server, which is shown as Fig. 1. Finally, we can see that the virtual machine has successfully connected to external network.

```
ubuntu@ubuntu:~$ ping google.com
PING google.com (142.251.42.238) 56(84) bytes of data.
64 bytes from tsa01s11-in-f14.1e100.net (142.251.42.238): icmp_seq=1 ttl=110 time=8.90 ms
64 bytes from tsa01s11-in-f14.1e100.net (142.251.42.238): icmp_seq=2 ttl=110 time=7.42 ms
64 bytes from tsa01s11-in-f14.1e100.net (142.251.42.238): icmp_seq=3 ttl=110 time=7.73 ms
^C
--- google.com ping statistics ---
3 packets transmitted, 3 received, 0% packet loss, time 2004ms
rtt min/avg/max/mdev = 7.420/8.017/8.900/0.637 ms
```

Fig. 1. Verify the status of network connection

After that, two virtual machines are created as a user equipment (UE) and a Next Generation Node B (gNB). Next, modify the host name and the local network IP address to facilitate instance identification during experiment process. The IP addresses of core network, user equipment and gNB are 192.168.56.101, 192.168.56.102 and 192.168.56.103, respectively. In addition, they are abbreviated as CN, UE102, and UE103 throughout experiments in this paper. Lastly, to validate the UE102 and UE103 are connected each other through core network rather than external network, the external network cards of them are disabled.

To construct the core network environment, the free5GC is downloaded from GitHub [20] with version 3.0.5. Note that some required packages should be downloaded and installed according to the readme document. The network parameters such as IP addresses of components in core network should be modified. Then, download the customized Linux kernel module as well as the gtp5g to handle packets. It is a core module developed for the Linux OS kernel to handle packets for the N4 interface which defined by 3GPP. After that, the core network can be executed and launched. Note that the connection status of core network and the configurations of network interface can be examined by using commands. Once the core network has been successfully established, we can set up the UE102 and UE103 in core network.

With respect to UE, the UERANISM project is downloaded and installed in the UE102. The UERANISM project is used to simulate UE and base stations (BSs). Before connecting to core network, execute the script in the virtual machine of core network and add set the IP address of UE102 as a subscriber. Then, execute the script of UERANISM project to boot user device and BS, and check the connection between cloud network and core network. A successful connection means that the overall core network environment has been successfully built up. Last, set up and check the UE103 by following the same above-mentioned steps to perform experiment environment.

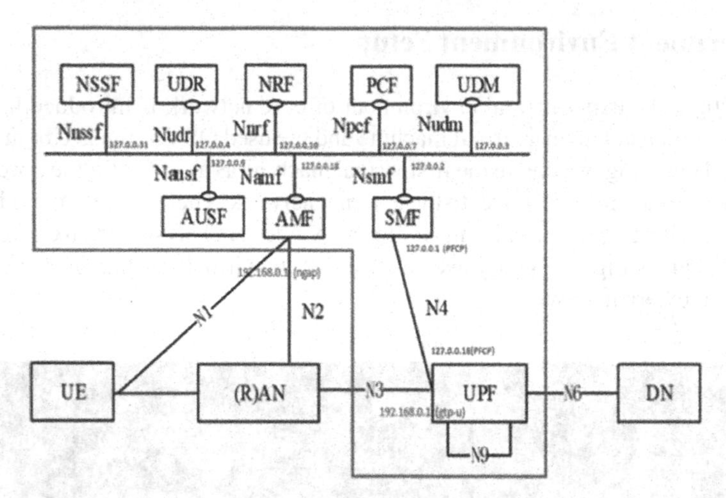

Fig. 2. The framework of free5GC

The framework and setting of the implemented free5GC is illustrated in Fig. 2. The default components of the free5GC are marked and framed in the orange lines. There are 9 components, i.e., the Network Slice Selection Function (NSSF), Unified Data Repository (UDR), Network Function Repository Function (NRF), Policy Control Function (PCF), Unified Data Management (UDM), Authentication Server Function (AUSF), Access and Mobility Management Function (AMF), Session Management Function (SMF), and User Plane Function (UPF). The functionality of these components is introduced as follows.

The NSSF is responsible for selecting a slice collection for serving users. On the basis of subscriber's data and location, the NSSF selects the AMF for users. The UDR aims to store subscription-related data, e.g., user contract, static contract, and policy data. The NRF is responsible for registering, managing, and checking the status of network. The PCF provides control and implementation of all network policies in control plane. The UDM is responsible for data management, such as identity authentication and certification repository. The UDM selects authentication method based on the subscriber identity and the configured policy, and calculates authentication data and key data as needed. The AUSF certificates the access authentication of subscribers from 3GPP and non-3GPP users.

In experiments, the settings of three components are modified, i.e., the AMF, SMF, and UPF. The AMF manages and authorizes users to access core network, such as the management of registration, connection, and reachability. The configuration of AMF is captured in Fig. 3. The IP address of the ngapIpList is changed from 127.0.0.1 to 192.168.56.101, which is marked with red line in Fig. 3. The SMF component establishes, modifies and release user sessions, and allocates and manages users' IPs. In addition, the SMF provides functions of charging and roaming. The configuration of SMF is captured in Fig. 4. The interface IP address is set to 192.168.56.101, which is marked with red line in Fig. 4. Lastly, the UPF component provides functions of routing, forwarding, packets transmission, and the users' QoS management. The configuration of UPF is captured in Fig. 5. The IP address of the gtpu is set to 192.168.56.101.

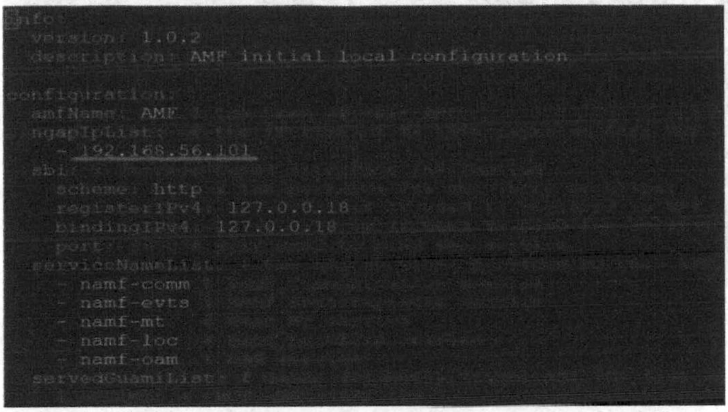

```
info:
  version: 1.0.2
  description: AMF initial local configuration

configuration:
  amfName: AMF
  ngapIpList:
    - 192.168.56.101
  sbi:
    scheme: http
    registerIPv4: 127.0.0.18
    bindingIPv4: 127.0.0.18
    port:
  serviceNameList:
    - namf-comm
    - namf-evts
    - namf-mt
    - namf-loc
    - namf-oam
  servedGuamiList:
```

Fig. 3. Configuration of AMF

```
    node_id: 127.0.0.8
    sNssaiUpfInfos:
      - sNssai:
          sst:
          sd:
        dnnUpfInfoList:
          - dnn: internet
            pools:
              - cidr: 60.60.0.0/16
      - sNssai:
          sst:
          sd:
        dnnUpfInfoList:
          - dnn: internet
            pools:
              - cidr: 60.61.0.0/16
    interfaces:
      - interfaceType: N3
        endpoints:
          - 192.168.56.101
        networkInstance: internet
links:
  - A: gNB1
    B: UPF
```

Fig. 4. Configuration of SMF

The network architecture of the experiment is captured in Fig. 6. We add the users as well as UE 102 and UE 103 and base stations to the free5GC framework. Note that to simplify the diagram, we illustrate the experiment architecture with core network and the UE102 only but there is another UE103 in practical experiment. The difference between UE102 and UE103 is IP addresses only.

```
debugLevel: info
ReportCaller:

pfcp:
  - addr: 127.0.0.8

gtpu:
  - addr: 192.168.56.101

dnn_list:
  - dnn: internet
    cidr: 60.60.0.0/24
```

Fig. 5. Configuration of UPF

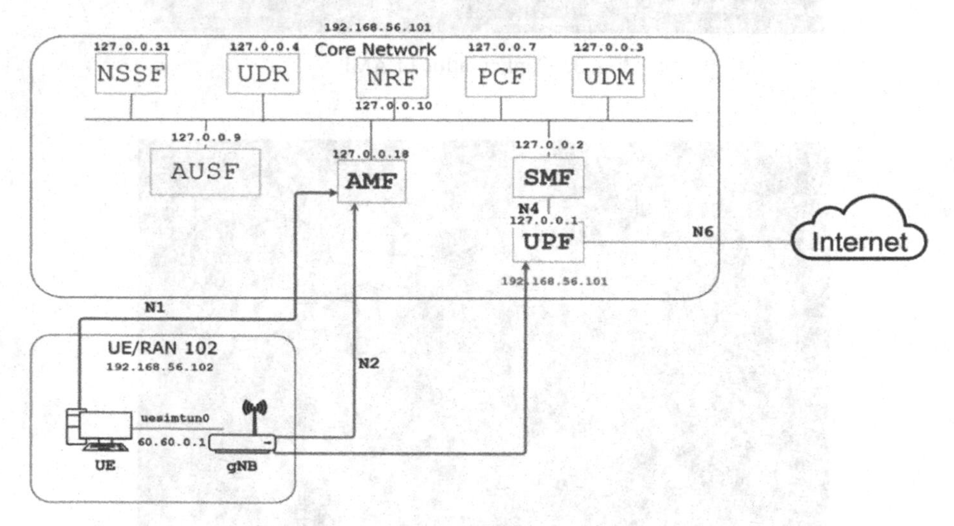

Fig. 6. The network architecture of experiment

4 Instant Messaging Application Deployment

In this section, the designed IM application is introduced. The IM application is developed by Python. The function of the program is to create a socket on both UEs so that they can communicate with each other through this socket communications. The operation steps of the IM application are described as follows.

1. Both UEs execute one after another and establish sockets on both UEs.
2. Enter one of the pre-defined mode codes on one side to activate IM types, e.g., "**m**" for text messages and "**f**" for file transmission. Then, the communication starts.

a). If text message mode "**m**" is selected, then UEs can start sending text messages. In Fig. 7, the top side is the sender and bottom side is the receiver. The sender sends "testing" message to receiver, then the receiver receives the message sent from sender with the beginning of "Incoming message".

b). If file transmission mode "**f**" is selected, then files can be transmitted through the designed IM application. In Fig. 8, the top side is the sender and bottom side is the receiver. The sender enters a file name "pi_200K.txt" but there is no such file, hence the file transmission is unsuccessful. The sender enters a file name "pi_204K.txt", and then the receiver receives a message "File transmission complete" represents that the file transmission is completed.

3. UEs can stop text or file mode by entering "stop", then the process of IM backs to step 2.
4. If any UE wants to terminate the IM application, they can enter "exit" while the IM application stage is in step 2.

```
waiting for connection
connection setup
m for message, f for file, exit to exit program:
m
Enter message: testing
Enter message:
```

```
waiting for connection
connection setup
m for message, f for file, exit to exit program:
Incoming message: testing
```

Fig. 7. Text mode of the IM

```
[sudo] password for ubuntu:
waiting for connection
connection setup
m for message, f for file, exit to exit program:
f
enter filename: pi_200K.txt
file does NOT exist, aborting
enter filename: pi_204K.txt
File transmission complete
enter filename:
```

```
[sudo] password for ubuntu:
waiting for connection
connection setup
m for message, f for file, exit to exit program:
File transmission complete
```

Fig. 8. File mode of the IM

In this paper, we utilized Tshark [21] to examine the packets of the proposed IM application pass through free5GC or not. The major goal is to confirm that the packets of IM application have passed through the 5G core network. Tshark is a packet analyzer tool in command line mode, which is similar to the tcpdump command. Tshark is developed by Wireshark without a Graphical User Interface (GUI).

The validation process is described as follows. Firstly, execute the IM application and make sure the transmission status works correctly. Then, access CN, UE102 and UE103 by using Secure Shell Protocol (SSH) and then monitor the network interfaces of three components by Tshark. Finally, execute IM application to transmit messages and then monitor the components through Tshark. The expected packet flow is illustrated in Fig. 9.

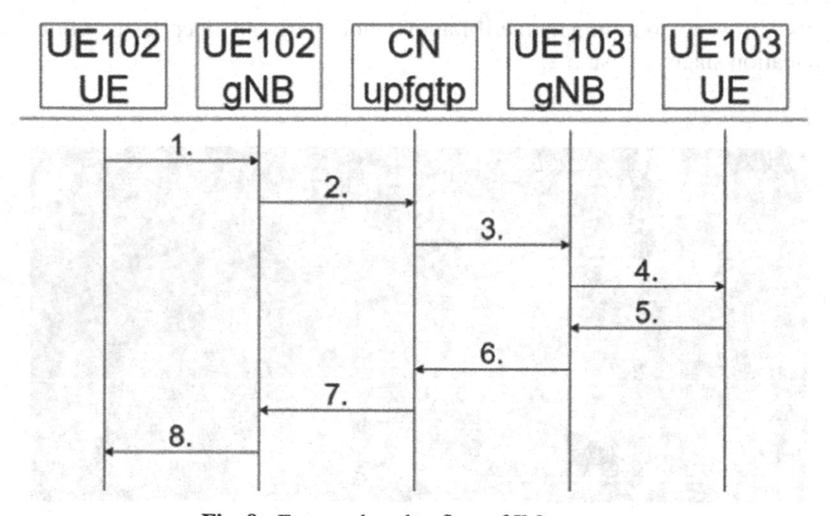

Fig. 9. Expected packet flow of IM message

Three network interfaces are monitored, i.e., the upfgtp in CN, the uesimtun0 tunnel of the UE102, and the uesimtun0 tunnel of the UE103. In all experiments, the IM communication initialized at UE102 sends a message to UE103. The packets captured in CN, UE102, and UE103 are shown as Figs. 10, 11, and 12, respectively.

The beginning field of the numbered sequences is the timestamp of a packet. According to these timestamps, it can be observed that the UE102 sends packets at the earliest on 07:50:44.338431664, which is shown in Fig. 10. Then, the UE 103 receives packets on 07:50:44.403228908, which is shown in Fig. 11. After that, the CN receives packets on 07:50:44.430627594, which is shown in Fig. 12. The results validate that the packets of the designed IM application passed through CN from UE102 to UE103 correctly. However, we found that the packets flow is not as expected.

The practical packet flow of the proposed IM is shown as Fig. 13. Note that the external network interfaces of UE102 and UE103 have been disabled to guarantee that the IM application works correctly in the free5GC. Although the practical packet flow does not match our expected packet flow, the packets of the IM application passed through CN based on the result of capturing packets in Fig. 12.

```
ubuntu@ue102:~$ sudo tshark -i uesimtun0 -t a
Running as user "root" and group "root". This could be dangerous.
Capturing on 'uesimtun0'
    1 07:50:44.338431664    60.60.0.1 → 60.60.0.2    TCP 59 60142 → 6000
0 [PSH, ACK] Seq=1 Ack=1 Win=510 Len=7 TSval=255852167 TSecr=2282608875
    2 07:50:44.339060211    10.0.2.15 → 60.60.0.1    ICMP 87 Redirect
         (Redirect for host)
    3 07:50:44.340207777    60.60.0.2 → 60.60.0.1    TCP 52 60000 → 6014
2 [ACK] Seq=1 Ack=8 Win=506 Len=0 TSval=2282631900 TSecr=255852167
```

Fig. 10. The monitored packets in UE102

```
ubuntu@ue103:~/socket$ sudo tshark -i uesimtun0 -t a
Running as user "root" and group "root". This could be dangerous.
Capturing on 'uesimtun0'
    1 07:50:44.403228908    60.60.0.1 → 60.60.0.2    TCP 59 60142 → 6000
0 [PSH, ACK] Seq=1 Ack=1 Win=510 Len=7 TSval=255852167 TSecr=2282608875
    2 07:50:44.403238973    60.60.0.2 → 60.60.0.1    TCP 52 60000 → 6014
2 [ACK] Seq=1 Ack=8 Win=506 Len=0 TSval=2282631900 TSecr=255852167
    3 07:50:44.404229436    10.0.2.15 → 60.60.0.2    ICMP 80 Redirect
         (Redirect for host)
```

Fig. 11. The monitored packets in UE103

```
ubuntu@free5gc:~$ sudo tshark -i upfgtp -t a
Running as user "root" and group "root". This could be dangerous.
Capturing on 'upfgtp'
    1 07:50:44.430627594    60.60.0.1 → 60.60.0.2    TCP 59 60142 → 60000
 [PSH, ACK] Seq=1 Ack=1 Win=510 Len=7 TSval=255852167 TSecr=2282608875
    2 07:50:44.430671043    10.0.2.15 → 60.60.0.1    ICMP 87 Redirect
         (Redirect for host)
    3 07:50:44.430687636    60.60.0.1 → 60.60.0.2    TCP 59 [TCP Retransm
ission] 60142 → 60000 [PSH, ACK] Seq=1 Ack=1 Win=510 Len=7 TSval=25585216
7 TSecr=2282608875
    4 07:50:44.431796082    60.60.0.2 → 60.60.0.1    TCP 52 60000 → 60142
 [ACK] Seq=1 Ack=8 Win=506 Len=0 TSval=2282631900 TSecr=255852167
    5 07:50:44.431822964    10.0.2.15 → 60.60.0.2    ICMP 80 Redirect
         (Redirect for host)
    6 07:50:44.431835915    60.60.0.2 → 60.60.0.1    TCP 52 [TCP Dup ACK
4#1] 60000 → 60142 [ACK] Seq=1 Ack=8 Win=506 Len=0 TSval=2282631900 TSecr
=255852167
```

Fig. 12. The monitored packets in CN

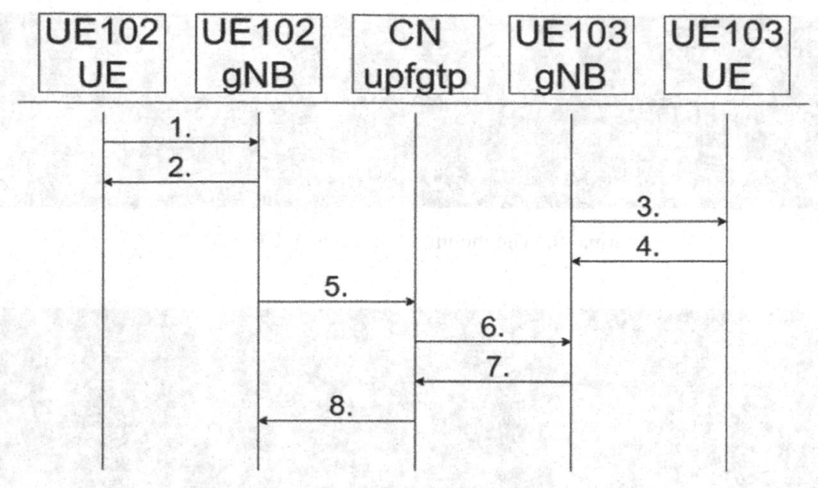

Fig. 13. Practical packet flow of IM message

5 Conclusion and Future Works

The 5G applications are on the fly based on the mobile network technology development. The paper presented an instant messaging application implemented by Python and realized in the free5GC framework. The experimental results showed that the packets of the IM application truly pass through the free5GC but the sequence of packets is not as expected. We presume that it may be attributed to the modification of free5GC's settings or the synchronization of core network and UEs. In the future works, we intend to fix the problem by synchronizing these components. Furthermore, we will try to propose VoIP and video conferencing applications.

References

1. Chen, C.-Y., Tseng, F.-H., Lai, C.-F., Chao, H.-C.: Network planning for mobile multi-hop relay networks: network planning for MMR networks. Wirel. Commun. Mob. Comput. **15**(7), 1142–1154 (2015). https://doi.org/10.1002/wcm.2396
2. Zhang, H., Wang, Y., Qiu, X., Li, W., Zhong, Q.: Network operation simulation platform for network virtualization environment. In: 17th Asia-Pacific Network Operations and Management Symposium (APNOMS), pp. 400–403. IEEE, Busan, South Korea (2015)
3. Tseng, C.-W., Huang, Y.-K., Tseng, F.-H., Yang, Y.-T., Liu, C.-C., Chou, L.-D.: Micro operator design pattern in 5G SDN/NFV network. Wirel. Commun. Mob. Comput. **2018**(3471610), 1–14 (2018)
4. Tseng, F.-H., Chang, K.-D., Liao, S.-C., Chao, H.-C., Leung, V.C.M.: sPing: a user-centred debugging mechanism for software defined networks. IET Networks **6**(2), 39–46 (2017)
5. Wollschlaeger, M., Sauter, T., Jasperneite, J.: The future of industrial communication: automation networks in the era of the internet of things and industry 4.0. IEEE Indust. Electron. Magaz. **11**(1), 17–27 (2017)
6. Sabuj, S.R., Ahmed, A., Cho, Y., Lee, K.-J., Lo, H.-S.: Cognitive UAV-aided URLLC and mMTC services: analyzing energy efficiency and latency. IEEE Access **9**, 5011–5027 (2021)

7. Dhir, A., Kaur, P., Rajala, R.: Continued use of mobile instant messaging apps: a new perspective on theories of consumption, flow, and planned behavior. Soc. Sci. Comput. Rev. **38**(2), 147–169 (2020)
8. Kaufmann, K., Peil, C.: The mobile instant messaging interview (MIMI): Using WhatsApp to enhance self-reporting and explore media usage in situ. Mobile Media Commun. **8**(2), 229–246 (2020)
9. free5GC. http://www.free5gc.org/. Accessed 21 Nov 2022
10. Open5GS. http://open5gs.org/. Accessed 21 Nov 2022
11. Open Mobile Evolved Core. http://opennetworking.org/omec/. Accessed 21 Nov 2022
12. Jang, H., Jeong, J., Kim, H., Park, J.: A survey on interfaces to network security functions in network virtualization. In: IEEE 29th International Conference on Advanced Information Networking and Applications Workshops, pp. 160–163. Gwangju, South Korea (2015)
13. Guo, Y., Zhang, Y.: Study on core network security enhancement strategies in 5G private networks. In: IEEE 21st International Conference on Communication Technology (ICCT), pp. 887–891. Tianjin, China (2021)
14. Lee, W., Na, T., Kim, J.: How to create a network slice? - a 5G core network perspective. In: 21st International Conference on Advanced Communication Technology (ICACT), pp. 616–619. PyeongChang, Korea (2019)
15. Kim, E., Choi, Y.: Traffic monitoring system for 5G core network. In: Eleventh International Conference on Ubiquitous and Future Networks (ICUFN), pp. 671–673. Zagreb, Croatia (2019)
16. Huang, L.: Instant messaging based on semantic P2P network. In: Fourth International Conference on Networking and Distributed Computing, pp. 33–35. Los Angeles, CA, USA (2013)
17. Fu, C., Tang, Y., Yuan, C., Xu, Y.: Cross-platform instant messaging system. In: 12th Web Information System and Application Conference (WISA), pp. 27–30. Jinan, China (2015)
18. Zhang, H., Wang, C., Lai, J.: Topic detection in instant messages. In: 13th International Conference on Machine Learning and Applications, pp. 219–224. Detroit, MI, USA (2014)
19. Su, K., Cheng, H., Wang, H., Lv, R.: Instant messaging application for the internet of things. In: IEEE International Conference Computer Science and Service System (CSSS), pp. 166–169. Nanjing, China (2012)
20. Open source 5G core network base on 3GPP R15. http://github.com/free5gc/free5gc. Accessed 21 Nov 2022
21. Tshark. http://tshark.dev/. Accessed 21 Nov 2022

Genetic Algorithm-Based Fair Order Assignment Optimization of Food Delivery Platform

Min-Yan Tsai[1], Guo-Yu Lin[2], Jiang-Yi Zeng[2], Chia-Mu Yu[1], Chi-Yuan Chen[2], and Hsin-Hung Cho[2(✉)]

[1] Department of Information Management and Finance, National Yang Ming Chiao, Tung University, Hsinchu, Taiwan
chiamuyu@nycu.edu.tw

[2] Department of Computer Science and Information Engineering, National Ilan University, Yilan, Taiwan
{n1043027,chency,hhcho}@niu.edu.tw, r1043002@ems.niu.edu.tw

Abstract. Most existing food delivery platforms lack responsibility when it comes to route planning. This often results in uneven assignment of orders or difficulty in arranging orders for delivery drivers. These issues have led to loss of consumer rights and reduced revenue for delivery platforms, as well as negative feedback and evaluations. To address this problem, it is necessary to first resolve the issue of uneven distribution of orders. In this paper, we propose using the Genetic Algorithm (GA) to solve the order assignment optimization problem. By utilizing GA's strong global search ability, we can achieve fair assignment of orders, optimize delivery routes, and balance revenue distribution. This approach creates a fair competition environment for delivery drivers and improves service quality, ultimately leading to positive feedback from consumers and creating a win-win situation.

Keywords: Artificial intelligence · genetic algorithm · traveling salesperson · delivery route planning

1 Introduction

In the era of the COVID-19 pandemic, people have increasingly been cutting back on eating out and staying at home, which has led to a rise in the use of delivery platforms for daily necessities. According to the Market Intelligence & Consulting Institute (MIC) in Taiwan, an investigation found that 72% of consumers have experience using delivery services in 2021. Among the most popular platforms, foodpanda has 78% and UberEats has 61% market share. However, the sudden influx of orders during peak hours can often lead to a misestimation of the delivery staff's ability to take and deliver orders, resulting in uneven distribution of orders and causing delays or disputes between consumers and couriers. This can ultimately harm the platform's reputation.

Y. Chen et al. (Eds.): BICT 2023, LNICST 512, pp. 122–132, 2023.
https://doi.org/10.1007/978-3-031-43135-7_12

This paper will discuss the use of metaheuristic algorithms for solving multi-objective optimization problems in delivery operations [1]. The approximate optimal solution can be effectively found by considering multiple factors such as the number of delivery personnel, the number of orders, service quality, delivery revenue, and delivery distance. The goal is to balance the rights of the delivery platform, delivery staff and consumers as much as possible. The algorithm will help assign suitable orders to delivery staff, ensuring timely delivery and reducing complaints from consumers. The multi-objective path optimization problem is an NP-hard problem [2, 3], which makes it difficult to solve using traditional algorithms. Therefore, this study proposes an architecture based on Genetic Algorithm (GA) to find a solution to this problem.

This paper is divided into five sections. Section 2 provides background information and related work on the Genetic Algorithm (GA) and Traveling Salesman Problem (TSP). Section 3 uses linear programming to define the delivery order problem. The simulation results of our proposed method are presented in Sect. 4, and the conclusion is given in the last section.

2 Related Works

2.1 Genetic Algorithm (GA)

The metaheuristic algorithm can give a combinatorial optimal solution at affordable cost through the objective observation and principle. The so call cost can be temporal, spatial or anything which can be defined [4]. Furthermore, the metaheuristic algorithm refers to the use of past experience or rules to define a random search methods in a population or individual. Compared with greedy-based algorithm, although the solution cannot be guaranteed that is definitely better, the concept of the conditional random process can avoid the local optimum and find the approximate optimal solution. Moreover, in a very hard problem, the cost will be less than the greedy approach. Therefore, these methods are widely used in much more applications which has the demand of optimization. The common and classical algorithm are Hill Climbing, (HC) [5], Simulated Annealing, (SA) [6] Ant Colony Optimization, (ACO) [7] and GA [8], etc.

GA was proposed in 1975 by John Henry Holland. It is according to the Nature selects, the fittest survives theory to choose the solutions. This mechanism is to reserve the fitness value better's solutions to weed out the poor fitness values solutions. GA major focus on the optimal solution search with combinatorial objectives. It has since been used in many fields such as scheduling optimization [9], data mining [10], finical prediction [11], process management [12], etc. The basic GA is the simulation of the gene evolutionary process the process including Selection, Crossover, and Mutation. The solution of the random combination is an individual that is generally represented as a set of sequences and named a chromosome. In the initialization of GA, the initial population is created that including several chromosomes. Then GA will choose two chromosomes to operate crossover and mutation. It can be getting new solutions and comparing which solutions need to eliminate. In general, the eliminating solutions have poor fitness values. The process of selection, crossover, and mutation are continuously iterated until the termination condition is reached.

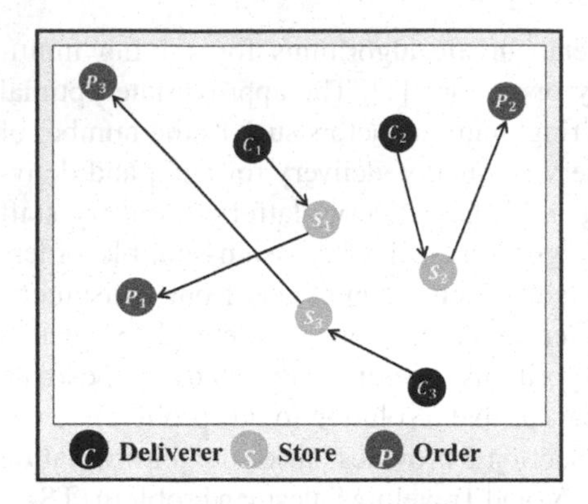

Fig. 1. Schematic diagram of the deliverer delivering the order

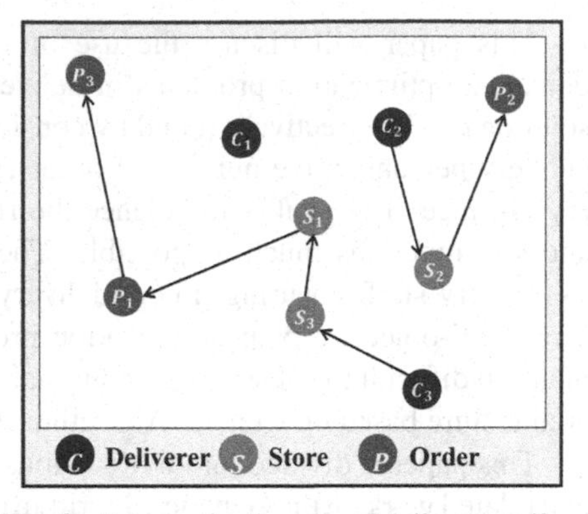

Fig. 2. Schematic diagram of delivery staff stacking orders

Some scholars have used GA to solve the Vehicle Routing Problem (VRP). This is a path planning problem for vehicles starting from warehouses and delivering goods. The similarities with this study are the characteristics of delivery from point A to point B and continuous delivery to different places and finding the shortest path [13]. Therefore, this study also GA will be taken as the main method.

2.2 Traveling Salesman Problem (TSP)

TSP is a classic NP-Hard problem proposed in the 19th century [14], which is still being studied in the field of discrete combinatorial optimization problems. The problem is defined as a set n cities where the distance between any 2 cities can be explicitly calculated. When the salesman's travel speed is at the same speed, find the shortest circuit to start from the designated city to visit each city without repeating it, and finally return to the starting city. If the exhaustive method is used to find the best solution for n cities, the time complexity is O(n!). Therefore, the more cities you visit, the less you can solve in a reasonable time. This paper wants to solve the problem of the deliverer's food delivery. However, this problem is more difficult than the original TSP that this problem considers the scenario which includes multi deliverers and the multi paths. Optimization problems with multiple variables and multiple objectives are practical problems for many complex engineering optimizations. The biggest difficulty is that when optimizing for a single goal or variable, it is necessary to exclude conflicts between other goals or variables, resulting in poor results for other goals or variables [14]. For example, the problem defines three salespeople and nine cities, to meet the conditions all 3 salesmen must travel to 3 cities, each city is visited by only 1 salesperson, and each salesperson does not repeatedly visit the cities that other people have visited at the same time.

3 Problem Definition

The delivery platform need to make order P divided the deliverer C. P and C is the sets $P = \{P_1, P_2, \ldots, P_m\}$ and $C = \{C_1, C_2, \ldots, C_n\}$. The coordination of the partner store S is the set $S = \{S_1, S_2, \ldots, S_o\}$ The delivery path assigned by the deliverer C_i is noted as R^{C_i} and R^{C_i} include the partner store S_k of the order P_j that will be passed through. It can be noted as $R^{C_i} = \{\{p_1, p_2, \ldots, p_x\}, \{s_1, s_2, \ldots, s_y\} | p_i \in P, s_i \in S\}$. When the platform divided the order to deliverer C_i is consider the between partner store s_j and the consumer p_k the Euclidean distance as $d^{C_i}_{s_j p_k}$ the next get order distance as $d^{C_i}_{S_x P_k}$ and find a deliverer that is closer to the partner store. The distance of that deliverer arrives at the first store is $d1^{C_i}_{p_1}$. When the deliverer finishes the job that can obtain the new order the flow is shown in Fig. 1. In addition, the platform also supplies the deliverer have much order to send at the same time its names as stacked orders that is shown in Fig. 2. The number of stacked orders varies according to the settings of each delivery platform. When the deliverer is not choose stacked orders, the deliverer unknown the next order or the store's location is near or far, which cause much more problem in delivery. If the deliverer chose stacked order method, the food delivery path will be shown after the platform accept this request. If the deliverer finds the path is not expected, the deliverer can abandon the order, but it may cause delays in overall delivery. The stacked orders have to be competed from the each deliverer, otherwise the meaning of fairness will disappear.

The delivery platform's Business Model is to build customer brand loyalty [15]. However, once the order is delayed, the consumer will lose trust in the platform. The optimization problem of the delivery path can be regarded as a TSP, and the optimal path can be obtained by a metaheuristic algorithm to shorten the delivery path of the deliverer. It's just that this only addresses a single TSP. Generally speaking, a delivery platform will have many delivery people delivering meals at the same time. It means that the Multiple Traveling Salesman Problem (MTSP) is still not solved. The deliverer can obtain enough orders and each order has a reasonable distance, Making every deliverer receive fair dispatch and remuneration, and consumers can also receive orders within a reasonable and fair time so the platform can conform to the reasonableness brought about by the principle of fair labor, and the consumer is reduced complain at the same time.

The distance of the deliverer path is $D^{C_i} = d1^{C_i}_{p_1} + \bigcup_{j=1}^{x} \sum_{k=1}^{y} d^{C_i}_{s_j p_k}$. This paper hope that each deliverer receive at least one order. Therefore we use any two D^{C_i} and D^{C_j} to find the standard deviation σ_{ij}. Our main purpose is to pursue the minimum total distance and at the same time consider the minimum difference between the individual distances of each courier, which means that each deliverer C has received orders, which means that this mechanism is fairer. The linear programming model is as follows:

Minimize $\sum_{i=1}^{n} \rho_i$

s.t.

$$\sigma_{ij} \sim 0, i \neq j, i, j \leq n,$$

ρ is the fitness function which means pursuit the lowest standard deviation and the shortest total distance that is shown in the following:

$$\rho = \frac{1}{\sum_{i=1}^{n} D^{C_i}} \times \frac{1}{\sum_{i=1}^{n} \sum_{j=1}^{n} \sigma_{ij}} \tag{1}$$

4 Proposed Method

This paper proposed the mechanism is use GA-based MTSP optimization. The first step is to randomly generate the orders P_i, stores S_i, and deliverers C_i to combine into a chromosome. The initialization phase defines a lot of chromosomes $Z = \{Z_1, Z_2, \ldots, Z_p\}$ and then operate the selection, crossover and mutation to create the new chromosome. Finally, the population will reserve the best chromosome via evaluation until the termination condition is reached. The process is shown in Fig. 3. The overall operation is as follows:

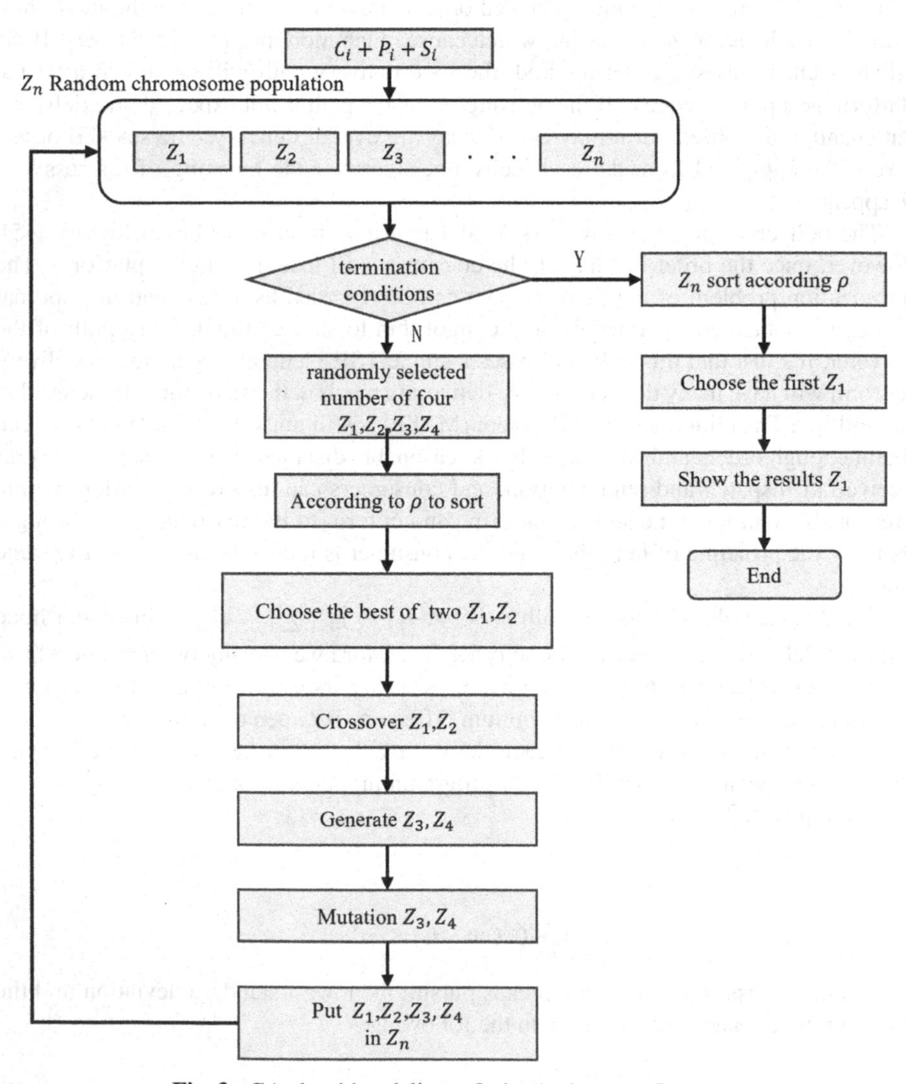

Fig. 3. GA algorithm delivery Order Assignment flowchart

1. Randomly generate P_iS_k and C_i to combine into chromosome.
2. Generate the multiple chromosomes to population Z.
3. Randomly chose 4 chromosomes from Z.
4. Define the fitness function via Eq. (1).
5. Choose the top two Z_1 and Z_2 depending on the sorting results of the fitness function.
6. Randomly invite 30% of Z_1 and Z_2 individually to exchange for crossover.
7. The new chromosomes Z_3andZ_4 will be created that is shown Fig. 4.
8. Take the random order between the non-deliverer position and the non-route last position in a chromosome for mutation such as the position of P_2S_2 and the position ofP_4S_4. Arrange and combine the store coordination (S_2, S_4) and consumer coordination (P_2,P_4) of P_2S_2 andP_4S_4, and select the shortest combination and put it back to the position of chromosome$P_2P_4S_4S_2$. The mutation rate is set 30%.
9. Put the Z_1, Z_2, Z_3, Z_4 into the Z that is shown in Fig. 5
10. Repeat the above steps until the termination condition is reached.
11. After terminating the iteration, sort Z_p according to the smallest standard deviation and the shortest total distance in ascending power, and select the first chromosome.
12. This chromosome is the best solution in the whole domain. The calculation is over.

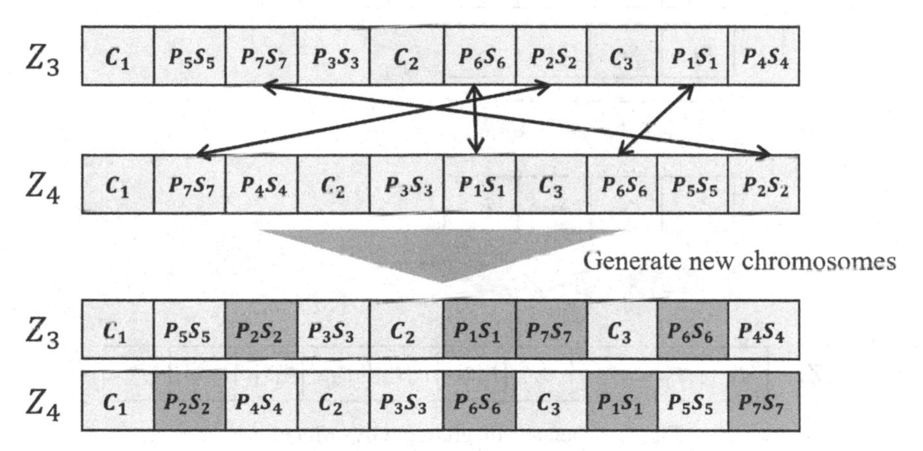

Fig. 4. Schematic diagram of GA's crossover

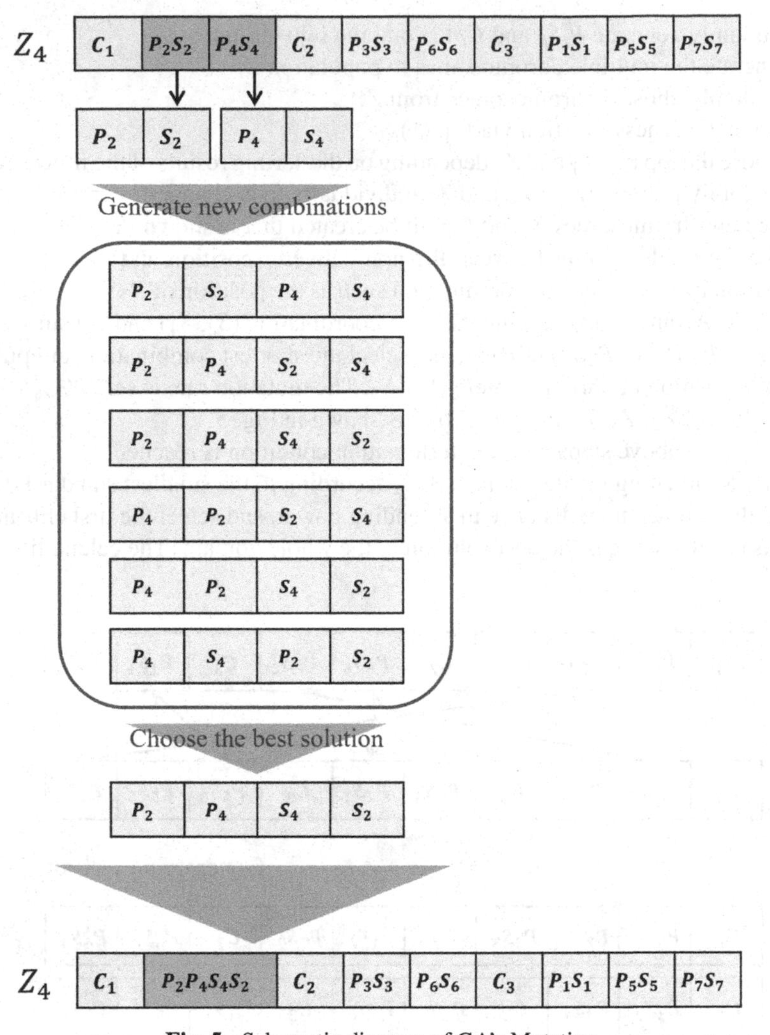

Fig. 5. Schematic diagram of GA's Mutation

The goal of this study is to make the deliverer who can use the system to obtain sufficient quantity and proper distance to reduce the occurrence of order grabbing by deliverer.

5 Simulation Results

This experiment uses C# to build and cooperate with winform's rapid development features, and then refers to Chart components for data visualization that is shown in Table 1. When the algorithm staring, an object name and random coordination should be created. The main thing is to randomize each deliverer and order and then combine them as evenly as possible that is shown in Table 2. The population initial total distance is 3157.35. The standard deviation is 431.94. We can find that the distance of SalesMan_003 is 1451.99 and the distance of SalesMan_001 is 258.66 that there is a significant difference between each other. It is not met the fairness of this paper. After many iterations, the resulting population is significantly improved that the distance of SalesMan_003 is reduced from 1451.99 to 629.3 and the distance of SalesMan_001 is increased from 258.66 to 628.80 that the difference is closer. Although the total distance raises to 3446.71 after the optimization, the standard deviation is reduced to 41.94. This result has met the fairness. This is because when each deliverer has an order and their standard deviation ate similar, it represents the delivery distance of each deliverer is also similar so that the waiting time of the consumers are also more balanced.

Table 1. Experimental environment

Type	Tool
Development	Visual Studio 2019
Platform	Winform
Language	C#
Framework	.Net Framework 4.8
Visualization	Chart
Data source	Randomly generated coordinates

This paper still has unresolved issues, such as the stack orders situation. Therefore how to use GA to optimize this problem is a major point in the future.Since the order of the store to the consumer cannot be reversed so that the path looks complex and messy that is shown in Fig. 6. For example, the coordination of SalesMan_002 is [28, 27] the former coordination represents the store coordination. The latter one represents the consumer coordination. The last index is the order code. We can see that the SalesMan_002 got three orders that the orders index are 002,008 and 011. This deliverer starts from [28, 27]. Firstly, this deliverer should go to coordination's [115, 184] of store of StoreCost 002 to take an order and then deliver this meal to the consumer [92, 193]. Then this deliverer continue to go to the coordination [26, 144] of StoreCost 008 to take the second order and go to the consumer [192, 44] to deliver the order. Then return to the coordination [16, 140] of StoreCost 011 to get the third order and finally go to deliver the meal to consumer [153, 68] to finish the task.

Table 2. Simulation results

Population initial value			Optimal distance		Distance difference
Delivery men coordinates	Distance	Path coordinates	Optimal distance	Optimal coordinates	Difference
SalesMan_000 [119,175]	340.26	SalesMan_000:[119,175]→ StoreCost_009:[146,122][122,4]→ StoreCost_010:[75,60][166,106]→ StoreCost_002:[115,184][92,193]	646.10	SalesMan_000:[119,175]→ StoreCost_017:[63,40][60,83]→ StoreCost_005:[53,107][100,54]→ StoreCost_014:[76,113][186,121]→ StoreCost_013:[176,114][19,12]→ StoreCost_001:[28,27][188,32]	+305.84
SalesMan_001 [166,30]	258.66	SalesMan_001:[166,30]→ StoreCost_000:[119,175][166,30]→ StoreCost_018:[136,74][32,191]	628.80	SalesMan_001:[166,30]→ StoreCost_004:[7,59][77,107]→ StoreCost_010:[75,60][166,106]→ StoreCost_015:[121,4][21,120]→ StoreCost_003:[51,56][83,7]	+370.14
SalesMan_002 [28,27]	444.64	SalesMan_002:[28,27]→ StoreCost_005:[53,107][100,54]→ StoreCost_011:[16,140][153,68]→ StoreCost_001:[28,27][188,32]	749.15	SalesMan_002:[28,27]→ StoreCost_002:[115,184][92,193]→ StoreCost_008:[26,144][192,44]→ StoreCost_011:[16,140][153,68]	+304.51
SalesMan_003 [188,32]	1451.99	SalesMan_003:[188,32]→ StoreCost_008:[26,144][192,44]→ StoreCost_017:[63,40][60,83]→ StoreCost_003:[51,56][83,7]→ StoreCost_012:[49,141][180,61]→ StoreCost_015:[121,4][21,120]→ StoreCost_006:[42,106][104,184]→ StoreCost_013:[176,114][19,12]→ StoreCost_004:[7,59][77,107]→ StoreCost_014:[76,113][186,121]	629.30	SalesMan_003:[188,32]→ StoreCost_016:[14,194][167,105]→ StoreCost_018:[136,74][32,191]→ StoreCost_012:[49,141][180,61]	-822.69
SalesMan_004 [115,184]	661.79	SalesMan_004:[115,184]→ StoreCost_019:[139,192][39,101]→ StoreCost_016:[14,194][167,105]→ StoreCost_007:[11,193][170,96]	793.33	SalesMan_004:[115,184]→ StoreCost_000:[119,175][166,30]→ StoreCost_006:[42,106][104,184]→ StoreCost_009:[146,122][122,4]→ StoreCost_019:[139,192][39,101]→ StoreCost_007:[11,193][170,96]	+131.54
Total distance: 3157.35			Total distance: 3446.71		+289.39
Standard deviation: 431.94			Standard deviatio: 68.59		-363.35

It can be found from the delivery route of SalesMan_002 that this research can improve the method of increasing stacking orders and optimize the route to be shorter. If StoreCost 008 and 011 are stacked for delivery, a new route can be generated:

StoreCost_008Join011
[26, 144][16, 140][153, 68][192, 44]

The SalesMan 002' starting coordination is [28, 27]. Firstly, the deliverer came to coordination [115, 184] of StoreCost 002 to obtain the meal and then deliver it to consumer [92, 193]. Second, the deliverer came to coordination [26, 144] of StoreCost 008 to get the meal for [16, 140]. Finally, the deliverer will go to coordination [153, 68] of StoreCost 01 to obtain the meal then deliver to coordination [192, 44] to consumer. The optimization paths between three consecutive order can be found.

In the future, in addition to adding the method of stacking orders, this research will continue to strengthen the data visualization part, so that the coordination of merchants and consumers are more clearly marked, and directional optimization reading is added to the behavior route.

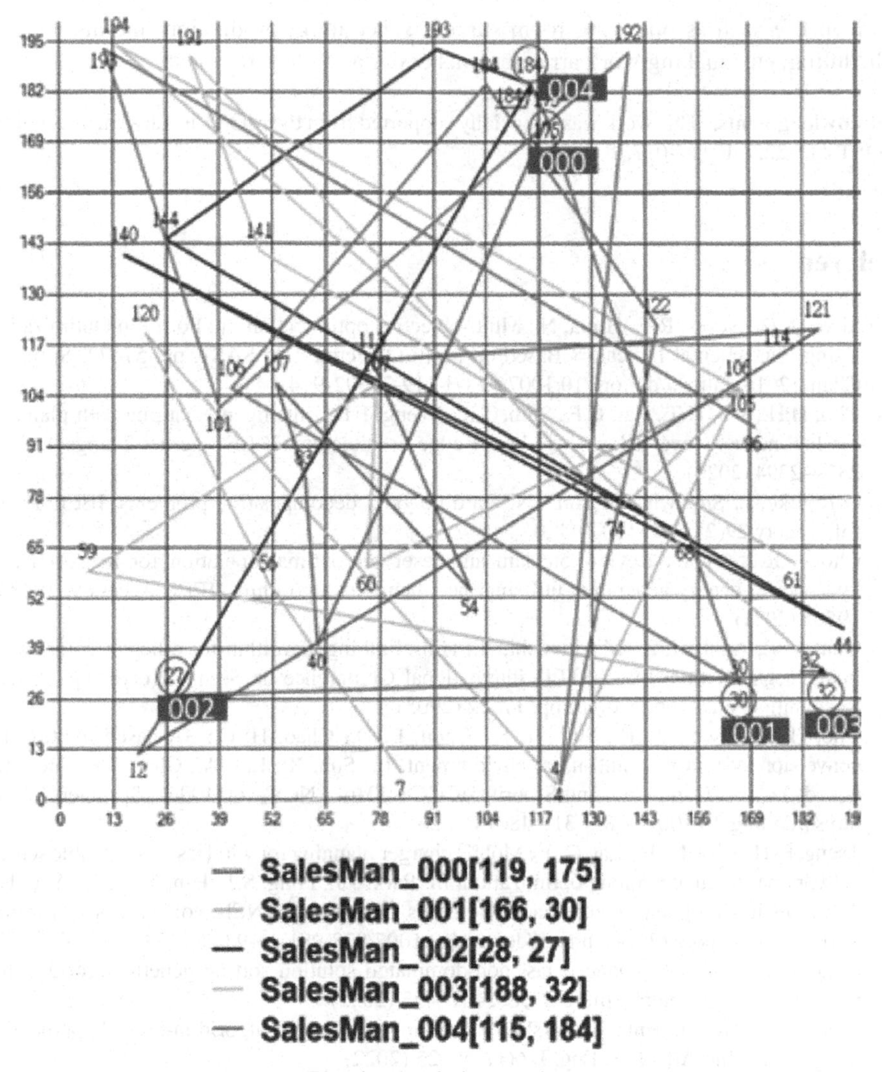

Fig. 6. Simulation topology

6 Conclusion

Food delivery platforms have become inseparable from human life. But the current popular platform just allows the deliveryman to grab the order by himself. This is not fair. Therefore, the main purpose of this study is to propose a fair delivery platform, which takes into account the ability of each delivery person and the time it takes for the meal to reach the consumer. It turns out that we will be able to set the delivery path and it can have fair and effective characteristics to achieve the highest net value for all. In the future, we will consider the situation of stacking orders to make the process closer to reality. Besides, In the future, we hope the proposed method can extend to annoying work in life such as operating room surgery scheduling, production line automation scheduling, etc., making work arrangements easier and more reasonable.

Acknowledgments. This work was financially supported from the MOST in Taiwan, under grant MOST 111-2221-E-197-017.

References

1. Silva, A.D., Neves, R.F., Horta, N.: Multi-objective optimization. In: Portfolio Optimization Using Fundamental Indicators Based on Multi-Objective EA. SAST, pp. 57–72. Springer, Cham (2016). https://doi.org/10.1007/978-3-319-29392-9_4
2. Cho, H.H., Wu, H.T., Lai, C.F., Shih, T.K., Tseng, F.H.: Intelligent charging path planning for IoT network over Blockchain-based edge architecture. IEEE Internet Things J. **8**(4), 2379–2394 (2020)
3. O'Rourke, J., Supowit, K.: Some NP-hard polygon decomposition problems. IEEE Trans. Inf. Theory **29**(2), 181–190 (1983)
4. Chong, K.L., et al.: Review on dam and reservoir optimal operation for irrigation and hydropower energy generation utilizing meta-heuristic algorithms. IEEE Access **9**, 19488–19505 (2021)
5. Ohashi, T., Aghbari, Z., Makinouchi, A.: Hill-climbing algorithm for efficient color-based image segmentation. In: IASTED International Conference on Signal Processing, Pattern Recognition, and Applications, pp. 17–22 (2003)
6. Cho, H. H., Tseng, F. H., Shih, T. K., Chou, L. D., Chao, H. C.: SA-based multimedia conversion system for multi-users environment. In: Sun, X., Liu, A., Chao, HC., Bertino, E. (eds.) Cloud Computing and Security. ICCCS 2016. LNCS, vol. 10040. Springer, Cham. https://doi.org/10.1007/978-3-319-48674-1_26
7. Tseng, F.-H., Cho, H.-H., Lai, C.-F.: Mobile charger planning for wireless rechargeable sensor network based on ant colony optimization. In: Park, J.J., Fong, S.J., Pan, Y., Sung, Y. (eds.) Advances in Computer Science and Ubiquitous Computing. LNEE, vol. 715, pp. 387–394. Springer, Singapore (2021). https://doi.org/10.1007/978-981-15-9343-7_53
8. Deng, W., et al.: An enhanced fast non-dominated solution sorting genetic algorithm for multi-objective problems. Inf. Sci. **585**, 441–453 (2022)
9. Akarsu, C.H., Küçükdeniz, T.: Job shop scheduling with genetic algorithm-based hyperheuristic approach. Int. Adv. Res. Eng. J. **6**(1), 16–25 (2022)
10. Dogan, A., Birant, D.: Machine learning and data mining in manufacturing. Expert Syst. Appl. **166**, 114060 (2021)
11. Elazouni, A.M., Metwally, F.G.: Finance-based scheduling: tool to maximize project profit using improved genetic algorithms. J. Constr. Eng. Manag. **131**(4), 400–412 (2005)
12. Lee, C.K.H.: A review of applications of genetic algorithms in operations management. Eng. Appl. Artif. Intell. **76**, 1–12 (2018)
13. Liu, N., Pan, J.S., Chu, S.C.: A competitive learning quasi affine transformation evolutionary for global optimization and its application in CVRP. J. Internet Technol. **21**(7), 1863–1883 (2020)
14. Konak, A., Coit, D.W., Smith, A.E.: Multi-objective optimization using genetic algorithms: a tutorial. Reliab. Eng. Syst. Saf. **91**(9), 992–1007 (2006)
15. Liberti, L., Lavor, C., Maculan, N., Mucherino, A.: Euclidean distance geometry and applications. SIAM Rev. **56**(1), 3–69 (2014)

Preliminary Considerations on Non-invasive Home-Based Bone Fracture Healing Monitoring

Roope Parviainen[1], Timo Kumpuniemi[2(✉)], Juha-Pekka Mäkelä[2], Matti Hämäläinen[2], Juha-Jaakko Sinikumpu[1], and Jari Iinatti[2]

[1] Unit of Pediatric Surgery, Oulu University Hospital, and Clinical Medicine Research Unit, University of Oulu, P.O. Box 23, 90029 Oulu, Finland
{roope.parviainen,Juha-Jaakko.Sinikumpu}@oulu.fi

[2] Centre for Wireless Communications, Network and Systems (CWC-NS), University of Oulu, P.O. Box 4500, 90014 Oulu, Finland
{timo.kumpuniemi,juha-pekka.makela,matti.hamalainen, jari.iinatti}@oulu.fi

Abstract. Fractures are common injuries causing pain and morbidity. Stable fractures with acceptable initial alignment are treated by immobilization. The alignment and angulation need to be controlled during the first 1–3 weeks (depending on the fracture), because the alignment may worsen. This is performed by taking X-ray images of the fracture site. Repetitive X-rays expose the patient to ionizing radiation repetitively. We have studied techniques to monitor the alignment of the fracture continuously and without the routinely taken control X-rays. The idea is to place sensors underneath the cast and on to the skin of the patient that would follow the angulation and alignment of the fracture and alert if a change is detected. Two approaches with radio technology are made: transmitter-receiver pairs and radar pairs. The challenges and their possible solutions are discussed.

Keywords: cast · children · UWB · radio technology · transmitter · receiver · radar

1 Introduction

Bone fractures are such common injuries that they are considered as a public health issue in global scale [1]. According to [1] the age-standardized incidence rate of fracture globally was 2296.2/100 000 in 2019. The fracture of hand, wrist or other distal part of arm are the most common sites of the fracture: Wu et.al. State that the age-standardized incidence ratio of these fractures in 2019 was 175.9/100 000 [1].

Fractures cause pain and morbidity to children as well. Literature implies that 27%-50% of children suffer at least one fracture before the age of 18 [2, 3]. The incidence of childhood fractures before the age of five has been shown to be 50–100/10 000 person years in European population [4–6].

Y. Chen et al. (Eds.): BICT 2023, LNICST 512, pp. 133–142, 2023.
https://doi.org/10.1007/978-3-031-43135-7_13

Distal radius fracture is a very common wrist injury, which is usually caused by a low-energy injury [7]. The usual mechanism of injury is falling on an outstretched arm from a standing height. The risk of suffering of a distal radius fracture is highest among post-menopausal women and adolescent males [7].

The most common method of treatment for any stable fracture is closed reduction (if needed) and immobilization in a cast. This is the accepted treatment also for stable distal radius fractures. A fracture is considered stable when the alignment of the bone is acceptable after reduction and the possibility of displacement is considered small [8]. Surgical treatment is considered if the proper alignment cannot be accomplished by closed reduction, the alignment fails in control x-rays taken at 1–2 weeks after injury or the features of the fracture predict poor functional outcome of the wrist.

Although, immobilization by casting is applied, the alignment of the fracture can still change into worse before the fractured bone is ossified. Typically, the alignment of the fracture is monitored by taking a control X-rays during the first weeks after the trauma. This exposes the patient to ionizing radiation on several occasions, which may have cumulative consequences (i.e., an increased cancer risk). The risk is higher among children due to their smaller size and immature tissue composition. Further, clinical and radiographic follow-ups have economic effects on the patient/family and the society as they require visits to the healthcare. Therefore, our research group is developing a method to monitor continuously the angulation and displacement of the fracture by sensors that are attached either directly to the skin or percutaneously to the bone by using extremely thin pins. These sensors will remain beneath the cast. The approach discussed in this paper utilizes radio-based sensors with sophisticated wireless communications solutions to measure and transfer constantly their reciprocal orientation using radio waves and alert if the distance or angle between them change. By using this method, it would be possible to monitor the position of the fracture constantly and thus avoid the regular control X-rays. The proposed monitoring can be done non-invasively. The system will alert if it detects significant change in the fracture alignment and then the patient will contact healthcare for a control X-ray. Based on our current knowledge, there are no existing radio-based displacement monitoring systems available for clinical use. In the first phase, we will concentrate on distal radius (i.e., wrist) fractures, because they are very common and there is no extensive amount of soft tissue surrounding the radius and ulna.

2 Anatomy of the Forearm

The bony structure of the forearm is formed by radius and ulna. They are connected distally by the Distal Radio Ulnar Joint (DRUJ) and proximally by the Proximal Radio Ulnar Joint (PRUJ). These joints enable the pronation-supination-movement of the forearm. In addition, the radius and ulna form a joint with the carpal bones enabling the flexion and extension of the wrist, as well as adduction and abduction. The distal radial articular surface is tilted in the volar direction 11° in average (variation 2°–20°) and the mean inclination angle is 22° (21°–25°) [9], see Fig. 1. Ulna is approximately 0.5 cm shorter than radius at the DRUJ, but there is large variation individually (1–5 mm).

The long bones can be divided into three distinct regions that are important to separate, especially in children: diaphysis, metaphysis and epiphysis. The diaphysis is the hollow shaft of a long bone, the epiphysis is the round and wide end of a long bone, the metaphysis resides between the epiphysis and diaphysis, and it is separated from the epiphysis by the growth plate during childhood (see Fig. 2). The longitudinal growth of a long bone happens mainly in the metaphysis and therefore the bone tissue is softer in this region. Thus, the pediatric bone fractures are most often located in the metaphysis.

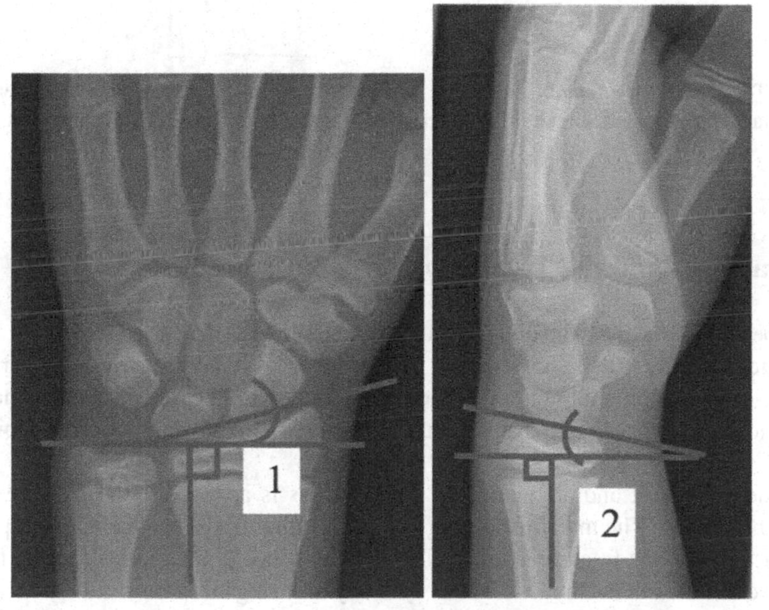

Fig. 1. The anatomy of the radius: 1) inclination, 2) volar tilt. Picture by Dr Ian Bickle, Radiopaedia.org, Creative Commons License, https://creativecommons.org/licenses/by-nc-sa/3.0/.

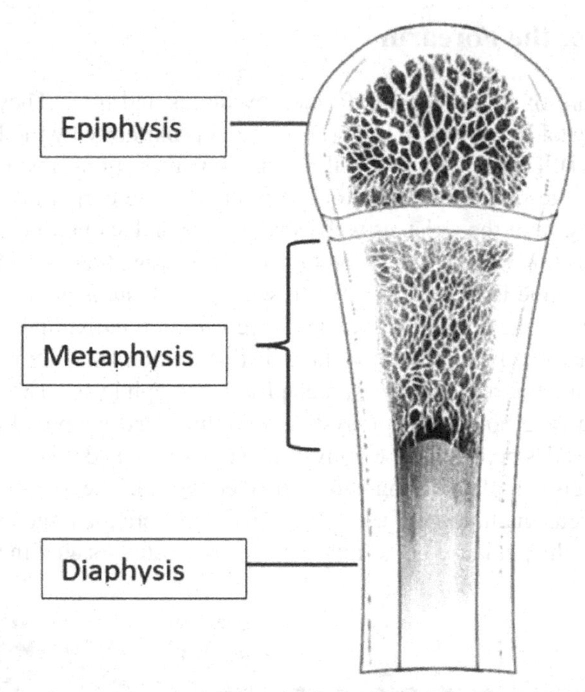

Fig. 2. The regions of a long bone. Picture modified from Parviainen, R. "The intrauterine and genetic factors associated with the childhood fracture risk", Acta Universitatis Ouluensis, D1591, 2020.

3 Fractures and the Treatment

The types of fractures can be divided roughly to stable torus fractures, bowing fractures, greenstick fractures (these three types happen only in children), transverse fractures with exact alignment and transverse fractures with varying degree of displacement and angulation. An example of a distal radius fracture with dorsal angulation is depicted in Fig. 3.

If the alignment and angulation of the fractures is acceptable the fracture can be treated by casting. The material used in the cast is either traditional calcium sulphate or fiberglass. When the fracture is immobilized using a cast, the alignment of the fracture must be controlled in the first 1–2 weeks by X-rays. During this period, the ossification of the fracture has just begun and the angulation or the displacement may worsen. The most important parameter that is followed in distal radius fracture is the dorsal of volar tilt seen in the lateral X-ray. In adults, the acceptable angulation is 15–20 degrees dorsally and under 20 degrees in the volar direction. In children, the acceptable angulation in the primary situation can be greater (20–30 degrees) depending on the age [10].

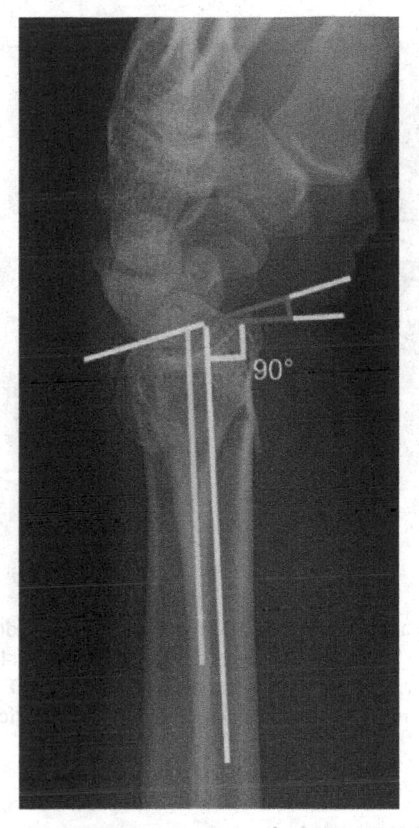

Fig. 3. Typical radius fractures with dorsal tilt. Picture by Mikael Häggström, Creative Commons License, https://creativecommons.org/licenses/by-sa/3.0/deed.en.

4 Example of the Needed Measurement Accuracy

In distal radius fractures 5–10-degree change compared to the starting point may be meaningful. If the width of the radius is 2 cm (in anterior-posterior dimension), it would mean that 5–10-degree change in the angulation would result in 17–35 mm change in vertical direction measured at the edge of the bone (see Fig. 4A). The younger the person with the fracture is, the smaller the bones are. Therefore, the vertical change may be smaller. The horizontal change is affected more by the location of the fracture (see Fig. 4B). From these calculations it is evident that the distances and angles that need to be measured are quite small. This places high demands on the accuracy of the monitoring system, especially as the system is planned to be placed on the skin. The skin sensor placing must be carefully considered and it is dependent on the type and location of the fracture. The sensors should be placed in such a way that there is as small amount of soft tissue as possible between the sensor and the bone.

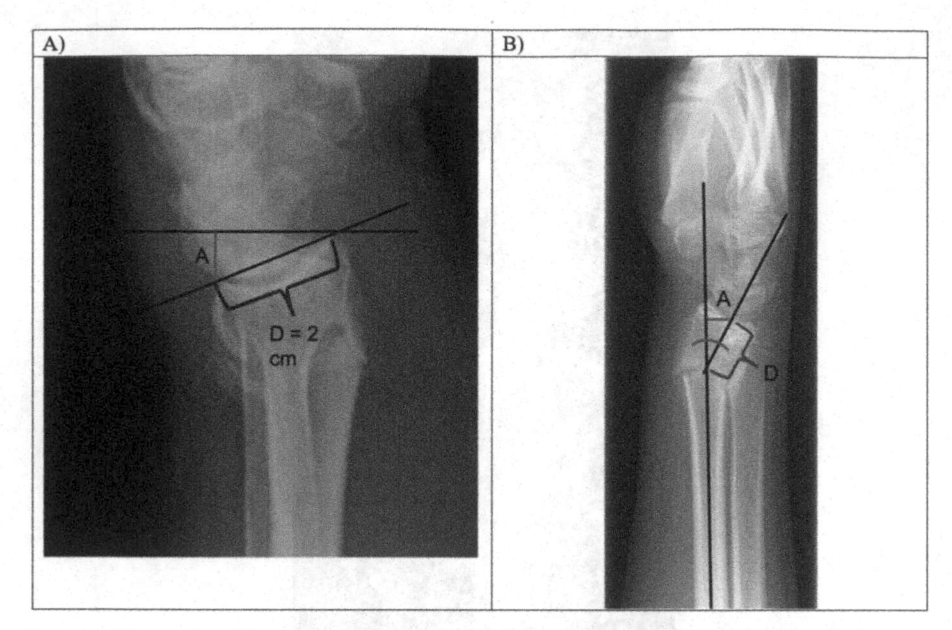

Fig. 4. A) Example of the amount of vertical displacement if the width of the bone is 2 cm and the dorsal tilt caused by the fracture is 10 degrees. The length A is 3.47 mm. B) Example of the amount of horizontal displacement if the fracture location is 3 cm (D = 3 cm) from the head of the bone and the dorsal tilt is 20 degrees. The length A is 1.02 cm. Picture by Roope Parviainen ©.

5 Possibilities and Challenges from Radio Technological Point of View

The scope of this work is to consider possible solutions for monitoring the bone fracture healing process at home and on a continuous basis during the whole healing process in an non-invasive manner. In the literature, there are corresponding solutions, which are using invasive methods [11, 12] requiring surgical operations. Non-invasive methods are proposed, e.g., in [13, 14] based on radio waves by measuring scattering parameters. They may however need quite sizeable measurement devices or are more suitable to detect the presence and size of a fracture instead of the healing process itself.

Two possible radio-based approaches are presented to detect an abnormal healing of a fracture. Besides being non-invasive, the solutions should be as small and light as possible, very reliable and dependable, easy to install and easy to monitor the results by the user, inexpensive, comfortable to wear during the healing process and have low power consumption. Furthermore, as described in Fig. 4, the accuracy or resolution demand for the solution is very high, up to less than 5 mm in distance, or 10 degrees in angle.

5.1 Several Transmitter/Receiver Pairs

The first approach to present is the usage of several transmitter (TX) and receiver (RX) pairs. The concept is depicted in Fig. 5. TX and RX are mounted under the cast directly on the skin of the patient on both sides of the bone fracture. For simplicity the skin is not drawn. Time of flight of the signal can be measured and based on the result the distance can be found out. On the right side of Fig. 5 the situation of an abnormal fracture healing process is illustrated. The bending of the bone causes a change in distance between the TX/RX pair indicating the need for more detailed examination carried out by a doctor. Since the bone may bend in different directions, several TX/RX pairs are needed.

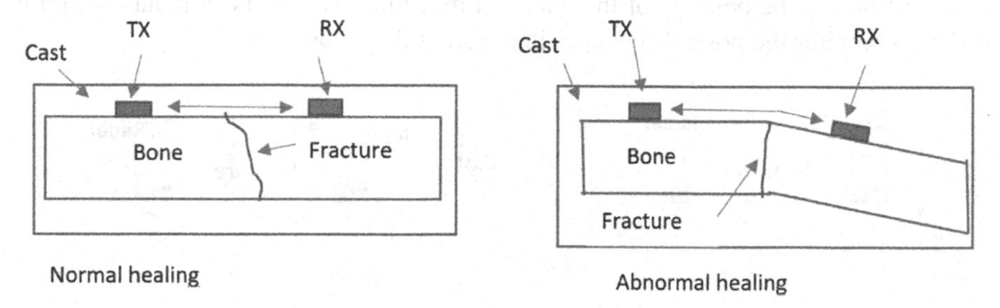

Fig. 5. TX/RX approach.

One serious challenge in this method is the demand of high resolution in distance. It is well known that the time resolution is the inverse of the signal bandwidth [15, 16]. Assuming the speed of light c in air this will lead to a distance resolution R stated as

$$R = c/B, \tag{1}$$

where B is the signal bandwidth. It is obvious that for a high distance resolution a wideband signal is needed. One technique possessing a wide bandwidth is impulse radio ultra wideband (IR-UWB) technology [15, 16].

However, assuming, e.g., a target resolution of 5 mm for the concept it would correspond approximately the need of a signal bandwidth of 60 GHz. This is impossible to accomplish in practice not only due to the bandwidths of the existing antennas but also due to the nonexistence of pulse generators to be able to produce such a short pulse in time, 0.17 ps.

By using advanced signal processing methods, the signal bandwidth demands could be relieved. In [17, 18], interpolation method is proposed. Based on the results, in theory, e.g., 1 mm resolution could be achieved with a bandwidth of approx. 10 GHz. The method assumes a correlation receiver structure. Furthermore, in [19] a pulse generator in presented being able to produce a pulse of approx. 120 ps in time, corresponding roughly 8 GHz in bandwidth.

Instead of using IR-UWB time of flight approach also the usage of received signal strength indication (RSSI) was considered. However, with RSSI the resolution problem may be seen from a different angle: since the resolution demand in distance is high, the power decay in just a few millimeters of a distance difference may be impossible to detect.

5.2 Radar Approach

The second approach considered is the use of radars. The concept is presented in Fig. 6. A radar pair is installed on or in the cast on both sides of the bone fracture as depicted. Radars are sensing the distance to the bone on both sides. At the installation phase the distances are recorded as a reference on both sides, as shown in the normal healing case. Should abnormal healing occur on either side of the fracture the distance changes leading to an alert due to the bending of the bone. At minimum two pairs of radars would be needed to monitor the possible changes in different directions.

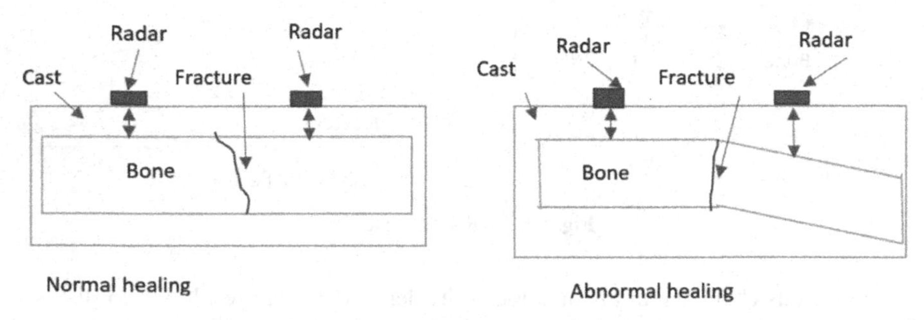

Fig. 6. Radar approach.

As with TX/RX approach in Sect. 5.1, one of the key questions in the radar approach is the distance resolution. Similar kind of methods as in [17, 18] could be applied here as well, assuming UWB communications in time domain. Regarding to signal processing methods, in [20, 21] also other signal processing methods are proposed for an UWB radar to increase accuracy. However, even though they are presented to be used as monitoring vital signals of a human, such as respiratory rate and heartbeat, they are not suggested to be used for bone fracture monitoring or to be used at the extreme vicinity of a human body, i.e., on-body.

Considering the concept in Fig. 6, another challenge can be noted. As the radars are thought to be on the cast or in it, the distance to be defined by the radars will be short, perhaps only 2–3 cm. Assuming pessimistic propagation speed of the signal in air, the echo will start to return already after approx. 0.02 ps. If the pulse generator operates as in [19] (pulse duration 120 ps), normal pulse radar does not have time to turn into the reception mode.

One solution to this could be to use full-duplex type of receivers [22] with pulse radars. Another solution could be to use a frequency modulated continuous wave (FMCW) radar where the previous mentioned problem does not, in theory, occur. FMCW radars are investigated, e.g., in [20, 21] at 77 GHz center frequency area but the minimum measurement distance is noted to be on the level of 3.5 cm.

Finally, in theory with radar approach, several echoes will arrive to reception from skin, fat, muscle and bone as noted in [23].

6 Conclusion and Future Work

This paper presents preliminary considerations leading to two possible approahes on non-invasive bone fracture healing monitoring. The monitoring is planned to be conducted at home continuously. The background on bone fractures and their healing, as well as the rationale for this approach is given. Furthermore, detailed exemplary cases from real life showing the demands for the question under scope are given.

The resolution demands are high, as the distances are very short. In here, our team discusses on two different approaches based on radio signals. Firstly the use of several transmitter-receiver-pairs by using either time of flight technique or RSSI measurement. In this case, the resolution can be seen as the main problem. Possible solutions related to digital signal processing are proposed to ease up the problem.

The second presented approach is based on radars at short distance. Also here, the resolution problem is present, even though similar kind of solutions as with transmitter-receiver pairs could be applied. However, in this case very short distance to be measured by radars sets limits to the technology to be used.

As future work, the two presented approaches should be examined in more detail also beyond theoretical investigations. Furthermore, other technologies outside the scope of the presented ones in this paper will be under research.

Acknowledgements. The authors wish to thank Adjunct Professor Harri Saarnisaari from Centre for Wireless Communications-Networks and Systems (CWC-NS) for fruitful and enlightening discussions on the technological matters.

This research has been financially supported in part by Academy of Finland 6Genesis Flagship (grant 346208), and in part by the University of Oulu's Infotech New Research Initiative project Fracture and FracturePoC project by the University of Oulu.

References

1. Wu, A.-M.: Global, regional, and national burden of bone fractures in 204 countries and territories, 1990¬2019: a systematic analysis from the global burden of disease study 2019. Lancet Healthy Longev **2**, 580–592 (2021). https://doi.org/10.1016/S2666-7568(21)00172-0
2. Jones, I.E., Williams, S.M., Dow, N., Goulding, A.: How many children remain fracture-free during growth? A longitudinal study of children and adolescents participating in the dunedin multidisciplinary health and development study. Osteoporos. Int. **13**(12), 990–995 (2002). https://doi.org/10.1007/s001980200137
3. Manias, K., McCabe, D., Bishop, N.: Fractures and recurrent fractures in children; varying effects of environmental factors as well as bone size and mass. Bone **39**(3), 652–657 (2006). https://doi.org/10.1016/j.bone.2006.03.018
4. Goulding, A., et al.: First fracture is associated with increased risk of new fractures during growth. J Pediatr **146**(2), 286–288 (2005). https://doi.org/10.1016/j.jpeds.2004.09.029
5. Orton, E., Kendrick, D., West, J., Tata, L.J.: Persistence of health inequalities in childhood injury in the UK; a population-based cohort study of children under 5. PLoS ONE **9**(10), e111631 (2014). https://doi.org/10.1371/journal.pone.0111631
6. Mäyränpää, M.K., Mäkitie, O., Kallio, P.E.: Decreasing incidence and changing pattern of childhood fractures: a population-based study. J. Bone Miner. Res. **25**(12), 2752–2759 (2010). https://doi.org/10.1002/jbmr.155

7. MacIntyre, N.J., Dewan, N.: Epidemiology of distal radius fractures and factors predicting risk and prognosis. J. Hand Ther. **29**(2), 136–145 (2016). https://doi.org/10.1016/j.jht.2016.03.003
8. Schneppendahl, J., Windolf, J., Kaufmann, R.A.: Distal radius fractures: current concepts. J. Hand. Surg. Am. **37**(8), 1718–1725 (2012). https://doi.org/10.1016/j.jhsa.2012.06.001
9. Obert, L., Loisel, F., Gasse, N., Lepage, D.: Distal radius anatomy applied to the treatment of wrist fractures by plate: a review of recent literature. SICOT J. **1**, 14 (2015). https://doi.org/10.1051/sicotj/2015012
10. Lynch, K.A., Wesolowski, M., Cappello, T.: Coronal remodeling potential of pediatric distal radius fractures. J. Pediatric Orthopaed. **40**(10), 556–561 (2020). https://doi.org/10.1097/BPO.0000000000001580
11. Bhavsar Mit, B., Moll, J., Barker John H.: Bone fracture sensing using ultrasound pitch–catch measurements: a proof-of-principle study. Ultrasound Med. Biol. **46**(3), 855–860 (2020)
12. Mattei L., Di Puccio F., Marchetti S.: Fracture healing monitoring by impact tests: single case study of a fractured tibia with external fixator. IEEE J. Transl. Eng. Health Med. **7**, 1–6, Art no. 2100206 (2019). https://doi.org/10.1109/JTEHM.2019.2901455
13. Kiriş, S., İncesu, A., Karaaslan, M., Akgöl, O., Ünal, E.: Study of helical antenna as a bone fracture sensor. In: 2020 4th International Symposium on Multidisciplinary Studies and Innovative Technologies (ISMSIT), pp. 1–4 (2020). https://doi.org/10.1109/ISMSIT50672.2020.9254610
14. Riaz, M., Tiberi, G., Asani, H., Ghavami, M., Dudley, S.: A non-invasive bone fracture monitoring analysis using an UHF antenna. In: 2020 12th International Symposium on Communication Systems, Networks and Digital Signal Processing (CSNDSP), pp. 1–5 (2020). https://doi.org/10.1109/CSNDSP49049.2020.9249570
15. Oppermann, I., Hämäläinen, M., Iinatti, J. (eds.): UWB Theory and Applications. Wiley, West Sussex, England (2004)
16. Ghawami, M., Michael, L.B., Kohno, R.: Ultra Wideband Signals and Systems In Communication Engineering, 2nd edn. Wiley, West Sussex, England (2007)
17. Saarnisaari, H., Tapio V.: A simple multipath delay estimator based on alternating projection algorithm. In: 2006 IEEE/ION Position, Location, and Navigation Symposium, pp. 1086–1093 (2006). https://doi.org/10.1109/PLANS.2006.1650714
18. Saarnisaari, H.: Some design aspects of mobile local positioning systems. PLANS 2004. In: Position Location and Navigation Symposium (IEEE Cat. No.04CH37556), 2004, pp. 300–309. https://doi.org/10.1109/PLANS.2004.1309009
19. Malajner, M., Šipoš, D., Gleich, D.: Design of a low-cost ultra-wide-band radar platform. Sensors **20**, 2867 (2020). https://doi.org/10.3390/s20102867
20. Alizadeh, M., Shaker, G., Almeida, J.C.M.D., Morita, P.P., Safavi-Naeini, S.: Remote monitoring of human vital signs using mm-wave FMCW radar. IEEE Access **7**, 54958–54968 (2019). https://doi.org/10.1109/ACCESS.2019.2912956
21. Wang, Y., Wang, W., Zhou, M., Ren, A., Tian, Z.: Remote monitoring of human vital signs based on 77-GHz mm-Wave FMCW radar. Sensors **20**, 2999 (2020). https://doi.org/10.3390/s20102999
22. Tapio, V., Juntti, M.: Non-linear self-interference cancelation for full-duplex transceivers based on Hammerstein-Wiener model. IEEE Commun. Lett. **25**(11), 3684–3688 (2021). https://doi.org/10.1109/LCOMM.2021.3109669
23. Lee D., Shaker G., Augustine R.: Preliminary study: monitoring of healing stages of bone fracture utilizing UWB pulsed radar technique. In: 2018 18th International Symposium on Antenna Technology and Applied Electromagnetics (ANTEM), pp. 1–2 (2018). https://doi.org/10.1109/ANTEM.2018.8572872

Features of Audio Frequency Content of Respiration to Distinguish Inhalation from Exhalation

Souhail Katti, Federica Aveta, Saurav Basnet, and Douglas E. Dow(✉)

Electrical and Computer Engineering, Wentworth Institute of Technology, Boston, MA 02115, USA
{kattis,avetaf,basnets,dowd}@wit.edu

Abstract. The life-sustaining function of respiration becomes impaired by diseases that occur more with old. A system that monitors the inhalations and exhalations of the respiratory cycle could raise an alert when abnormal patterns or prolonged disruptions are detected. Noninvasive methods are suitable to chronically monitor respiration. Methods include analyzing audio sounds generated during respirations and analyzing changes in the volume of the thorax or abdomen. In casual observations of eupneic breathing, inhalation often sounds different from exhalation, though may be quite similar. One of the challenges for signal processing is to distinguish inhalation from exhalation based on only the audio. The purpose of this study was to find a method of analyzing the audio frequency content that could differentiate the inhalation and exhalation. Volunteer subjects were recruited to record audio during eupneic respiration for analysis. To classify the timing of each inhalation and exhalation, both respiratory sounds and volume changes of the thorax were simultaneously recorded. The audio files were analyzed by Fast Fourier Transform (FFT) to determine the frequency content. Features of the frequency power spectrum were found that appear promising for distinguishing inhalation and exhalation. Such differences could be used to characterize audio respiratory signals and improve the monitoring of individuals at risk for impaired respiratory function.

Keywords: Fourier · FFT · LabView · MATLAB · eupneic · breathing · chronic monitoring

1 Introduction

The population in the world is aging, and the number of people aged over 65 years is growing [1]. Old age is associated with an increased risk of chronic health conditions that may degrade respiration including sleep apnea, respiratory disease and cardiovascular diseases [2, 3]. Sleep disorder breathing (SDB) includes impairments that involve obstructions or narrowing of the upper airway during sleep. SDB is responsible for numerous problems, including fragmented sleep, hypertension and traffic accidents [4].

Y. Chen et al. (Eds.): BICT 2023, LNICST 512, pp. 143–155, 2023.
https://doi.org/10.1007/978-3-031-43135-7_14

Episodes of sleep apnea results in intermittent deregulation of oxygen saturation, which have short and long-term consequences [5]. Sleep apnea increases the risk for high blood pressure, heart problems, type 2 diabetes, metabolic syndrome, liver problems [6].

The diagnosis of SDB typically takes place in clinical settings using intrusive instrumentation, such as spirometer or polysomnography. Due to the clinical setting and procedures, patients may not exhibit their normal pattern of sleep and respiration, thus hampering diagnosis.

Efforts have been made to develop less intrusive methods to monitor respiration during activities of daily living or sleep [7–11]. One method involves recording and analyzing the audio sounds that occur during respiration [12–14]. In one of these studies, audio sounds recorded during respiration were analyzed to classify as normal, wheezes or crackles [12]. Another study analyzed respiratory audio to determine the respiration rate and then determine the duration of exhalations [13]. Another study analyzed respiratory audio to find spectral features to characterize the respiration as normal or abnormal [13]. These studies did not analyze the audio to find features toward distinguishing the behavior of inhalation from the exhalation, which was the purpose of our study.

Factors that may influence the audio pattern include the rate of airflow, occurrence of turbulence in airways, the respiratory rate, individual differences in the anatomy, differences in behavior or condition of the airway, and whether the person is breathing through their nose or mouth. The audio profile of an inhalation may be similar to an exhalation, but may have distinguishing features [14]. For respiratory function to be derived from only analyzing audio signals, the analysis would need to distinguish whether a burst of sound is for an inhalation or an exhalation. Analysis for the timing of respiration would need to distinguish and identify each burst of sound generated during an inhalation and each burst generated during an exhalation.

The purpose of our study was to find features of the audio frequency content of respiration that could be used to distinguish inhalations from exhalations. The findings of this study would be useful for monitoring respiration and health.

2 Methods and Materials

2.1 Recording of Respiratory Activity

In this study audio recordings were made of volunteer subjects while they underwent eupneic cycles of breathing. Of the 13 volunteer subjects for the recording sessions 54% identified as male and 46% as female. Six subjects were aged between 20 to 30 years old, six were between 30 to 60 years old and one subject was 98 years old. Figures 1 and 2 show the demographics of the volunteers.

Since audio sounds generated during inhalations or exhalations may vary in amplitude and frequency content due to the respiration rate, depth of breathing, mode of breathing through the nose or mouth. The subjects were asked to breathe in several different ways during the recording sessions.

Each subject had 3 recordings of 1 min each while they underwent eupneic breathing at a moderate level of intensity. Of the 3 recordings for each subject, one recording was made for each of 3 modes. The first mode had the subject inhale and exhale through

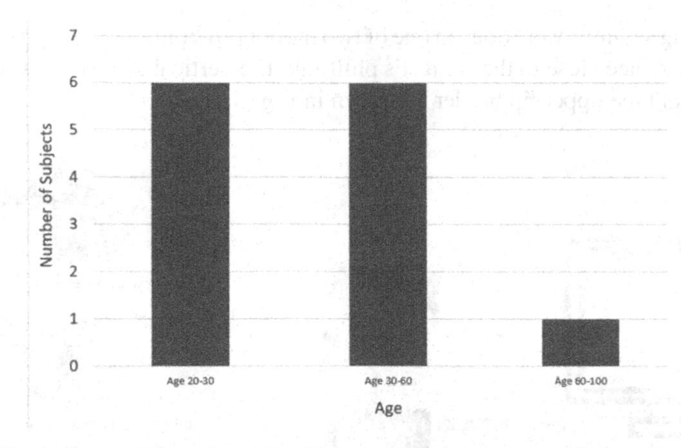

Fig. 1. Ages of the volunteer subjects who underwent recording sessions.

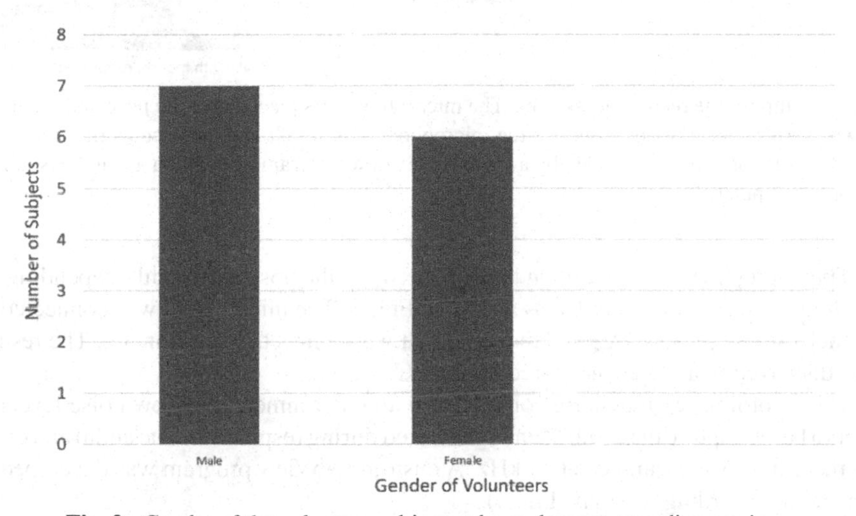

Fig. 2. Gender of the volunteer subjects who underwent recording sessions.

their nose. The second mode had the subject inhale and exhale through their mouth. The third mode had the subject inhale through nose and exhale through mouth.

During the recording sessions, every subject had to undergo 3 recordings while doing eupneic breathing and sitting in a chair. All the subjects had a break of 10 min at the start before they began their recordings to allow for relaxation of their breathing rate prior. Between each of the 3 recordings, the subject had a break of 1 min.

The recording sessions utilized Vernier (Vernier Software and Technology LLC, Beaverton, OR, USA) devices for the microphone, chest belt, and data acquisition module (Fig. 3). The data acquisition module was controlled by a custom LabView (National Instruments, Austin, TX, USA) program running on a Windows (Microsoft, Redmond, WA, USA) laptop personal computer (PC).

Recording sessions were done in one of two indoor environments using a microphone placed at a distance close to the subject's philtrum, the vertical groove between the base of the nose and the upper lip border as shown in Fig. 3.

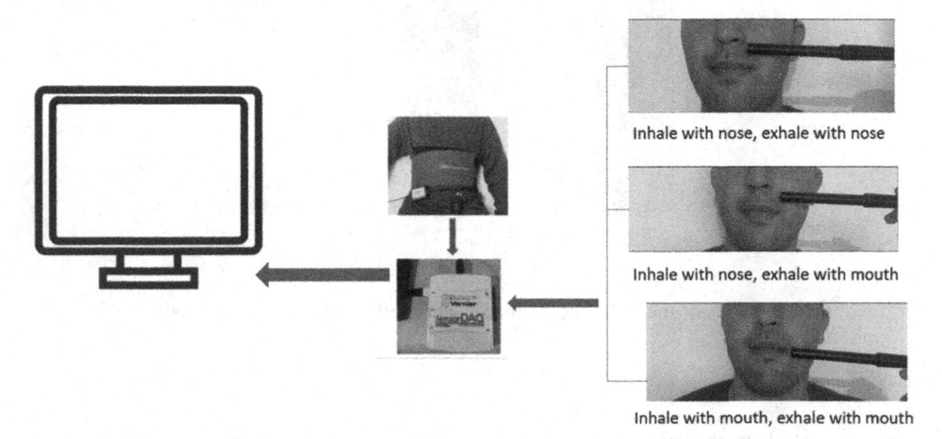

Fig. 3. Setup for the recording sessions. The microphone was placed near the nose and mouth as shown. Vernier devices were used for the microphone, chest belt and data acquisition unit. The recording sessions were controlled by a custom LabView program running on a Windows laptop personal computer (PC).

The microphone was placed on the side between the nose and mouth, depending on which mode was being recorded as shown in Fig. 3. The microphone was connected to channel 1 of the sensor-DAQ and the chest belt was connected to channel 2. The results were displayed in a graph and stored on the PC.

The recording sessions were conducted in an environment with low noise levels in order to better capture the audio sounds generated during respiration. The audio and chest belt recordings were sampled at 24 kHz. A custom LabView program was developed to manage the recording sessions (Fig. 4).

An example of a recording in the LabView program is shown in Fig. 4. The plots show the 60-s recorded signal that was sampled at 24 kHz. The top plot shows the audio signal, and the bottom plot shows the chest belt signals. Cursors in yellow were manually added to this image to help visually correlate the signals for the respiration cycle. The chest belt signal rose during inhalations as the volume of the thorax increased. Then the signal values decreased during exhalations as the volume of the thorax decreased. The peak of the chest belt signal (where the vertical yellow cursors were placed) indicated the transition from inhalation to exhalation. The yellow vertical cursors were placed at approximately the same time in the upper plot showing the audio signal. The audio signal appeared as an almost flat plot line centered at zero of during the quiet periods of no airflow, in between the bursts of inhalation or exhalation. The airflow during inhalations or exhalations appeared as a burst of activity in the audio signal still centered at an amplitude of 0. The smaller audio burst before each yellow cursor corresponded with an inhalation, being at the same time as the rise of the chest belt signal. The larger audio burst after each yellow cursor corresponded with an exhalation, as the chest belt signal fell to lower values.

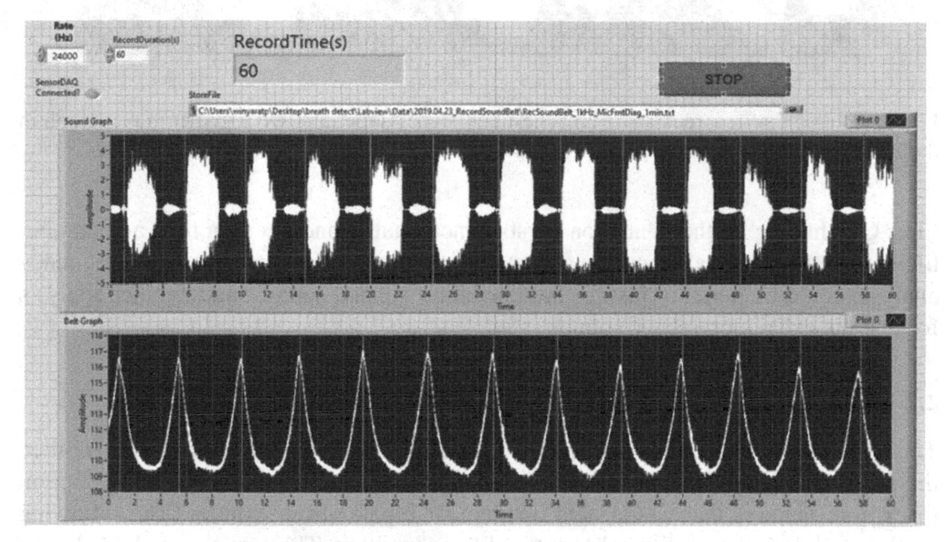

Fig. 4. Recording of breathing in the LabView program that is sampled at 24 kHz with duration of 60 s.

As the purpose of the study was to explore features in the frequency content that could distinguish inhalation from exhalation, the inhalation audio segments were isolated from the exhalation audio segments prior to analysis by FFT. Another Custom LabView program was developed to manually isolate each inhalation and exhalation audio segment. In the example shown in Fig. 5, the white trace was the audio bursts during inhalation or exhalation with a short quiet period in between. The red trace in the background was the chest belt signal with rises indicating inhalations (expanding chest circumference) and falls indicating expirations.

According to Nyquist's theorem, a periodic signal must be sampled at a frequency that is more than twice as high as the signal's highest frequency component to be observed. Thus, the frequency content to be analyzed would be less than 12 kHz. The range of frequencies used in this study was further reduced to the range of 0 to 6 kHz.

The data from the microphone and the chest belt were displayed within the Graphical Interface of the LabView program. The recorded audio and chest belt data were transferred to an Excel sheet and stored as a file.

The cursors were used to manually isolate each inhale or exhale. Cursors were used in the chest belt signal to mark the beginning or end of an inspiration. Then the corresponding audio signal was copied as a list of numbers of the sampled audio values to a growing list of all the inhalations. Each burst of inhalation was concatenated to the end of the growing list. The same was done for each burst of exhalation. Figure 5 shows an example of this. Figure 6A shows an example of concatenated exhales, and Fig. 6B of concatenated inhales. Thus, each audio recording for a volunteer doing eupneic breathing for one of their 3 recordings having different modes would be separated into two audio

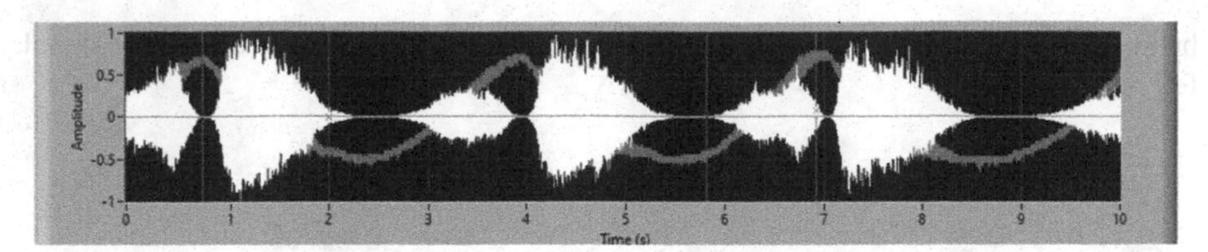

Fig. 5. Example of recorded audio and chest belt signal superimposed together in one graph in the LabView program. The respiration signals were loaded from previously recorded files.

files. One having all the inhalation bursts concatenated one after another, and another file having all the exhalations concatenated one after another. In this way, a frequency analysis could be done for just for the inhalation-related sounds, and another analysis for just the exhalation-related sounds.

2.2 Frequency Domain Analysis

The algorithm developed in this study did a series of steps to form a signal that was called the Smoothed Normalized Spectrum. This signal was the basis for feature extraction to explore features that would help distinguish inhales from exhales. The following section will describe the following steps of the algorithm. The concatenated inhale and exhale signals underwent FFT to form the power spectrum. The power spectrum was normalized for amplitude. The normalized spectrum was smoothed out to form the Smoothed Normalized Spectrum.

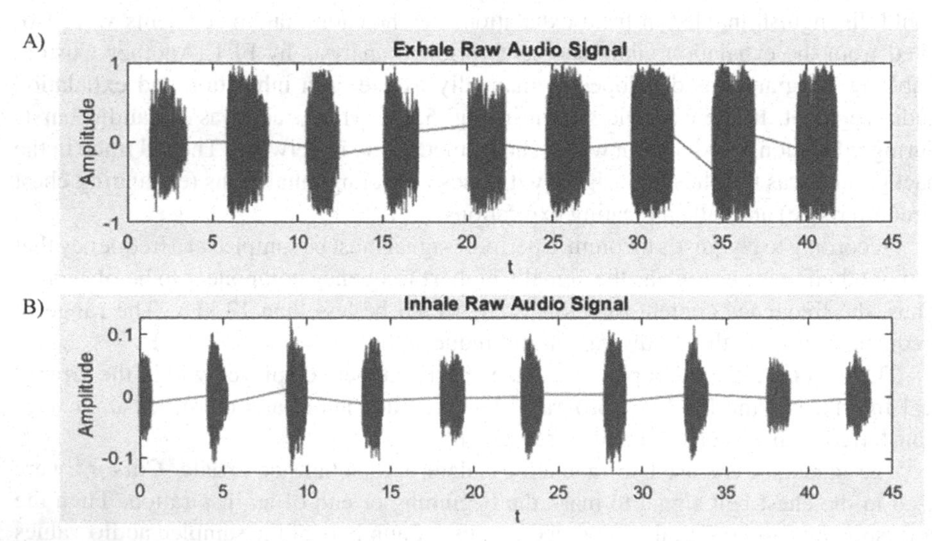

Fig. 6. Examples of isolated audio bursts of only exhalations (A) or only inhalations (B). Chest belt recordings of thorax volume changes were used to help move cursors to manually select each exhalation or inhalation. These isolated signals then underwent FFT for analysis of frequency content.

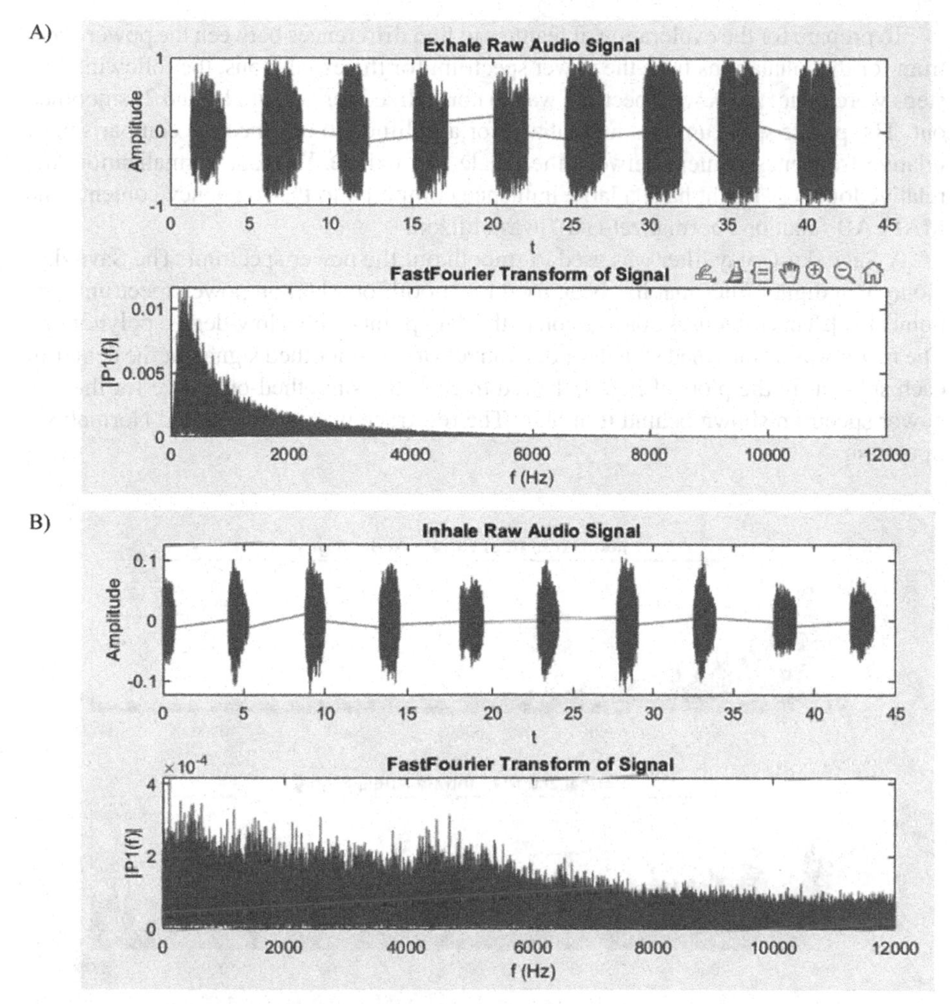

Fig. 7. Example of isolated audio segments that underwent FFT, with the resulting power spectrum. A) shows plots exhalation, and B) shows plots for inhalation. The power spectrum of the exhalation had a different profile than the one for inhalation.

FFT separates a time-domain signal into its distinct spectral components, giving frequency information about the signal. The FFT implements the Discrete Fourier Transformation. The FFT algorithm was implemented within a custom-made MATLAB program on a Windows laptop PC.

The FFT converted the waveform signal in the time domain into the frequency domain. The FFT disassociated the time-based waveform into a number of sinusoidal terms, each with a distinct magnitude, frequency, and phase. The resulting power spectrum was plotted with amplitude of each sinusoidal phase against its frequency as shown in Fig. 7.

To prepare for the exploration of features to find differences between the power spectrum for the inhalations from the power spectrum for the exhalations, the following two steps were done: the power spectrum was 1) normalized for amplitude, and 2) smoothed out. The power spectrum was normalized for amplitude to enhance the comparison of relative frequency content between the inhale and exhale. Without normalization, the relative loudness might have a large influence compared to the frequency content. The MATLAB function "normalize(data)" was utilized.

A Savitzky-Golay filter was used to smooth out the power spectrum. The Savitzky-Golay is a digital filter that has been used to smooth out data on power spectrum data points [15]. This filter uses convolution to the data points with a low-degree polynomial. The result was a smoothed signal (or derivatives of the smoothed signal) at the center of each sub-set. In the plots of Fig. 8, the red trace is the smoothed-out curve for the raw power spectrum shown behind it in blue. The red trace was the Smoothed Normalized Spectrum.

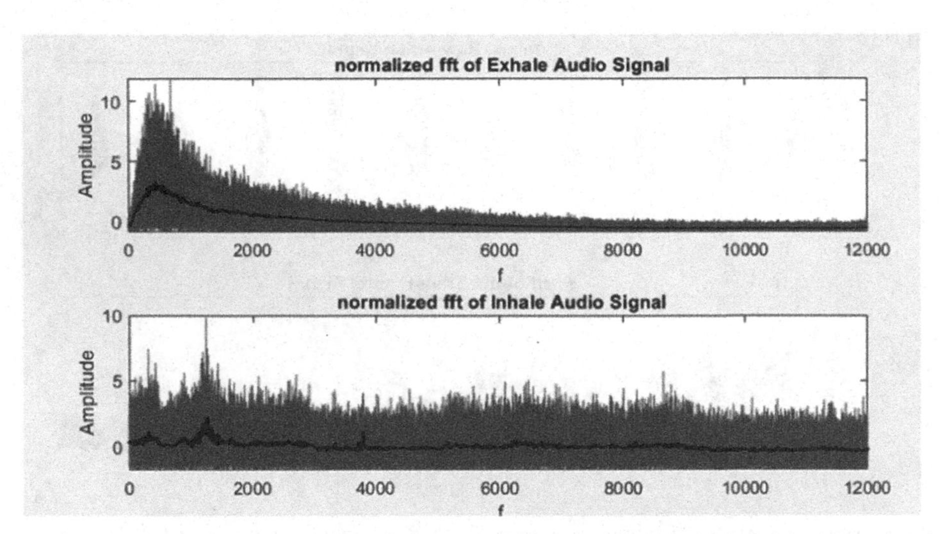

Fig. 8. Normalized FFT of Exhales and Inhales. The plots show the results of the isolated and concatenated exhale audio bursts having undergone FFT, then being normalized for amplitude, followed by being smoothed out with the Savitzky-Golay digital filter. The blue plot shows the normalized power spectrum, and the red plot shows the smoothed-out trace, called the Smoothed Normalized Spectrum. The top plot is for exhalation and the bottom plot is for inhalation.

The Smoothed Normalized Spectrum was used in this study as the basis for feature extraction. Thus, each audio recording of the volunteer subjects breathing resulted in two Smoothed Normalized Spectrums, one for the concatenated inhales and one for the concatenated exhales.

2.3 Feature Extraction

To form features, a curve was fit to each Smoothed Normalized Spectrum, one for inhale and one for exhale. Several types of curves were tried. The cubic polynomial curve appeared able to fit the curve in ways that could distinguish the exhale from the inhale smoothed normalized spectrum. The cubic polynomial was form of this form.

$$Y = Ax^3 + Bx^2 + Cx + D \tag{1}$$

The determined coefficients (A, B, C, D) became the features.

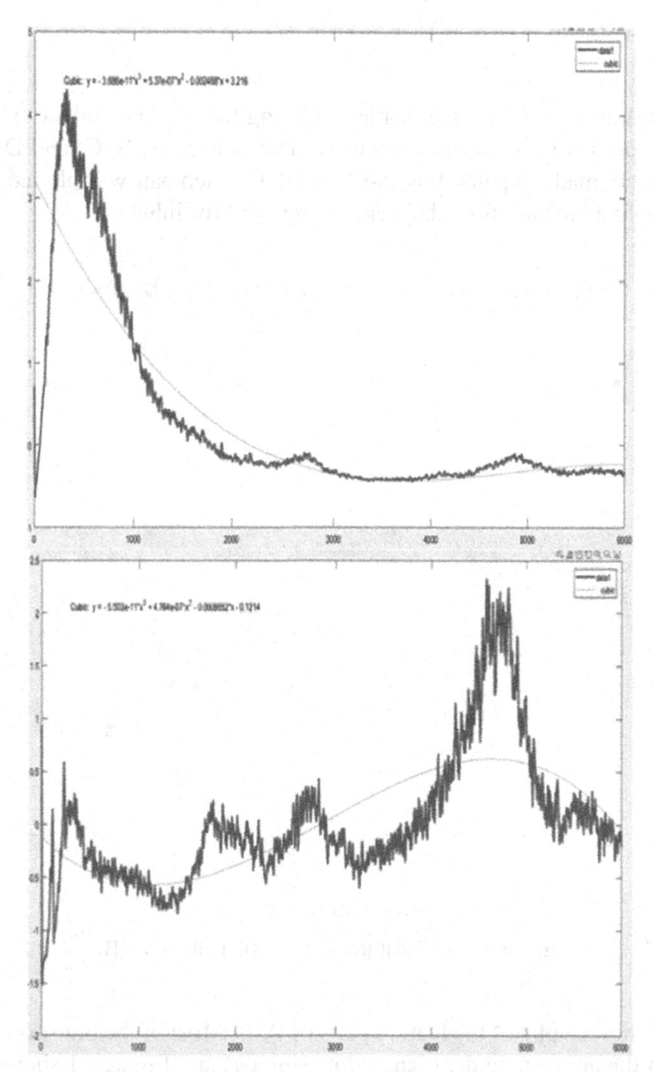

Fig. 9. Fit cubic polynomial to find the features. The red plot shows the normalized, smoothed-out power spectrum of 0–6 kHz. The green plot shows the fitted cubic polynomial. The coefficients (A, B, C, D) of the equation for the fitted cubic polynomial became the features to use for further analysis. The top graph (A) shows the plots for exhalation, and the bottom graph (B) shows the plots for inhalation.

Figure 9 shows an example of the red trace was the normalized, smoothed-out power spectrum. The top graph (Fig. 9A) was derived from the concatenated exhales, and the bottom graph (Fig. 9B) was derived from the concatenated inhales. The green curve was the fitted curve. The coefficients (A, B, C, D) of the fitted polynomial equation were the features used for further analysis. In general, the fitted curve and resulting features for the exhales were quite different than those for the inhales.

3 Results

For each recording of the volunteer subjects having the mode of Inhale from nose and exhale from mouth with a moderate intensity. The feature A, B, C, and D were determined. Plots were made of pairs A-B, A-C, and B-C. Each pair was plotted as a colored dot on the graph: a blue dot for exhale and orange dot for inhale.

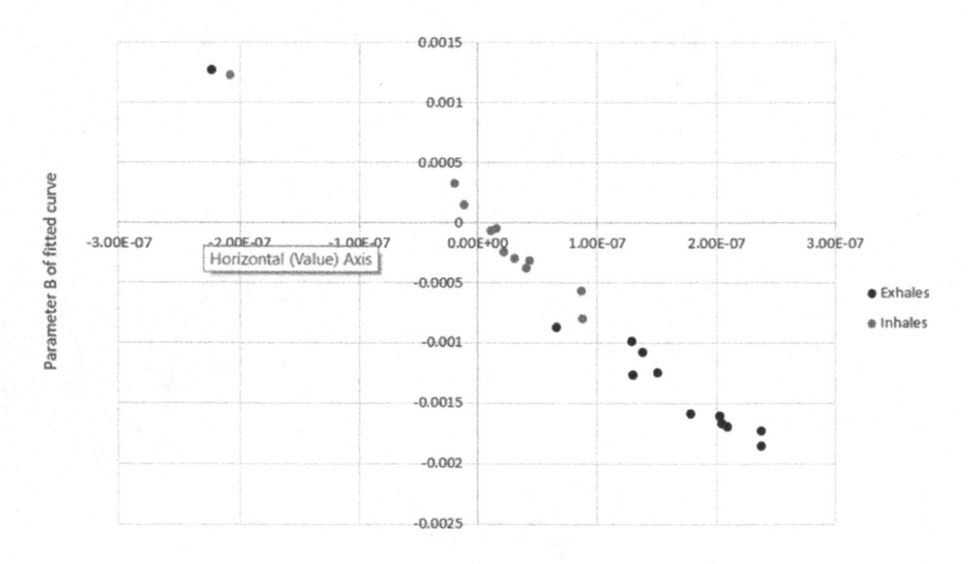

Fig. 10. Plot of features for Coefficients A vs. B.

Figure 10 shows a plot of the feature pairs of A-B. Most of the blue dots (exhale) are separate from the area where the orange dots (inhale) are. Figure 11 shows the feature pairs for A–C. Figure 12 show the feature pairs for B-C. In general, the region of the blue dots (exhales) is separate from the region of the orange dots (inhales).

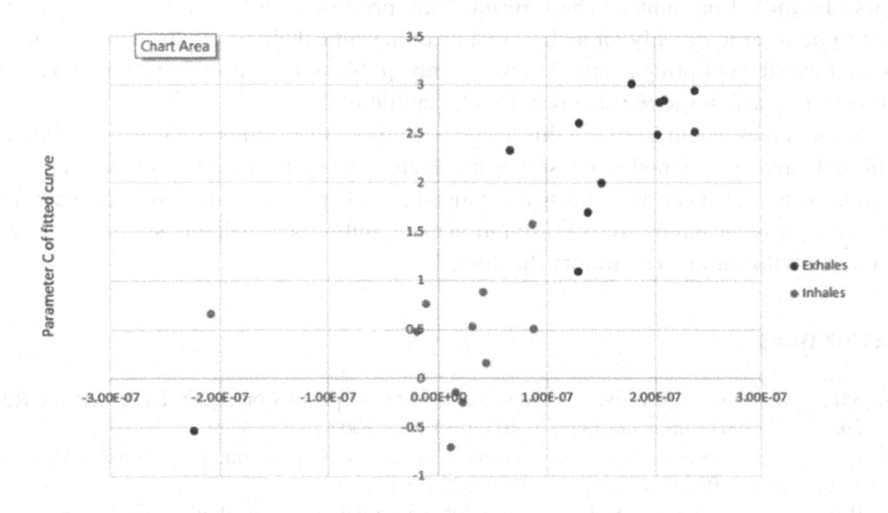

Fig. 11. Plot of features for Coefficients A vs. C.

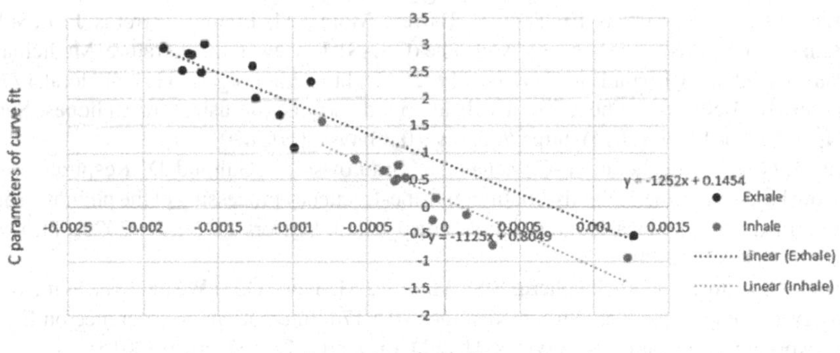

Fig. 12. Plot of features for Coefficients B vs. C.

4 Discussion and Future Directions

An algorithm was developed to determine features for the frequency content of the audio sounds during inhalations and exhalations of breathing. The main steps were converting the breathing signal of concatenated inhales and exhales from the time domain to the frequency domain by applying FFT. The resulting power spectrum was normalized and smoothed to form the Smoothed Normalized Spectrum, which was then fit with a cubic polynomial. The features were the coefficients of the polynomial.

The features were plotted in pairs, and appear to be in separate into regions for the pairs of exhale from inhale. These results look promising in that the features may be able to be used to classify an audio burst as either an exhale or inhale. This would be useful toward monitoring respiration using only audio signals. Further testing would be required to verify these results toward wider application.

Future steps would be to run this analysis on more recordings, such as those having different intensities or modes of respiration. Machine learning could be applied to classify an unknown sound as an exhalation, inhalation or neither. Development of such a method would improve the monitoring of respiration that would enhance diagnosis and treatment of people with chronic respiratory disorders.

References

1. Sanderson, W.C., Scherbov, S.: A new perspective on population aging. Demographic Res. **16**, 27–58 (2007). https://doi.org/10.4054/DemRes.2007.16.2
2. Bloom, D.E., Boersch-Supan, A., McGee, P., Seike, A.: Population aging: facts, challenges, and responses. Benefits Compens. Int. **41**, 22 (2011)
3. Bloom, D.E., Canning, D., Lubet, A.: Global Population Aging: Facts, Challenges, Solutions & Perspectives, vol. 144, pp. 80–92. MIT Press, Daedalus (2015)
4. Zhao, J., Wang, W., Wei, S., Yang, L., Wu, Y., Yan, B.: Fragmented sleep and the prevalence of hypertension in middle-aged and older individuals. Nature Sci. Sleep **13**, 2273–2280 (2021). https://doi.org/10.2147/NSS.S337932
5. Gottlieb, D.J., Punjabi, N.M.: Diagnosis and management of obstructive sleep apnea: a review. JAMA **323**(14), 1389 (2020). https://doi.org/10.1001/jama.2020.3514
6. Azagra-Calero, E., Espinar-Escalona, E., Barrera-Mora, J.M., Llamas-Carreras, J.M., Solano-Reina, E.: Obstructive sleep apnea syndrome (OSAS). Review of the literature. Medicina Oral Patología Oral y Cirugia Bucal e925–e929 (2012). https://doi.org/10.4317/medoral.17706
7. Reyes, B., Reljin, N., Chon, Ki.: Tracheal sounds acquisition using smartphones. Sensors **14**(8), 13830–13850 (2014). https://doi.org/10.3390/s140813830
8. Lu, X., Coste, C.A., Nierat, M.-C., Renaux, S., Similowski, T., Guiraud, D.: Respiratory monitoring based on tracheal sounds: continuous time-frequency processing of the phonospirogram combined with phonocardiogram-derived respiration. Sensors **21**(1), 99 (2020). https://doi.org/10.3390/s21010099
9. Bethel, S.J., Joslin, C.T., Shepherd, B.S., Martel, J.M., Dow, D.E.: Wearable at-home recording system for sleep apnea. Presented at the 2015 17th International Conference on E-health Networking, Application & Services (HealthCom), pp. 149–152. IEEE (2015)
10. Dow, D.E., Petrilli, A.M., Mantilla, C.B., Zhan, W.-Z., Sieck, G.C.: Electromyogram-triggered inspiratory event detection algorithm. Presented at the The 6th International Conference on Soft Computing and Intelligent Systems, and The 13th International Symposium on Advanced Intelligence Systems, pp. 789–794. IEEE (2012)
11. Dow, D.E., Horiguchi, Y., Hirai, Y., Hayashi, I.: Monitoring of respiratory cycles utilizing sensors on sleeping mat. Presented at the ASME International Mechanical Engineering Congress and Exposition, vol. 58363:V003T04A011 (2017)
12. Meza, C.A.G., del Hoyo Ontiveros, J.A., Lopez-Meyer, P.: Classification of respiration sounds using deep pre-trained audio embeddings. Presented at the 2021 IEEE Latin American Conference on Computational Intelligence (LA-CCI), pp. 1–5. IEEE (2021)
13. Doheny, E.P., et al..: Estimation of respiratory rate and exhale duration using audio signals recorded by smartphone microphones. Biomed. Signal Process. Control **80**, 104318 (2023). https://doi.org/10.1016/j.bspc.2022.104318

14. Gavriely, N., Palti, Y., Alroy, G.: Spectral characteristics of normal breath sounds. J. Appl. Physiol. **50**, 307–314 (1981)
15. Xu, P., Jia, Y., Jiang, M.: Blind audio source separation based on a new system model and the savitzky-golay filter. J. Electric. Eng. **72**(3), 208–212 (2021). https://doi.org/10.2478/jee-2021-0029

Sharing Health Records in Senegal Using Blockchain

Mouhamadou Moustapha Mbaye(✉) and Abdourahime Gaye

Department of Computer Engineering and Communication,
University Alioune DIOP of Bambey, Bambey, Senegal
m3moustapha@gmail.com, abdourahime.gaye@uadb.edu.sn

Abstract. Electronic Health Records offer real advantages for accessing and storing patient health information, which can improve the management of patient care. However, the attractive features of electronic records (accessibility, portability, and portability of patient health information) also present privacy risks. Organizations need to share person-specific health data without disclosing the privacy of their subjects.

Current mechanisms for effective management and protection of health records in Senegal have proven insufficient. In this paper, we propose a system that addresses the issue of sharing health data between hospitals in a trustless environment based on the Consortium Blockchain to improve the quality of care and the efficiency of the health system in Senegal. After a brief introduction, we present some characteristics of Blockchain as well as the different types and securing of Blockchain using cryptographic algorithms. Then an overview of related work is conducted. Finally, we presented the preliminaries of sharing health records with the Blockchain in Senegal. Our work ends with the description of the functioning of our medical record management mechanism with the Blockchain in Senegal and its implementation.

CCS Concepts: Security and privacy · Distributed systems security · Authorization

Keywords: hospitals · blockchain · data sharing · privacy · security · electroNic health records

1 Introduction

In Senegal's hospitals, there is a concern for good management practices to assist health professionals in their care process, so it is necessary to find solutions to the problems essentially posed by the management of health records in public health establishments. To do this, it is essential to make a representation of computer domain knowledge easily interpretable and exploitable, the organization of medical data collection considering the context of the patient and the exploitation of good practice guides and shared clinical experience [18, 19].

Y. Chen et al. (Eds.): BICT 2023, LNICST 512, pp. 156–165, 2023.
https://doi.org/10.1007/978-3-031-43135-7_15

We must therefore trust‘ this enterprise to manage this shared data within the parameters of confidentiality. We can imagine a professional network operating based on a decentralized application. The application could link all users to connect them to each other to facilitate information exchange without the transaction being secured and validated by a central authority. Instead of a central authority, the Blockchain uses a consensus mechanism to reconcile discrepancies between nodes of a distributed application [1, 2].

To solve the above problems, we propose to build a consortium blockchain for security and privacy preserving EHR sharing. This health files could assist the doctors and other authorized health and social services professionals responsible for your care to access to the files in any hospital in order to offer better care and more efficient follow-up.

2 Presentation of the Blockchain

2.1 Characteristics of a Blockchain

It consists of a register composed of a series of time-stamped blocks of transactions. It is this precise aspect of the Blockchain technology, which is the object of the present development, that has led to its name being given, by metonymy, to all these protocols.

A block generally consists of the hash value of the previous block, the payload, the signature of the contributor and the timestamp. A block consists of a format that uniquely identifies the block. This is followed by the block size, which contains the entire size of the block [2]. Once the block is validated, on average every ten minutes for the bitcoin example, the transaction becomes visible to all the holders of the register, potentially all the users, who will then add it to their block chain. The blockchain is defined as a technology that allows the storage and transmission of information in a decentralized manner from one individual to another.

The blocks, thus constituted of several transactions “signed” by public keys are then “time-stamped” by their author and constitute a basic unit to be verified. The Blockchain allows to timestamp digital documents impossible to backdate or to modify the content once data is recorded. This aspect, called timestamping, is essential because it allows the relative dating of the blocks, thus constituted, as all the blocks are chained, the order of the blocks is deterministic; therefore, each block can serve as a timestamp of the transactions included to solve the problem of double spending [3]. Krawiec et al. [4] presented several existing problems with current health information ex-change systems and the advantages offered by blockchain technologies. The protocol invented by Nakamoto proposes a solution to limit the risk of such a simultaneous pro-duction of two blocks, and to ensure that a valid block has time to spread throughout the network before a next one is created (Fig. 1).

2.2 Categories of Blockchain

Generally, blockchain can be classified into three categories:

Blockchain without authorization (Public Blockchain), this type of Blockchain implementation in which any node is free to join the network and participate as a miner without requiring authorization or access authorization.

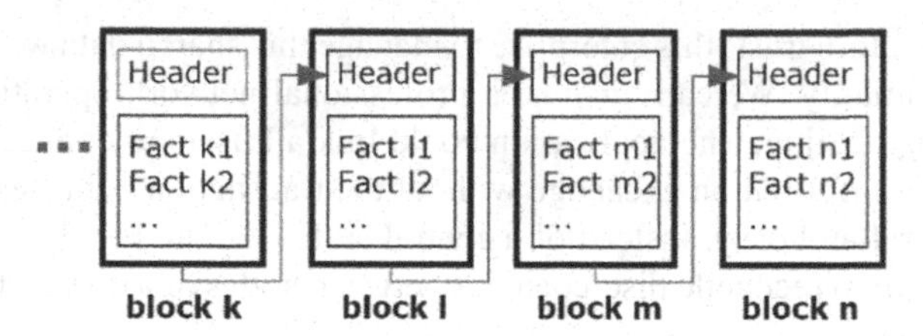

Fig. 1. Example of use of a Blockchain

Authorized blockchain: in this type participants must be authorized and have appropriate access permissions before they can join and participate in the network. In the Authorized Blockchain, only certain nodes can be authorized to participate in the mining process, this type is called the Private Blockchain or the Consortium Blockchain. The distinction between private and consortium blockchains is based on the number of nodes allowed to be miners. If only one node is allowed to be a miner, it is private, whereas if two or more nodes are allowed to participate in the mining process, it is a consortium blockchain [5]. For the consortium blockchain, all the members of its organizations will be able to read the transaction and verify that the sender was indeed the last owner of the transactions sent. Only the receiver will be able to sign the transaction with his private key to prove possession.

2.3 The Consensus Method in the Blockchain

Due to the incompressible latency of the network discussed above, multiple valid blocks could be created simultaneously by multiple nodes. The nodes would add one or another of these blocks and the network would then contain registers in different states.

The consensus mechanism is the core technology of the Blockchain, as it determines whether the new block is validated and who keeps the record [6]. This ensures that the most up-to-date and complete version' is the one used as a reference to validate transactions. Thus, this influences the security and reliability of the whole system. Therefore, it is necessary for the nodes to agree on the next block to be added to the chain, which is why Blockchain protocols provide a "consensus method". In practice, in a public blockchain such as Bitcoin, a mechanism for designating the validated block is used. Its author must provide proof of its designation to the other users of the network [7]. The simplest method of designation would be to draw lots for this validator, at a given interval of time (sufficient for a block to spread throughout the network.

3 Related Work

Xia, et al. [8] discusses a Blockchain-based data sharing framework that addresses the access control challenges associated with sensitive data stored in the cloud by using the built-in immutability and autonomy properties of the Blockchain. The proposed platform uses secure cryptographic techniques (encryption and digital signatures) to provide effective access control to sensitive shared data pools using an authorized Blockchain for enhanced security and a tightly monitored system. It allows users to access electronic medical records from a shared repository after their identities and cryptographic keys have been verified. Each block, in addition to the transactions and timestamp, has an identifier, which takes the form of a "hash" that links the blocks together. This hash is always the result of the "hash" of the previous block, the hash value of the previous block makes the blockchain unchangeable. The merkle root hash is part of the header, ensuring that none of the blocks in the Blockchain network can be modified without changing the header. This is achieved by taking the hashes of all events in the Blockchain network and adding of the output to the current block. The final output is a sha256 (sha256 ()) [7]. The proposed decentralized system consists of three entities, namely the user, system management and storage.

Zhang, et al. [9] Proposes a Blockchain-based secure and privacy-preserving PHI (BSPP) sharing scheme for diagnosis improvement in e-health systems. The patient's PHI and corresponding keywords are encrypted for data security while they are searchable by authorization for diagnostic improvements. Two types of Blockchain, Private Blockchain and Consortium Blockchain, are built by designing their data structures and consensus mechanisms. Lam Private Blockchain is responsible for storing PHI while the Consortium Blockchain keeps records of secure indexes of PHI.

Yue, et al. [10] propose a blockchain-based App architecture (called Healthcare Data Gateway (HGD)) to enable patients to own, control, and share their own data easily and securely without violating privacy, which offers a potential new way to improve the intelligence of healthcare systems while maintaining the privacy of patient data. The system is a smart smartphone application that allows patients to easily manage and control the sharing of their health data. The authors combine blockchain and off-blockchain storage to build a privacy-oriented personal data management platform, it manages personal electronic medical data on the blockchain storage system, evaluates all data requests by leveraging goal-centric access control, and uses secure multi-party computing to enable a third party to perform treatment on the given patient without risking patient privacy.

4 Sharing Medical Records with Blockchain for Health in Senegal

Given its characteristics, Blockchain technology could provide solutions to problems encountered in sensitive areas, particularly in the health sector. This leads us to reflect on the impact of the Blockchain in the health field. To put it plainly, what would be the contribution of Blockchain technology in the field of health? Several use cases are possible in the health sector. Blockchain could be used to [3, 11]:

- drug traceability;
- securing health data;
- managing patient data.

In recent years, Blockchain has been proposed as a promising solution to achieve personal health information (PHI) sharing with security and privacy preservation due to its immutability advantages [12]. Decentralization is an important feature of the Blockchain for health applications because it enables distributed health applications that do not rely on a centralized authority. Because the information in the Blockchain is replicated between all the nodes in the network, it creates an atmosphere of transparency and openness that allows healthcare stakeholders, and especially patients, to know how their data is being used, by whom, when and how.

The strength of the Blockchain lies in the fact that the compromise of one node in the Blockchain network does not affect the state of the ledger since the information in the ledger is replicated across multiple nodes in the network. Therefore, by its nature, the Blockchain can protect health data from potential data loss, corruption, or security attacks [13, 14].

5 Modeling the Health Record with the Blockchain in Senegal

In our field of application in this case that of health, the Blockchain would allow with effectiveness to justify the design and the setting in circulation of the medical files, i.e. to ensure its traceability, but also to fight against the attacks and usurpations of identity. In our project we implement our system composed of three entities, namely the hospital (doctor, nurse…), the system management (central authority) and the patient.

a. Node as hospital (doctors, nurses);
b. Minor as central authority (Ministry of Health and authorized physician);
c. Block as patient's medical record (EHR).

Each doctor will have his ID code at the central authority (Ministry of Health). Usually, each hospital has a server and many computers. Each computer is operated by a doctor to record the health information of his patients and then generate blocks for the health records of the patients and broadcast them to the Blockchain. In addition, the selected central authority is responsible for verifying the new blocks to come [9].

The use of cryptographic algorithms to encrypt the data stored on the Blockchain ensures that only users with legitimate permissions to access the data can decrypt it, thus improving the integrity and confidentiality of patient data (Fig. 2).

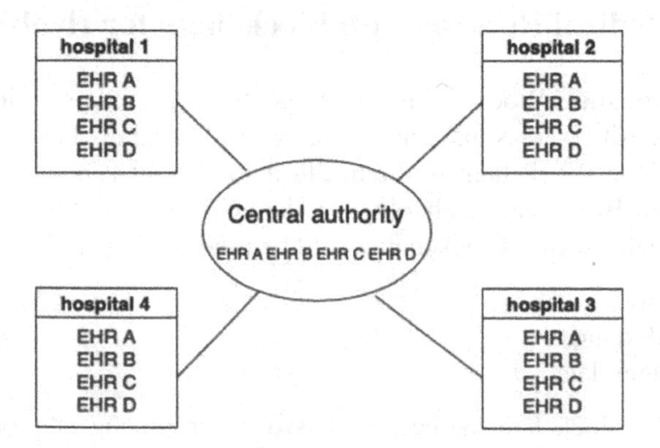

Fig. 2. Design of the medical record management mechanism

6 Operation of the Health Record Management Mechanism with the Blockchain in Senegal

We propose a data sharing mechanism based on the Blockchain consortium to secure electronic health records. The functioning of each entity can be described as follows:

1. Hospitals: are the users who are composed of doctor, nurse and patients who wish to register a patient record or access the data in the record. In the system, health records must be created by professional physicians and cannot be created by anyone else, including the patient himself. Personal health data belong to the patient and is used by requesters with the permission of the data owner. The doctor is part of the hospital layer by authenticating the patients to join the blockchain. He sends the block to the verifier (central authority) to accept patients who request to join the system. New blocks are accepted only after they are verified by the verifiers, who are responsible for checking the validity of the new blocks. The processes of generating, verifying, and adding new blocks to the blockchain is called mining.
2. The Central Authority or The Ministry of Health (verificator): The verifier is part of the data management layer by further authenticating the patient record and receives the physician's transaction key which is retained. The verifier subsequently approves the blocks that have been signed by the physician. This authenticates a block from the system into the Blockchain. To ensure the confidentiality and reliability of the mining processes, a consensus mechanism is essential in the Blockchain network. It determines who keeps the records and how to verify the validity of the new block. The consensus node is responsible for processing and verifying the authenticity and details of a block. Processed blocks are disseminated in the blockchain by the consensus node. An important role of the consensus node is to process and publish results based on irregularities in the system. The consensus node is the only entity allowed to access the system of unprocessed requests.

We have proposed consensus building for the validation of a new medical record as presented´ in the figure below:

- A hospital creates a new block (record) with patient information;
- The central authority verifies that the request has been´ issued in the network by a physician authorized to create a record through his or her identifier (Physician Id);
- If the central authority approves the record, and more than half of the randomly drawn hospital servers verify the new transactions are correct, they are accepted as a new validation block in the blockchain. The file can be shared to the whole network by meeting all the necessary requirements;
- If the consensus is successful, then a new record is entered into the Blockchain.

Since the Blockchain is immutable, it is impossible for a person to be able to open a record already' stored to add content without the presence of the patient who holds the private key for example. Indeed, by encrypting the data using asymmetric cryptography, there is no antagonism to sharing private data insofar as it will not be read by other users. Only the person with the private key will be able to decrypt and enjoy the data stored on the blockchain [15] (Fig. 3).

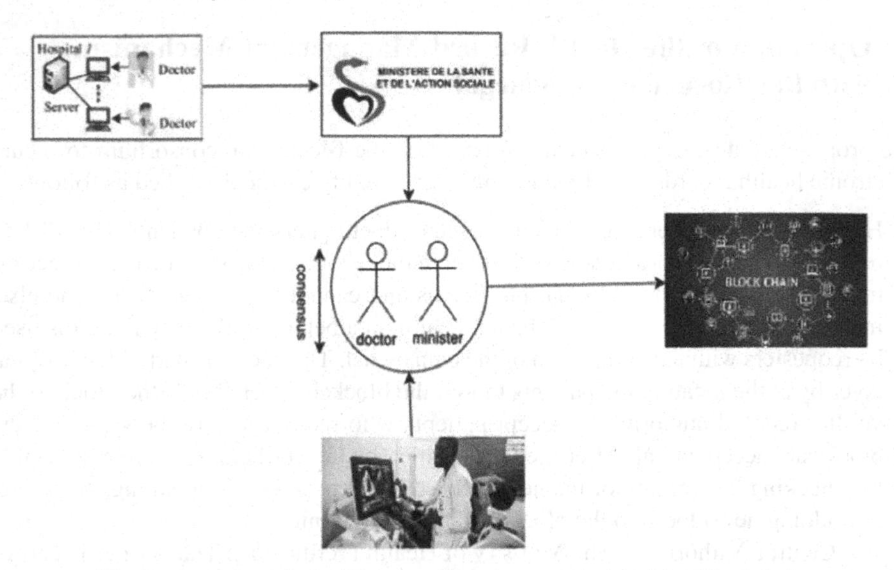

Fig. 3. Overview of a healthcare transaction associated with a Blockchain

7 Implementation of the Health Record Management Mechanism with Blockchain in Senegal

The purpose of this section is to provide a framework for building a system while analyzing the secure structures implemented to facilitate Blockchain-based data sharing for the logic of the electronic medical record system between hospitals in Senegal. We describe designed structures that achieve data sharing by presenting our data access system that aims to provide an appropriate sharing scheme while preserving the required security properties of the Blockchain.

The data structure of the Blockchain consortium is shown in Fig. 4. It consists of the block header, payload, contributor signature and timestamp.

1. The block header is hashed with sha256 (sha256()) as it is done in Bitcoin headers. The block header plays an important role in the Blockchain network in guaranteeing immutability. By changing a block header, an attacker should be able to change all block headers from the genesis block to forge a block record. The block header concludes three components:
   a. the block ID: a block consists of a format that uniquely identifies the block;
   b. the block size: the block size that contains the entire size of the block;
   c. the hash value of the previous block: which is a sha256 hash (sha256()) whose function is to ensure that no previous block header can be modified without changing that block header. If one of the transactions in a block is modified, even slightly, the corresponding hash output will change drastically, which will break the chain to the following blocks of the Blockchain.
2. The payload: is composed of four parts: The identity of the PHI generator (physician) which is the identity of the physician who created or accessed the record, the identity of the PHI owner (patient's first and last name), blood type and keyword which may

include test results for a patient and a corresponding diagnosis from a physician which are encrypted. Notably, not all PHI information is stored in plain text format. Since all hospitals can view the data transmitted by a node, the confidentiality of the data will not be intact, but physicians cannot open it without the patient's permission. The patient can select a representative who can access his information and/or medical history on his behalf, in case of emergency or the doctor contacts the central authority to access the file.

3. The contributor's signature allows to follow the generator (physician) of the block, for each node having contributed to the block, a digital signature is required.
4. The timestamp: indicates the time of generation of the block. When the conditions for this field are met, the block is ready to be distributed in the Blockchain network. The block lock time generally means the time the block enters the Blockchain. An update of the Blockchain by adding the new block into the Blockchain to all participants.

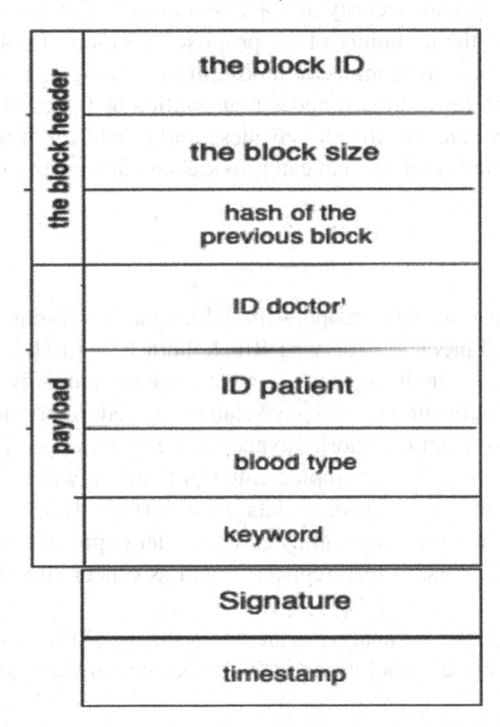

Fig. 4. Structure of a medical record block in the Blockchain consortium

8 Discussion

The fact that the information on the blockchain is replicated among all hospitals in the country creates an atmosphere of transparency and openness, allowing patients to know how their data is being used. When a transaction is performed, the corresponding healthcare parameters are sent to the validation devices which are the central authority (Ministry of Health) and the hospitals with the use of the Blockchain consortium that is' to say that the validation of the data and thus the registration of a new record would only be authorized by the central authority as well as the selected hospitals... Sharing electronic health records can help improve diagnostic accuracy, where security and privacy preservation are critical issues in systems.

Hospital's store, health records in the Consortium Blockchain, which has the advantages of faster transaction, better privacy preservation, low cost, and better security performance. However, access to the data could be public, at least in part. Patients can access' all the data related to the file thanks to their private key, others could for example, just see the name, first name and blood type of the patient. The blood type is in clear in the file because it allows in case of an accident to solve emergency problems without resorting to decryption protocols.

Each system adopts a different algorithm that meets the requirements. The proposed protocol ensures data security and access control. The essential features of the Blockchain guarantee the immunity of the proposed protocol. In other words, the data stored in the Blockchain are immutable unless there is a significant attack (51%) that occurs when there are fewer honest nodes than malicious nodes in the whole network [16, 17]. This protocol ensures the effectiveness and reliability of the proposed system to work in different environments and can provide satisfactory security protection.

9 Conclusion

In this paper, we illustrate how to apply the Blockchain technique in healthcare. We proposed a secure and privacy-preserving Blockchain-based EHR sharing protocol for diagnostic improvements in the Senegalese healthcare system. Advantages of this strategy include easy authentication since a physician only needs to provide his ID associated with his identity to the central authority to access the system.

The digital health record, once implemented in Senegal, will be used to make available, in real time, all of a user's medical data, from birth to death, to share data with all health professionals involved in providing care in order to provide management, quality and performance indicators, to feed registers, such as cancer, and the daily emergency situation report.

In our future work, we are analyzing the performance of our system and comparing it with current state-of-the-art solutions for data sharing between hospitals.

References

1. Swan, M.: Blockchain: Blueprint for a new economy. O'Reilly Media, Inc. (2015)
2. Angraal, S., Krumholz, H.M., Schulz, W.L.: Blockchain technology: applications in health care. Circul. Cardiovascul. Qual. Outcomes **10**(9), e003800 ((2017))
3. Kuo, T.-T., Kim, H.-E., Ohno-Machado, L.: Blockchain distributed ledger technolo-gies for biomedical and health care applications. J. Am. Med. Inform. Assoc. **24**(6), 1211–1220 (2017)
4. Transaction, C.P., MPI, M.P.I.: Blockchain: opportunities for health care, CP Transaction (2016)

5. Helliar, C.V., Crawford, L., Rocca, L., Teodori, C., Veneziani, M.: Permission-less and permissioned blockchain diffusion. Int. J. Inf. Manage. **54**, 102136 (2020)
6. Bentov, I., Lee, C., Mizrahi, A., Rosenfeld, M.: Proof of activity: Extending bit-coin's proof of work via proof of stake [extended abstract] y. ACM SIGMETRICS Perform. Eval. Rev. **42**(3), 34–37 (2014)
7. Zhou, T., Li, X., Zhao, H.: Med-ppphis: blockchain-based personal healthcare in-formation system for national physique monitoring and scientific exercise guid-ing. J. Med. Syst. **43**(9), 1–23 (2019)
8. Xia, Q., Sifah, E.B., Smahi, A., Amofa, S., Zhang, X.: Bbds: blockchain-based data sharing for electronic medical records in cloud environments. Information **8**(2), 44 (2017)
9. Zhang, A., Lin, X.: Towards secure and privacy-preserving data sharing in e-health systems via consortium blockchain. J. Med. Syst. **42**(8), 1–18 (2018)
10. Yue, X., Wang, H., Jin, D., Li, M., Jiang, W.: Healthcare data gateways: found health-care intelligence on blockchain with novel privacy risk control. J. Med. Syst. **40**(10), 1–8 (2016)
11. Siyal, A.A., Junejo, A.Z., Zawish, M., Ahmed, K., Khalil, A., Soursou, G.: Applications of blockchain technology in medicine and healthcare: challenges and future perspectives. Cryptography **3**(1), 3 (2019)
12. Esposito, C., De Santis, A., Tortora, G., Chang, H., Choo, K.-K.R.: Blockchain: a panacea for healthcare cloud-based data security and privacy? IEEE Cloud Comput. **5**(1), 31–37 (2018)
13. Liu, W., Zhu, S., Mundie, T., Krieger, U.: Advanced block-chain architecture for e-health systems. In: 2017 IEEE 19th International Conference on e-Health Net-working, Applications and Services (Healthcom), pp. 1–6. IEEE (2017)
14. Azaria, A., Ekblaw, A., Vieira, T., Lippman, A.: Medrec: using blockchain for medical data access and permission management. In: 2016 2nd International Conference on Open and Big Data (OBD), pp. 25–30. IEEE (2016)
15. Burniske, C., Vaughn, E., Cahana, A., Shelton, J.: How Blockchain Technology can Enhance Electronic Health Record Operability. Ark Invest: New York, NY, USA (2016)
16. Buterin, V., et al.: A next-generation smart contract and decentralized application platform. White Paper **3**(37) (2014)
17. Greenspan, G.: Multichain private blockchain-white paper, pp. 57–60 (2015). http://www.mul-tichain.com/download/MultiChain-White-Paper.Pdf
18. Aly Konte, M. : Secteur de la santé au sénégal: malaises actuels et perspectivesfutures; rapport de conférence (2006)
19. Hane, F.: Production des statistiques sanitaires au sénégal: entre enjeux politiques et jeux d'acteurs. Santé publique **29**(6), 879–886 (2017)

A Novel Durable Fat Tissue Phantom for Microwave Based Medical Monitoring Applications

Mariella Särestöniemi[1,2](✉), Rakshita Dessai[3], Sami Myllymäki[3], and Teemu Myllylä[1,4]

[1] Health Sciences and Technology, Faculty of Medicine, University of Oulu, Oulu, Finland
mariella.sarestoniemi@oulu.fi

[2] Centre for Wireless Communications, Faculty of Information Technology and Electrical Engineering, University of Oulu, Oulu, Finland

[3] Microelectronics laboratory, Faculty of Information Technology and Electrical Engineering, University of Oulu, Oulu, Finland

[4] Optoelectronics and Measurement Techniques Research Unit, Faculty of Information Technology and Electrical Engineering, University of Oulu, Oulu, Finland

Abstract. Human tissue mimicking phantoms allow development of realistic emulations platforms which are essential for design of several biomedical monitoring and diagnosis systems. This first aim of this paper is to present a novel and durable fat tissue phantom for lower microwave frequency ranges 2.5–10 GHz. The phantom is developed from the liquid propylene glycol (pure) which we found to have similar dielectric properties as the fat tissue and hence, it is suitable to be used as liquid fat phantom. Development steps of solid fat phantoms with different trials are presented to provide insight how each ingredient affect on the dielelctric properties of the mixture. Additionally, phantom's stability over time in terms of dielectric and physical properties are evaluated. The second main aim of this paper is to present a novel approach to verify the feasibility and reliability of phantoms in practical scenarios with tissue layer model simulations. In the simulations, the antenna reflection coefficients are calculated with tissue layer models in which the dielectric properties of the fat tissue layer is varied between the proposed prolyne glycol -based fat phantoms as well as real human fat tissue values. Our goal is to show how small differences in the dielectric properties of the phantoms affect on a practical scenario which is based on antenna impedance measurements. The dielectric properties of the proposed fat phantom have very good correspondence with real fat tissue especially in the range of 5 GHz-10 GHz. Also, at lower ultrawide band (3.1–5 GHz), the difference in dielectric properties is minor. The layer model simulations show that the differences in dielectric properties do not have significant effect when modelling the practical scenarios in the frequency ranges targeted for medical applications. Hence the proposed liquid and solid fat phantoms are suitable to be used in the emulation platforms of biomedical applications.

Keywords: Adipose phantom · Biomedical applications · Dielectric properties · Microwave propagation · Tissue mimicking phantoms

Y. Chen et al. (Eds.): BICT 2023, LNICST 512, pp. 166–177, 2023.
https://doi.org/10.1007/978-3-031-43135-7_16

1 Introduction

Interest on development of microwave -based medical monitoring applications has increased significantly recently due to their non-ionized radiation, low-cost, and possibility for portability [1–6]. Development of new biomedical technology products requires precise modelling of the human body effects. Commonly, this involves large amount of experiments and measurements that need to be carried out with humans and animals, which in general are time consuming, complex, and expensive to perform. Thus, realistic tissue mimicking 3D phantom emulation platforms are suggested to be used when evaluating new concepts of medical applications [6].

The development of human tissue phantoms for microwaves has been an actively studied topic in recent years [7–15]. Different recipes for solution mixtures have been proposed for solid and liquid phantoms for different medical monitoring and imaging applications. Solid phantoms have the advantage of possibility for using realistic shaped 3D molds whereas liquid phantoms have the advantage of easy adjustability in terms of size.

Development of adipose, i.e., fat tissue phantoms for microwave ranges has also been studied actively [10–15]. Numerous of the proposed fat phantoms are targeted for breast cancer detection studies. However, fat phantom development has shown to be challenging due to several reasons: a) some of the proposed recipes contain ingredients which are toxic (e.g. formaldehyde), thereby requiring specific laboratory equipment to be handled, b) incredients are not easily accessible or are very costly, c) the fat phantoms aimed to be solid become oily in the room temperature and hence challenging to be used especially with 3D phantom models, d) or the dieletric properties of the presented fat phantoms have clear differences compared to the dielectric properties of the human fat tissue, e) phantoms are not durable: either the dielectric properties change significantly with the time or mildew appears on phantoms even after short time of storage in the refridgerator.

The first aim of this paper is to present for the first time a novel, easy-to-produce, non-toxic, and durable mixture for fat phantom which is based on pure propylene glycol. The developement procudure with different recipe trials are explained to give insight how different ingredients affect on the dieletric properties. *The second main aim* is to present a novel idea for phantom verifications with tissue layer model electromagnetic simulations. In the simulations, the antenna reflection coefficient is calculated with tissue layer models in which the dielectric properties of the fat tissue layer is varied between the proposed prolyne glycol -based fat phantoms as well as real human fat tissue values. Our goal is to show how small differences in the dielectric properties of the phantoms affect on a practical scenario which is based on antenna impedance measurements.

This paper is organized as follows: Sect. 2 presents the liquid and solid propylene glycol -based fat tissue phantoms. The development prodecure of the solid fat phantom and dielectric properties of different recipe trials are presented. Section 3 verifies the usability of the proposed fat phantoms using layer model simulations. Additionally, stable-of-time properties are evaluated. Summary and Conclusions are given in Sect. 4.

2 Propylene Glycol Based Fat Tissue Phantoms, Liquid and Solid

The development of propylene glycol-based phantoms started from authors' observation that the dielectric properties of the pure proplylene glycol (98%) were close to those of real fat tissue. The relative permittivity and conductivity values for fat tissue and liquid propylene glycol are shown in Fig. 1. Dielectric properties of the fat tissue are retrieved from [16]. As it can be seen, the relative permittivity of the propylene glycol is close to the dielectric properties of fat tissue from 2.5 GHz onwards. The relative permittivity of propylene glycol is slightly higher than the real fat tissue at 2.5–3.6 GHz, whereas from 3.6 GHz onwards, the relative permittivity is slighlty lower. However, the difference at these ranges is maximum 0.2. Also the difference in the conductivity values is minor 0.1–0.2 dB. Hence, the pure propylene glycol can be used as fat tissue phantom in the liquid form especially at lower microwave frequencies.

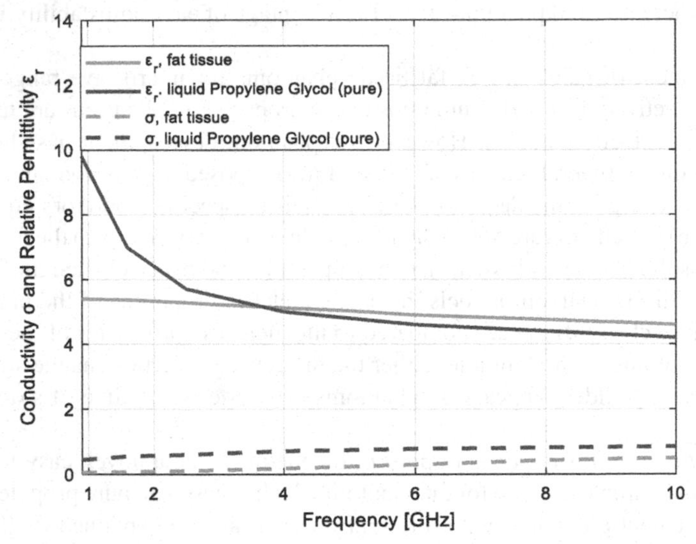

Fig. 1. Relative permittivity and conductivity values for fat tissue and liquid propylene glycol (pure) presented in the same linear scale.

Next, we started investigations for developing propylene glycol -based *solid* phantoms which enable the use of 3D fat phantom models. The aim was to solidify propylene glycol by adding gelatin and xanthan in the mixture. Since gelatin does not get dissolved directly with propylene glycol, small amount of water is needed to be mixed with gelatin first. However, water increases both relative permittivity as well as conductivity and therefore, the amount of water is intended to be minimized. Dishwashing liquid is added to enable smooth mixing of all the ingredients. In this study, we investigate the impact of different amounts of water in the solution mixtures. Additionally, the impact of the amount of gelatin and xanthan is also studied. Altogether 14 different mixture trials are investigated to find a solution with provides best match to a real fat tissue taken into the account of physical characteristics (solidness, possibility to be used in 3D molds). The different mixture solution trials FT1-FT14 with amount of ingredients are summarized in Table 1. Additionally, the corresponding measured dielectric properties of each sample are also presented in Table 1.

The phantom preparation is described briefly in the following: On a hot plate stirrer, presented in Fig. 2a, the distilled water is warmed in a beaker to 65 °C. Then, while keeping the temperature at 65 °C, the gelatin is gently added. The mixture is allowed to be stirred for 5 min. The propylene glycol is heated to around 50 °C and added to the gelatin water-based mixture and stirred continuously till the solution is heated to 65 °C. Then the xanthan is thoroughly combined with the solution. Finally, dishwashing liquid is added and well mixed into the solution. The solution is placed in a small petri dish and refrigerated for 24 h. Before taking any measurements, the phantoms rest at room temperature for about an hour to achieve room temperature 22°.

Firstly, the dielectric properties of the phantoms are measured using and Vector Network Analyzer (VNA) Keysight P9375A connected to a SPEAG's Dielectric Assessment Kit (DAK 3.5) [17]. The DAK software converts the measured complex S_{11} of the phantom sample into the complex permittivity and conductivity. The operation frequency range is 900 MHz to 10 GHz with a sweep of 117 points. The calibration was performed by applying the standard Open Source Load (OSL) calibration. The *open* was measured by holding the probe in the ambient air. The *short* was measured by connecting the probe to the shorting block and the measurement for the *load* was performed by setting the probe to DAK's official calibration liquid "Head" [17]. The success of the calibration was verified by measuring the dielectric properties of the calibration liquid Head and comparing results with the data sheet.

After the calibration, all of the fat phantom mixture solution trials (FT) are measured at room temperature. The dielectric properties of the samples are measured twice at 2–3 different locations and are given as an average of all. The measurement setup of dielectric properties of the phantom is presented in Fig. 2b.

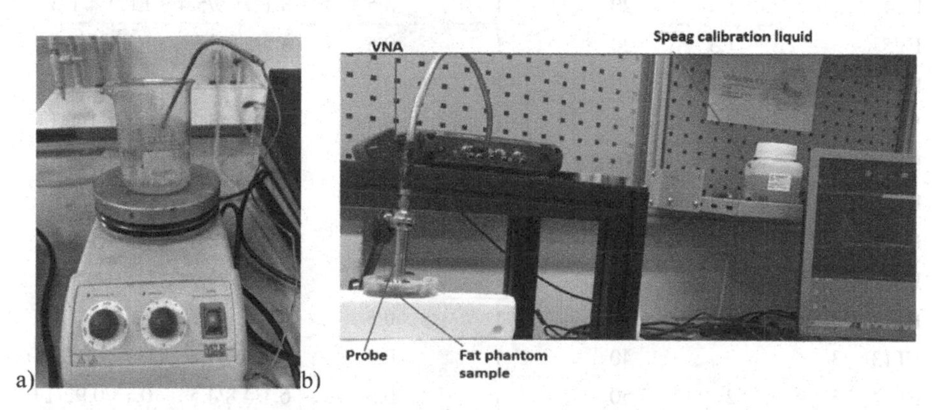

Fig. 2. The setup for measuring dielectric properties of the phantom samples.

The first fat-phantom trial (FT1) is developed only from propylene glycol (20 ml), distilled water (5 ml), and gelatine (3 g). As seen from Table 1, both relative permittivity and the conductivity values are too high compared to the real fat tissue. Therefore, the next trial FT2 includes less water and gelatine, but has incluision of xanthan and dishwashing liquid. FT2 has clearly lower relative permittivity and conductive than FT1, though still excessively high compared to the target values. FT3-FT5 are the trials where amount of water and gelatine are further decreased, both reductions yielding in lower relative permittivity and conductivity. In FT6-FT9, the amount of gelatine is drastically reduced to 0.75 g which however does not lower the relative permittivity to below 5 and also conductivity remains too high. Trials FT10-FT14 evaluates the impact of the increasing the amount of the propylene glycol in the mixture. The addition of propylene glycol requires addition of the water as well and thus, the decrease of the permittivity and conductivity values is more moderate. When using propylene glycol 40 ml and 50 ml (FT13 and FT14), the relative permittivity and conductivity values are already very close to those of the real fat tissue, especially with FT14. However, FT14 is not fully solidified even after several days and therefore is not suitable in its current form for 3D emulation platforms requiring fully solid phantoms. Thus, we chose the fully solid FT13 for the fat tissue phantom mixture solutions since the relative permittivity and conductivity are only 0.2 and 0.3 at higher level than those of the measured fat tissues.

Table 1. Different phantom mixture trials and their dielectric properties.

	Distilled water [ml]	Gelatine [g]	Propylene glycol [ml]	Xanthan [g]	Dish-washing liquid [ml]	Relative permittivity	Conductivity [S/m]
FT1	5	3	20	–	–	10.3/6.9/6.2	1.3/ 1.7/ 2.1
FT2	3	2	20	1	0.5	9.3/6.5 /5.9	1.1/ 1.5 /1.7
FT3	2	1.5	20	1	0.5	8.6/6.1/5.6	1.1/1.5/1.7
FT4	2	1	20	1	0.5	8.15/5.9/5.4	1.0/1.4/1.6
FT5	3	1.5	20	1	0.5	8.9/5.9/5.7	1.1/1.5/1.8
FT6	2	0.75	20	1	0.5	7.7/5.73/5.3	0.9/1.3/1.5
FT7	1.5	0.75	20	1	0.5	7.0/5.4/5.1	0.8/1.1/1.3
FT8	2	0.75	20	2	0.5	7.8/5.8/5.4	0.9/1.3/1.6
FT9	1.5	0.75	20	2	0.5	6.9/5.3/5.0	0.8/1.0/1.2
FT10	3	2	25	1	0.5	7.0/5.2/4.8	0.8/1.1/1.2
FT11	3	2	30	1	0.5	7.1/5.2/4.9	0.8/1.0/1.2
FT12	3	2	35	1	0.5	6.8/5.1/4.8	0.8/1.1/1.1
FT13	3	2	40	1	0.5	6.4/5.0/.4.7	0.75/0.95/1.0
FT14	3	2	50	1	0.5	6.1/4.8/4.5	0.73/0.92/1.0
Final = FT13	**3**	**2**	**40**	**1**	**0.5**	**6.4/5.0/.4.7**	**0.75/0.95/1.0**

The relative permittivity and conductivity values for fat tissue trials and real fat tissue at frequency range 0.9–10 GHz are presented in Fig. 3a-b, respectively.

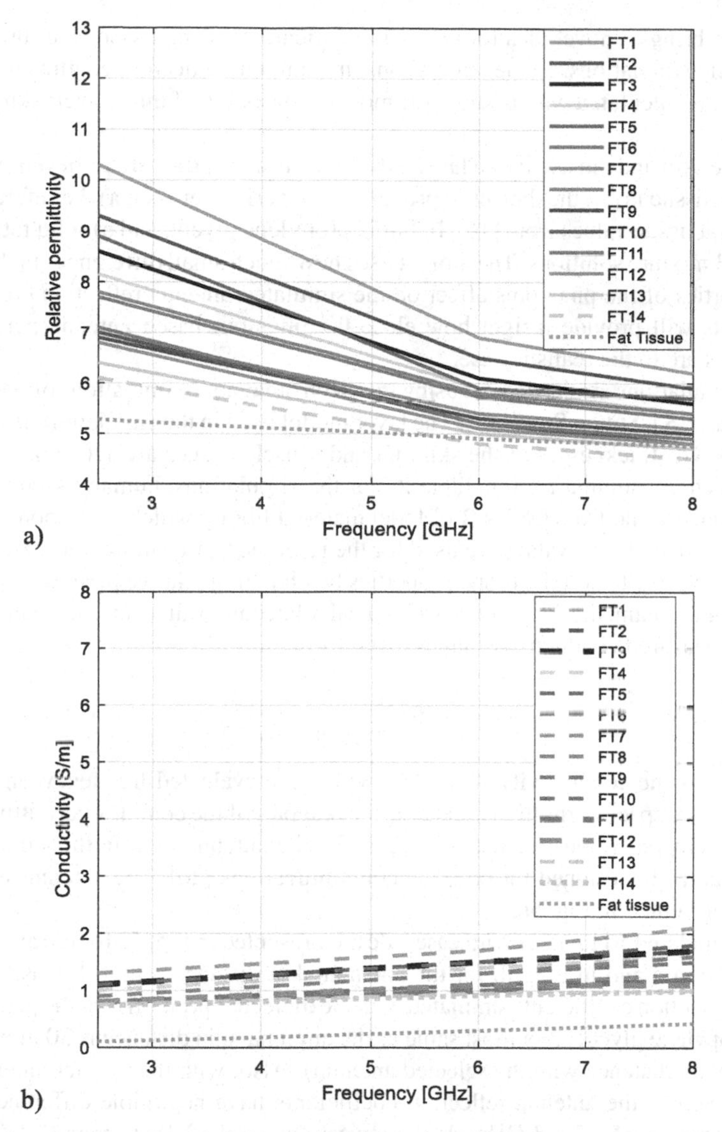

Fig. 3. Dielectric properties of different fat phantom trials (FTs) a) relative permittivity, b) conductivity.

3 Fat Phantom Evaluations

3.1 S11 Parameter Comparison with Layer Models

This paper brings a novel idea for phantom verifications using tissue layer model electromagnetic simulations. In the simulations, the antenna reflection coefficient, i.e. S11 parameter, is calculated with tissue layer models consisting of three layer: skin, fat, and muscle.

For the skin and muscle tissue layers, the dielectric proprties are retrieved from [16]. For the fat tissue layer, the dielectric properties are varied between a) the reference case with real fat tissue values from [16], b) liquid propylene glycol, and c) solid fat phantom with FT13 mixture solutions. The aim is to see how much small differences in the dielectric properties of the phantoms affect on the simulated antenna reflection coefficients. The results will provide insight how close the phantom based antenna performance evaluations are to the realistic case.

The simulations are carried out using the electromagnetic simulation software Simulia Dassault CST Studio Suite [18]. The layer model used in the simulations is presented in Fig. 4a. The thicknesses of the skin, fat and muscle tissues are 1.1 mm, 7 mm, and 8 mm which are summarized in Table 2. For the simulations, human tissue values are automatically found from CST's BioModel material library which correspond to values retrieved in [16]. Those values are used for the reference case simulations. However, in CST, it is possible to edit the tissue properties by changing relative permittivity and loss tangent values manually. Therefore, we first calculate tanδ values for the phantom cases from the measured conductivity values using formula:

$$\tan\delta = \frac{\sigma}{\omega\varepsilon_0\varepsilon_r}$$

in which σ is the conductivity, $\omega = 2\pi f$ with f the evaluated frequency, $\varepsilon_0 = 8.854$ e^{-12} is the free space permittivity and ε_r is the real part of the complex permittivity value [19]. Loss tangent values are listed in Table 2. The antenna used in the simulations is an UWB antenna designed for on-body communications [20]. The antenna simulation model is presented in Fig. 4b.

The simulated S11 values with cases a-c are presented in Fig. 5a for the antenna-skin distance 30 mm and in Fig. 5b for the antenna-skin distance 8 mm. It was found that antenna reflection coefficients simulated with the dielectric properties of the phantom and liquid propylene glycol are almost same as the antenna-skin distance is 30 mm (optimal antenna-body distance with the selected antenna). Also, with the smaller antenna-body distance (8mm), the antenna reflection coefficients have negligible differences at the frequency range 3.5 GHz-8 GHz. At the lowest part of the UWB range (3.1 GHz), the difference is maximum 4.5 dB, whereas at ISM frequency band 2.5 Ghz, the difference is maximum 2 dB. The difference is at largest, 6 dB, at 2.6 GHz but that frequency range is out of the interest for medical applications. These results are in line with the dielectric property differences presented in Fig. 3 since their differences are also larger at lower frequencies.

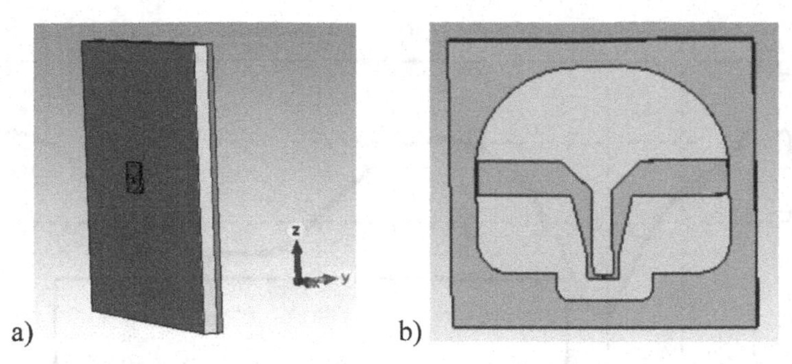

Fig. 4. a) The layer model used in the phantom verifications with S11 parameter simulations, b) UWB loop antenna used in the measurements.

Table 2. Loss tangent values for liquid and solid fat phantoms and real fat tissue

Fat Tissue/Phantom	2.5 GHz	6 GHz	8 GHz
Real fat tissue	0.14	0.19	0.21
Solid phantom	0.25	0.36	0.43
Liquid phantom	0.21	0.34	0.43

3.2 Fat Phantom Stability over Time

Normally, several gelatin-based phantoms last only a limited time especially if no preservatives are used. Mildew appears even in couple of weeks although the phantoms are stored in the refrigerator. Additionally, dielectric properties change remarkably as the water slowly evaporates from the phantoms with the time and hence, the phantom dries slowly.

Next, the proposed fat phantom's stability over time is evaluated measuring dielectric properties of the FT13 after 1, 7, 10, 11, 16, and 67 days. The phantom was stored in the refrigerator and measured at room temperature. After 10- and 67-days of storage, phantom was reheated and resolidified again, and the measurements were taken. Dielectric properties of the phantom after 1, 7, 10, 11, 16, and 67 days are presented in Fig. 6. It is found that dielectric properties change only slightly within the time for the first 16 days. Especially at the frequency range 6–8 GHz, the differences in relative permittivity are negligible: maximum 0.2. In conductivity, the variation is 0.1 S/m. At lower frequency range, the relative permittivity difference is 0.5 and conductivity difference 0.45. The relative permittivity decreases slightly for the first 10 days. Reheating arises the dielectric properties slightly: relative permittivity arises 0.4 units at lower part of the simulated frequency range and 0.1 units at higher frequency range. Differences in the conductivity values are 0.1 S/m over the whole simulated frequency range. After 67 days, the dielectric properties of the phantoms have changes more significantly: relative permittivity is increased 0.5–1 units and conductivity 0.1–0.45 S/m.

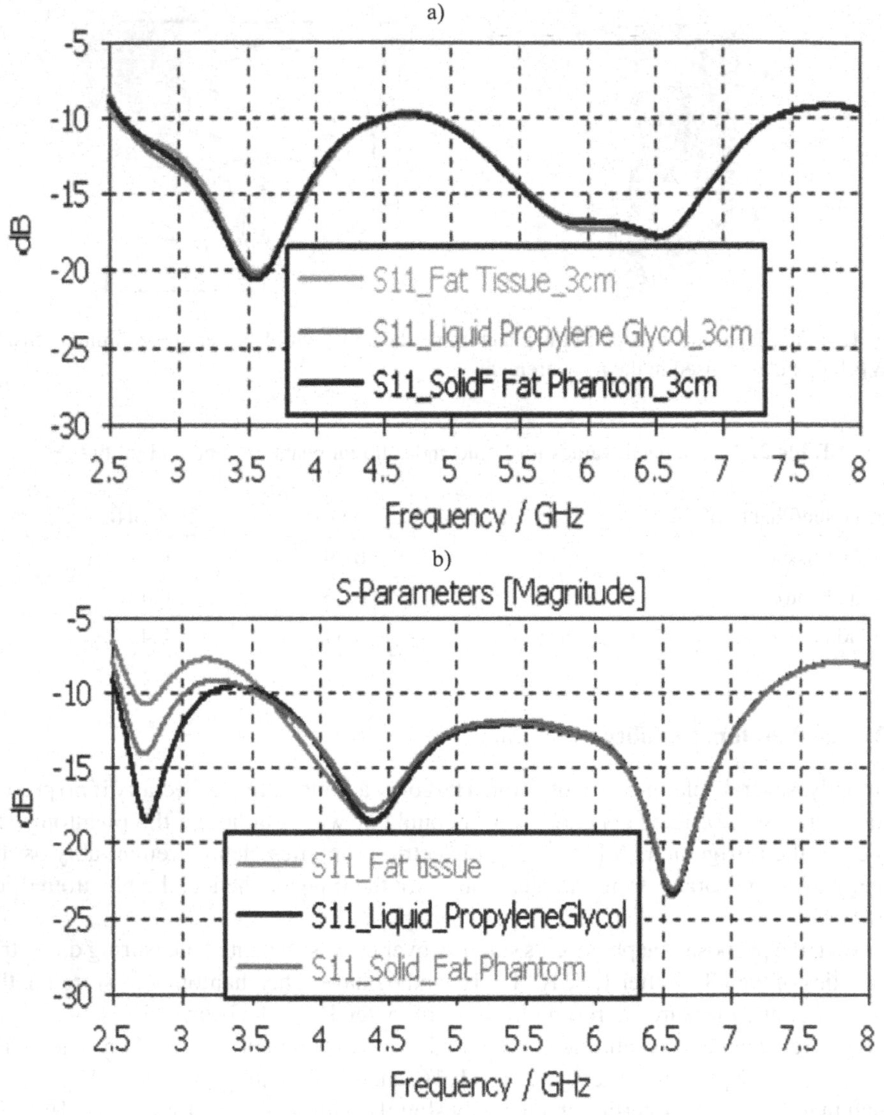

Fig. 5. S11-parameters obtained using dielectric properties of real fat tissue, liquid propylene glycol and solid fat phantom in the fat tissue layer: a) antenna-skin distance 30 mm and b) 8 mm.

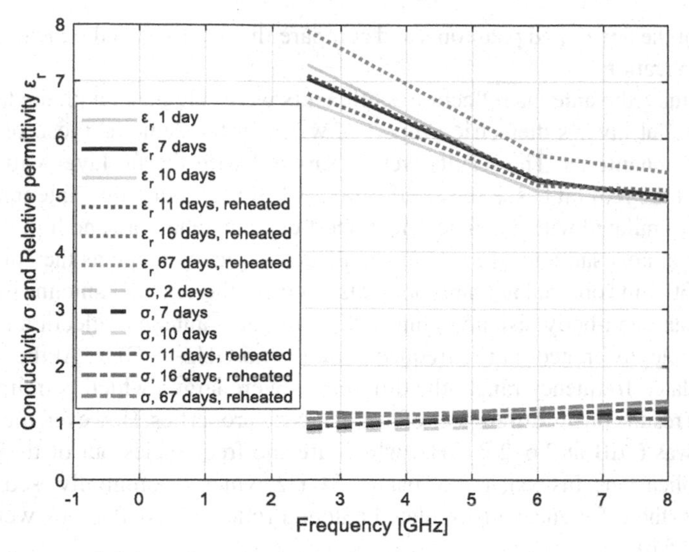

Fig. 6. Relative permittivity and conductivity values for phantom after 1,7,10, 11, 16, and 67 days after preparation.

4 Summary and Conclusions

This paper presented a novel and durable fat tissue phantom for lower microwave frequency ranges. The proposed phantom was developed from the liquid propylene glycol (pure) that we proved to have corresponding dielectric properties as the fat tissue. Development steps of the solid fat phantoms were described to provide insight how each ingredient affects on the dielectric properties of the mixture. The challenge in the development process is that jellying agents gelatin and xanthan do not get mixed directly with liquid propylene glycol and therefore, small amount of distilled water has to be added. However, distilled water and jellying agents affect on the dielectric properties and thus, optimal mixture, which yields to fully solid solution after polymerization, has to be developed taken into the account that the dielectric properties should be enough close to real fat tissue. The dielectric properties of the proposed solid phantom was shown to have very good correspondence with real fat tissue especially from 5 GHz-10GHz: the differences were almost negligible. Also, at lower ultrawide band (3.1–5 GHz), the difference was minor: differences in relative permittivity and conductivity values were 0.4–0.8 and 0.1–0.2, respectively. The differences were at largest around 2.5–2.7 GHz.

This paper also proposed for the first time the idea of evaluating the feasibility and reliability of the new phantoms for practical scenarions with human tissue layer model simulations. The idea is that the dielectric properties of the developed fat phantom is used in the simulation model as dielectric properties of the corresponding fat tissue layer. For other tissue layers, the dielectric properties are set same as in the realistic case. Simulations, e.g. the antenna reflection coefficients simulations, are carried out and compared with the antenna reflection coefficients obtained with the reference case in which also the fat layer has realistic dielectric properties. The proposed method provides the possibility to investigate smoothly the impact of small differences in the dielectric properties of the developed phantoms and compare the results with the ideal case for the selected application.

In this study, the antenna reflection coefficients were calculated with the layer model in which the fat layer's dielectric properties were set the same as those of the liquid and solid fat phantoms. The results were compared with the fat layer with dielectric properties of the real fat tissue retrieved from [16]. It was found that antenna reflection coefficients simulated with the dielectric properties of the phantom and liquid propylene glycol were almost same for the whole simulated frequency range as the antenna-body distance is 30 mm (optimal antenna-body distance for the selected antenna). Also, with the smaller antenna-body distance (8mm), the simulated antenna reflection coefficients have negligible differences at the frequency range 3.5 GHz–8 GHz. At the lowest part of the simulated frequency range, the differences were larger, which is in line with the comparison results on the differences of the dielectric properties. However, the maximum difference was 6 dB at 2.6–2-7 GHz, which are the frequencies out of the interest of medical applications. Instead, at ISM band 2.5 GHz, which is commonly used in medical applications, the differences in simulated antenna reflection coefficients were smaller: maximum 2.5 dB.

Based on the presented results, the pure propylene glycol can be considered as an excellent fat phantom in liquid form for the frequency range 2.5–10 GHz with only minor differences in the dielectric properties compared to those of the real fat tissue. The developed solid fat phantom is also very good for the frequency range 3–10 GHz and also good at the ISM band 2.5 GHz. Hence, it can be concluded that the proposed propylene glycol -based liquid and solid fat phantoms are suitable to be used in the emulation platforms of biomedical applications.

As a future work, we plan to test usability of novel fat phantoms with different antennas, both on-body and implant antennas. Additionally, we will use novel fat phantoms (both solid and liquid) to different medical monitoring application studies for which we already have realistic simulation -based results available: e.g. for realistic capsule endoscopy radio channel modelling [21] and breast cancer detection studies [22].

Acknowledgement. This research is funded by Academy of Finland Profi6 funding, 6G-Enabling Sustainable Society (University of Oulu, Finland), which is greatly acknowledged.

References

1. Vrba, D., Vrba, J., Fiser, O., Cumana, J., Babak, M., Vrba Senior, J.: Applications of microwaves in medicine and biology. Recent Microwave Technol. (2022). https://doi.org/10.5772/intechopen.105492
2. Khan, S., Saied, I.M., Ratnarajah, T., Arslan, T.: Evaluation of unobtrusive microwave sensors in healthcare 4.0-toward the creation of digital-twin model. Sensors (Basel). **22**(21), 8519 (2022). https://doi.org/10.3390/s22218519. PMID: 36366218; PMCID: PMC9657877
3. Li, C., Tofighi, M., Schreurs, D., Horng, T.: Principles and Applications of RF/Microwave in Healthcare and Biosensing, Elsevier,1st Edition (2016)
4. Kiourti, A., et al.: Next-generation healthcare: enabling technologies for emerging bioelectromagnetics applications. IEEE Open J. Antennas Propag. **3**, 363–390 (2022). https://doi.org/10.1109/OJAP.2022.3162110

5. Rafique, U., Pisa, S., Cicchetti, R., Testa, O., Cavagnaro, M.: Ultra-wideband antennas for biomedical imaging applications: a survey. Sensors. **22**(9), 3230 (2022). https://doi.org/10.3390/s22093230
6. Särestöniemi, M., Reponen, J., Myllymäki, S., Myllylä, T.: Remote diagnostics and monitoring using microwave technique – improving healthcare in rural areas and in exceptional situations, under review for FinJeHew journal (2023)
7. Costanzo, S., Cioffi, V., Qureshi, A.M., Borgia, A.: Gel-Like human mimicking phantoms: realization procedure, dielectric characterization and experimental validations on microwave wearable body sensors. Biosensors (Basel). **11**(4), 111 (2021). https://doi.org/10.3390/bios11040111.PMID:33917777;PMCID:PMC8068187
8. Garrett, J., Fear, E.: Stable and flexible materials to mimic the dielectric properties of human soft tissues. IEEE Antennas Wirel. Propag. Lett. **13**, 599–602 (2014). https://doi.org/10.1109/LAWP.2014.2312925
9. Castelló-Palacios, S., Garcia-Pardo, C., Fornes-Leal, A., Cardona, N., Vallés-Lluch, A.: Tailor-made tissue phantoms based on acetonitrile solutions for microwave applications up to 18 GHz. IEEE Trans. Microw. Theory Tech. **64**(11), 3987–3994 (2016). https://doi.org/10.1109/TMTT.2016.2608890
10. Pollacco, D.A., Conti, M.C., Farrugia, L., Wismayer, P.S., Farina, L., Sammut, C.V.: Dielectric properties of muscle and adipose tissue-mimicking solutions for microwave medical imaging applications. Phys. Med. Biol. **64**(9), 095009 (2019). https://doi.org/10.1088/1361-6560/ab0dda. PMID: 30844769
11. Di Meo, S., et al.: Tissue-mimicking materials for breast phantoms up to 50 GHz. Phys Med Biol. **64**(5), 055006 (2019). https://doi.org/10.1088/1361-6560/aafeec. PMID: 30650384
12. Lazebnik, M., Madsen, E.L., Frank, G.R., Hagness, S.C.: Tissuemimicking phantom materials for narrowband and ultrawideband microwave applications. Phys. Med. Biol. **50**(18), 4245–4258 (2005)
13. Porter, E., Fakhoury, J., Oprisor, R., Coates, M., Popovic, M.: Improved tissue phantoms for experimental validation of microwave breast cancer detection. In: Proceedings of the Fourth European Conference on antennas and Propagation (EuCAP 2010), pp. 1-5. Barcelona, Spain (2010)
14. Martellosio, A., et al.: Dielectric properties characterization from 0.5 to 50 GHz of breast cancer tissues. IEEE Trans. Microw. Theory Tech. **65**(3), 998–1011 (2017). https://doi.org/10.1109/TMTT.2016.2631162
15. Di Meo, S., et al.: Realization of tissue mimicking materials for breast phantoms using waste oil hardeners. In: 13th European Conference on Antennas and Propagation (EuCAP 2019), 31 March-5 April (2019)
16. IT IS dielectric properties (2022). https://www.itis.ethz.ch/virtual-population/tissue-properties/databaseM
17. Speag DAK SPEAG, Schmid & Partner Engineering AG
18. Dassault Simulia CST Suite. https://www.3ds.com/
19. Orfanidis, S.J.: Electromagnetic Waves and Antennas (2002) 2016. http://www.ece.rutgers.edu/~orfanidi/ewa/
20. Tuovinen, T., Yekeh Yazdandoost, K., Iinatti, J.: Comparison of the performance of two different UWB antennas for the use in WBAN on-body coimmunications. In: European Conference on Antennas and Propagation (EUCAP2012), pp. 2271-3374 (2012)
21. Särestöniemi, M., Wisanmongkol, J., Taparugsanagorn, A., Hämäläinen, M., Iinatti, J.: Radio channel model for WBAN capsule endoscopy with anatomical voxel models. IEEE Access (2023)
22. Särestöniemi, M., Reponen, J., Sonkki, M., Myllymäki, S., Pomalaza-Ráez, C., Tervonen O., Myllylä T.: Breast cancer detection feasibility with UWB flexible antennas on wearable monitoring vest. In: TELMED2022, pp. 751–756. Italy (2022)

ISI Mitigation with Molecular Degradation in Molecular Communication

Dongliang Jing[1,2,3](✉), Linjuan Li[1], and Jingjing Wang[1]

[1] College of Mechanical and Electronic Engineering, Northwest A&F University, Yangling 712100, China
{dljing,1765799773,2021012610}@nwafu.edu.cn

[2] Key Laboratory of Agricultural Internet of Things, Ministry of Agriculture and Rural Affairs, Yangling 712100, China

[3] Shaanxi Key Laboratory of Agricultural Information Perception and Intelligent Service, Yangling 712100, China

Abstract. Inter-symbol interference (ISI) decreases the performance of diffusion based molecular communication (MC) significantly. Especially, considering the molecular degradation during the propagation, the ISI mitigation becomes more tricky as the received molecules vary greatly. To tackle this problem, in this paper, we propose an optimal detection method based on maximum likelihood detection by minimizing the error probability. To characterize the proposed detection method, the optimal detection threshold and bit error rate (BER) is derived. Simulation results verified the effectiveness of the proposed ISI mitigation method in the considered MC system with molecular degradation.

Keywords: Inter-symbol interference · Molecular degradation · Molecular communication · Maximum likelihood

1 Introduction

Inspired by the nature communication between biological cells, a new communication paradigm named molecular communication (MC) is proposed which enables the communication between the nanomachines [1,2]. In the MC, biochemical molecules are employed as the information carriers to transmit the information. MC has broad various promising applications such as in-body health monitoring, drug delivery, etc. [3]. Especially, during the outbreak of COVID-19, MC can also be employed to model the propagation of the virus [4–6].

In MC, the information can be encoded in the molecular concentrations, molecular types, and the release time of molecules. Then the encoded molecules are released into the medium and propagate to the receiver by diffusion, active transport, and others. At the receiver, the nanomachine senses the received molecules and decodes the information.

Diffusion-based molecular communication (DBMC) attracts more attention to its energy efficiency. In DBMC, the released molecules diffuse to the receiver by

Y. Chen et al. (Eds.): BICT 2023, LNICST 512, pp. 178–189, 2023.
https://doi.org/10.1007/978-3-031-43135-7_17

Brownian motion making the molecules follow different trajectories and making the channel memory. Therefore, the DBMC suffers serve intersymbol interference (ISI) due to the previously released molecules arriving at the receiver in the subsequent time slots.

The ISI decreases the performance of the DBMC greatly. Therefore, various methods have been proposed to mitigate the ISI. In [2], a chemical reaction based method is proposed where acids, bases, and the concentration of hydrogen ions are employed to transmit the information. In [7], a pre-equalization scheme where the difference between the number of received two types of molecules as the actual signal is proposed to mitigate the ISI. In [8], to mitigate the ISI, the increase of the received molecular concentration rather than the absolute concentration is considered. By employing the K-means clustering algorithm, in [9], the detection thresholds can be reformulated and better bit error rate performance is achieved. An Extended Kalman filter is proposed for detection in [10] to mitigate the ISI. Deep learning schemes are also proposed to demodulate the received molecules, in [11], convolutional neural network is studied, and in [12], deep neural network is considered.

In DBMC, during the propagation of the released molecules, the molecules may be degraded due to the chemical reaction. In [13–15], the exponential degradation of the molecules during the propagation is considered. In [13], considering the molecular degradation, modulation schemes are proposed. In [16], the molecular degradation during the propagation and the molecular reaction with receiver receptor proteins is considered, and the expected received signal is analyzed. The results in [15] indicate that the degradation improves the system performance once appropriate select the degradation rate. Efficient deployment of the limited amount of enzymes in the channel to mitigate the ISI is studied in [17]. In [18], to mitigate the ISI, an enzymatic reaction is introduced by degrading the molecules in the channel.

In this paper, considering the degradation of the molecules in the channel, we propose a detection scheme based on maximum likelihood (ML). Though the ML detection scheme has been employed in the MC, however, in these schemes, molecular degradation is not considered, and molecular degradation is common in more practical MC systems. Therefore, in this paper, we first analyze the received signal considering molecular degradation during the propagation. Then, based on the received signal, a detection scheme based on the ML is conducted to mitigate the ISI. Finally, the simulations are performed to verify the effectiveness of the proposed scheme in the considered MC system.

The remainder of this paper is organized as follows. In Sect. 2, we discuss the considered MC system model with molecular degradation. In Sect. 3, we analyze the received signal and introduce the detection method based on the ML. In Sect. 4, we validate the ML detection method in the ISI mitigation with molecular degradation during the propagation. Finally, the conclusion of this paper is presented in Sect. 5.

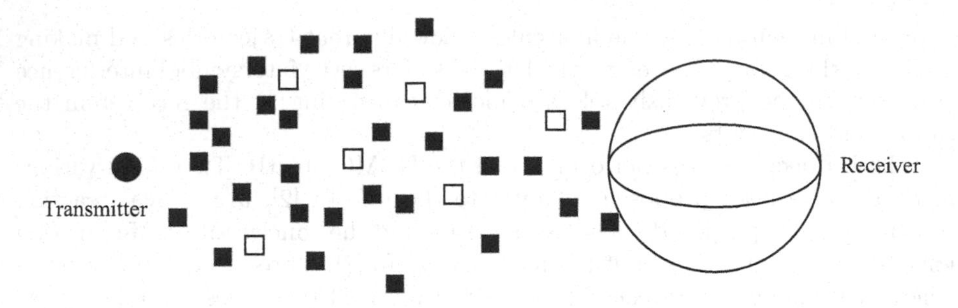

Fig. 1. The considered molecular communication system model.

2 System Model

In this paper, a three-dimensional (3D) MC system with a point transmitter and a sphere absorb receiver is considered. At the transmitter, the concentration shift keying (CSK) modulation scheme is employed, which indicates that to transmit bit 0, no molecule is released, while to transmit bit 1, N_m molecules are released. After the molecules are released from the transmitter, they propagate to the receiver by diffusion. Moreover, the exponential degradation of the molecules is taken into account during the propagation. The receiver detects and counts the number of received molecules. Then, based on the received signal, the received bits are demodulated by the ML detection method. The overall MC system model is shown in Fig. 1.

As shown in Fig. 1, the released molecules can be divided into two parts, one is the signaling molecules, which propagate in the channel and can be detected by the receiver. The other part is the degraded molecules, which are degraded in the channel before arriving at the receiver.

Considering the exponential decay degradation of the released molecules, assuming the initial concentration of released molecules is C_0, then, at time t, the molecular concentration can be expressed as [15]

$$C(t) = C_0 e^{(-\lambda t)}, \tag{1}$$

where λ can be achieved by

$$\lambda = \frac{\ln 2}{\Lambda_{1/2}}, \tag{2}$$

where $\Lambda_{1/2}$ is the half lifetime of molecules.

Then, at time t, the fraction of absorbed molecules which are released at time $t = 0$ can be expressed as [15]

$$F(\lambda, t|r_0) = \frac{r_{rx}}{d + r_{rx}} \exp\left[-\sqrt{\frac{\lambda}{D}}d\right] - \frac{r_{rx}}{2(d + r_{rx})} e^{-\sqrt{\frac{\lambda}{D}}d} \times \left\{\mathrm{erf}\left(\frac{d}{\sqrt{4Dt}} - \sqrt{\lambda t}\right) + e^{2\sqrt{\frac{\lambda}{D}}d} \times \left[\mathrm{erf}\left(\frac{d}{\sqrt{4Dt}} + \sqrt{\lambda t}\right) - 1\right] + 1\right\}, \tag{3}$$

where r_{rx} is the radius of the receiver, d is the distance between the transmitter and the receiver, and D is the diffusion coefficient of the molecules. During a bit interval T, the hitting probability can be expressed as

$$F_{\mathrm{hit},1,\lambda} = F(\lambda, t + T|r_0) - F(\lambda, t|r_0). \tag{4}$$

For simple the notation, $F_{\mathrm{hit},k-i+1,\lambda}$ denotes the hitting probability of molecules released at the beginning of ith bit interval and observed during the kth bit interval and considering the molecular degradation during the propagation. $F_{\mathrm{hit},1,\lambda}$ denotes the hitting probability of molecules released at the beginning of the kth bit interval and observed during the kth bit interval.

Then, after N_m are released from the transmitter at time t_0, the expected number of received molecules during a bit interval can be expressed as

$$\mathrm{E}[N_{rx,\lambda}] = N_m F_{\mathrm{hit},1,\lambda}. \tag{5}$$

In DBMC, due to the channel memory, making the absorbed molecules at the receiver in a bit interval not only from the at the current bit interval but also from molecules released at the previous bit interval. Therefore, considering the channel memory, the received molecules in the kth bit interval can be expressed as

$$N_{rx,k,\lambda} = N_{rx,c,k,\lambda} + N_{rx,\mathrm{ISI},k,\lambda} + N_{rx,n,k,\lambda}, \tag{6}$$

where $N_{rx,c,k,\lambda}$ denotes the molecules released at the beginning of kth bit interval and received during the kth bit interval; $N_{rx,\mathrm{ISI},k,\lambda}$ is the molecules released from the previous bit interval but received during the kth bit interval; and $N_{rx,n,k,\lambda}$ is the counting noise.

After N_m molecules are released from the transmitter at the beginning of kth bit interval, the received molecules $N_{rx,c,k,\lambda}$ during the kth bit interval can be expressed as

$$N_{rx,c,k,\lambda} = N_m F_{\mathrm{hit},1,\lambda}. \tag{7}$$

The $N_{rx,c,k,\lambda}$ can be approximated by the normal distribution and expressed as

$$N_{rx,c,k,\lambda} \sim \mathcal{N}\left(N_m F_{\mathrm{hit},1,\lambda}, N_m F_{\mathrm{hit},1,\lambda}\left(1 - F_{\mathrm{hit},1,\lambda}\right)\right). \tag{8}$$

The interference molecules $N_{rx,\mathrm{ISI},k,\lambda}$ can be expressed as

$$N_{rx,\mathrm{ISI},k,\lambda} = \sum_{i=1}^{k-1} \mathbb{N}_{rx,\mathrm{ISI},i,\lambda}, \tag{9}$$

where $\mathbb{N}_{rx,\mathrm{ISI},i,\lambda}$ denotes the molecules released at the beginning of ith bit interval but received during the kth bit interval and can be expressed as

$$\mathbb{N}_{rx,\mathrm{ISI},i,\lambda} = N_{tx,i} F_{\mathrm{hit},k-i+1,\lambda}, \tag{10}$$

where $N_{tx,i}$ denotes the number of transmitted molecules at the beginning of ith bit interval, and for bit 0, no molecule is released, while for bit 1, N_m molecules are released. The $\mathbb{N}_{rx,\mathrm{ISI},i,\lambda}$ can be approximated by the normal distribution

$$\mathbb{N}_{rx,\mathrm{ISI},i,\lambda} \sim \mathcal{N}\left(N_{tx,i} F_{\mathrm{hit},k-i+1,\lambda}, N_{tx,i} F_{\mathrm{hit},k-i+1,\lambda}\left(1 - F_{\mathrm{hit},k-i+1,\lambda}\right)\right). \tag{11}$$

In DBMC, the counting noise $N_{rx,n,k,\lambda}$ is a random process of molecules entering/leaving the receptor space of the receiver and is usually assumed to follow a Gaussian distribution with 0 mean and the variance depends on the received molecules and can be expressed as

$$N_{rx,n,k,\lambda} \sim \mathcal{N}\left(0, \sigma_n^2\right). \tag{12}$$

3 Maximum Likelihood Detection Method

In this section, the ML detection scheme is employed at the receiver to detect the received bits and mitigate the ISI. Without loss of generality, we assume all the transmission bits are random and independent. Then, based on (7)–(12), the received molecules can be approximated by the normal distribution $N_{rx,k,\lambda} \sim \mathcal{N}\left(\mu_{rx,k,\lambda}, \sigma_{rx,k,\lambda}^2\right)$. Let H_0 be the hypothesis that bit 0 is transmitted; therefore, under H_0, the mean $\mu_{0,k,\lambda}$ and variance $\sigma_{0,k,\lambda}^2$ of the received molecules can be expressed as

$$\mu_{0,k,\lambda} = \mu_{I,k,\lambda} + \mu_{n,k,\lambda} = \frac{1}{2} \sum_{i=1}^{k-1} N_{tx,i} F_{\mathrm{hit},k-i+1,\lambda}, \tag{13}$$

$$\begin{aligned} \sigma_{0,k,\lambda}^2 &= \sum_{i=1}^{k-1} \sigma_{I,j,\lambda}^2 + \sigma_{n,k,\lambda}^2 \\ &= \sum_{i=1}^{k-1} \left[\frac{1}{2} N_{tx,i} F_{\mathrm{hit},k-i+1,\lambda}\left(1 - F_{\mathrm{hit},k-i+1,\lambda}\right) + \frac{1}{4}\left(N_{tx,i} F_{\mathrm{hit},k-i+1,\lambda}\right)^2\right] + \mu_{0,k}. \end{aligned} \tag{14}$$

Let H_1 be the hypothesis that bit 1 is transmitted, therefore, under H_1, the mean $\mu_{1,k,\lambda}$ and variance $\sigma_{1,k,\lambda}^2$ of the received molecules can be expressed as

$$\begin{aligned}\mu_{1,k,\lambda} &= \mu_{c,k,\lambda} + \mu_{I,k,\lambda} + \mu_{n,k,\lambda} \\ &= N_{tx,k}F_{\text{hit},1,\lambda} + \frac{1}{2}\sum_{i=1}^{k-1} N_{tx,i}F_{\text{hit},k-i+1,\lambda},\end{aligned} \tag{15}$$

$$\begin{aligned}\sigma_{1,k,\lambda}^2 &= \sigma_{c,k,\lambda}^2 + \sigma_{I,k,\lambda}^2 + \sigma_{n,k,\lambda}^2 \\ &= N_{tx,k}F_{\text{hit},1,\lambda}\left(1 - F_{\text{hit},1,\lambda}\right) + \sum_{i=1}^{k-1}\left[\frac{1}{2}N_{tx,i}F_{\text{hit},k-i+1,\lambda}\left(1 - F_{\text{hit},k-i+1,\lambda}\right)\right] \\ &+ \sum_{i=1}^{k-1}\left[\frac{1}{4}\left(N_{tx,i}F_{\text{hit},k-i+1,\lambda}\right)^2\right] + \mu_{1,k,\lambda}.\end{aligned} \tag{16}$$

The bit detection at the receiver in the kth bit interval by employing the ML decision rule can be expressed as

$$\hat{b}_{rx,k} = \arg\max_{b_{tx,k}} f\left(N_{rx,k,\lambda}|H_x\right), \tag{17}$$

where $b_{tx,k}$ is the transmitted bits sequence and $b_{tx,k} \in \{0,1\}$, $f\left(N_{rx,k,\lambda}|H_x\right)$ denotes the conditional PDF of $N_{rx,k,\lambda}$ under the hypothesis H_x. Based on (13) and (14), the conditional PDF of $N_{rx,k,\lambda}$ under the hypothesis H_0 which assumes bit 0 is transmitted can be expressed as

$$f\left(N_{rx,k,\lambda}|H_0\right) = \frac{1}{\sqrt{2\pi\sigma_{0,k,\lambda}^2}}\exp\left(-\frac{\left(N_{\text{thr}} - \mu_{0,k,\lambda}\right)^2}{2\sigma_{0,k,\lambda}^2}\right). \tag{18}$$

And Based on (15) and (16), the conditional PDF of $N_{rx,k,\lambda}$ under the hypothesis H_1 which assumes bit 1 is transmitted can be expressed as

$$f\left(N_{rx,k,\lambda}|H_1\right) = \frac{1}{\sqrt{2\pi\sigma_{1,k,\lambda}^2}}\exp\left(-\frac{\left(N_{\text{thr}} - \mu_{1,k,\lambda}\right)^2}{2\sigma_{1,k,\lambda}^2}\right). \tag{19}$$

Therefore, the optimal detection threshold N_{thr} can be achieved by setting

$$\frac{1}{\sqrt{2\pi\sigma_{0,k,\lambda}^2}}\exp\left(-\frac{\left(N_{\text{thr}} - \mu_{0,k,\lambda}\right)^2}{2\sigma_{0,k,\lambda}^2}\right) = \frac{1}{\sqrt{2\pi\sigma_{1,k,\lambda}^2}}\exp\left(-\frac{\left(N_{\text{thr}} - \mu_{1,k,\lambda}\right)^2}{2\sigma_{1,k,\lambda}^2}\right), \tag{20}$$

Then the optimal detection threshold N_{thr} can be achieved as

$$N_{\text{thr}} = \frac{1}{\sigma_{1,k,\lambda}^2 - \sigma_{0,k,\lambda}^2}\left[\left(\mu_{0,k,\lambda}\sigma_{1,k,\lambda}^2 - \mu_{1,k,\lambda}\sigma_{0,k,\lambda}^2\right)\right.$$
$$\left.+\sigma_{0,k,\lambda}\sigma_{1,k,\lambda}\sqrt{\left(\mu_{0,k,\lambda} - \mu_{1,k,\lambda}\right)^2 + 2\left(\sigma_{1,k,\lambda}^2 - \sigma_{0,k,\lambda}^2\right)\ln\frac{\sigma_{1,k,\lambda}}{\sigma_{0,k,\lambda}}}\right]. \tag{21}$$

At the receiver, the bits are decoded according to the following rule

$$\hat{b}_{rx,k} = \begin{cases} 1, \ N_{rx,k,\lambda} \geq N_{\text{thr}} \\ 0 \ \ N_{rx,k,\lambda} < N_{\text{thr}}. \end{cases} \tag{22}$$

In DBMC, considering the channel memory, the error in the kth bit interval is not only related to the molecules transmitted in the current bit interval but also the previous bit interval. Therefore, the bit error rate can be expressed as

$$P_{e,k,\lambda} = \sum_{(b_{tx,1},\ldots,b_{tx,k})\in\{0,1\}^k} \Pr\left(b_{tx,1},\ldots,b_{tx,k}\right)\Pr\left(\text{error} \mid b_{tx,1},\ldots,b_{tx,k}\right). \tag{23}$$

In DBMC, an error is occurred when the decoded bit not equal to the transmitted bit, namely $\hat{b}_{rx,k} \neq b_{tx,k}$. The error for transmitting bit 0 and bit 1 can be expressed as

$$\Pr\left(N_{rx,k,\lambda} > N_{\text{thr}} | b_{tx,k} = 0\right) = \mathrm{Q}\left(\sqrt{\frac{\left(N_{\text{thr}} - \mu_{0,k,\lambda}\right)^2}{\sigma_{0,k,\lambda}^2}}\right), \tag{24}$$

$$\Pr\left(N_{rx,k,\lambda} > N_{\text{thr}} | b_{tx,k} = 1\right) = \mathrm{Q}\left(\sqrt{\frac{\left(N_{\text{thr}} - \mu_{1,k,\lambda}\right)^2}{\sigma_{1,k,\lambda}^2}}\right). \tag{25}$$

Thus, the average bit error probability in the kth time slot for the same probability to transmit bit 0 and bit 1 can be expressed as

$$P_{e,k,\lambda} = \frac{1}{2}\left(1 - Q\left(\frac{N_{\text{thr}} - \mu_{1,k,\lambda}}{\sigma_{1,k,\lambda}}\right)\right) + \frac{1}{2}Q\left(\frac{N_{\text{thr}} - \mu_{0,k,\lambda}}{\sigma_{0,k,\lambda}}\right). \tag{26}$$

4 Numerical and Simulation Results

In this section, numerical and simulations are conducted to verify the effectiveness of the ML detection scheme in mitigating ISI in the DBMC system considering the degradation of the molecules during the propagation. The parameters are listed in Table 1.

In Fig. 2, the molecular concentration varies with time under the different half lifetime of molecules $\Lambda_{1/2}$ are compared. As shown in Fig. 2, the molecular concentration decreases with time, however, the decrease rate is affected by the half lifetime of molecules $\Lambda_{1/2}$. For the larger half lifetime of molecules $\Lambda_{1/2}$, the longer time the molecules propagate, therefore, more molecules are probability to be absorbed by the receiver, making the higher molecular concentration.

Table 1. Simulation parameters.

Simulation parameters	Symbol	Value
Distance between transmitter and receiver	d	$4\,\mu m$
Radius of receiver	r_{rx}	$6\,\mu m$
Diffusion coefficient	D	$79.4\,\mu m^2/s$
Number of transmitter molecules for bit 1	N_{tx}	10000
Half-lifetime of released molecules	$\Lambda_{1/2}$	0.128 s,0.064 s,0.032 s
Bit interval	T	0.5 s

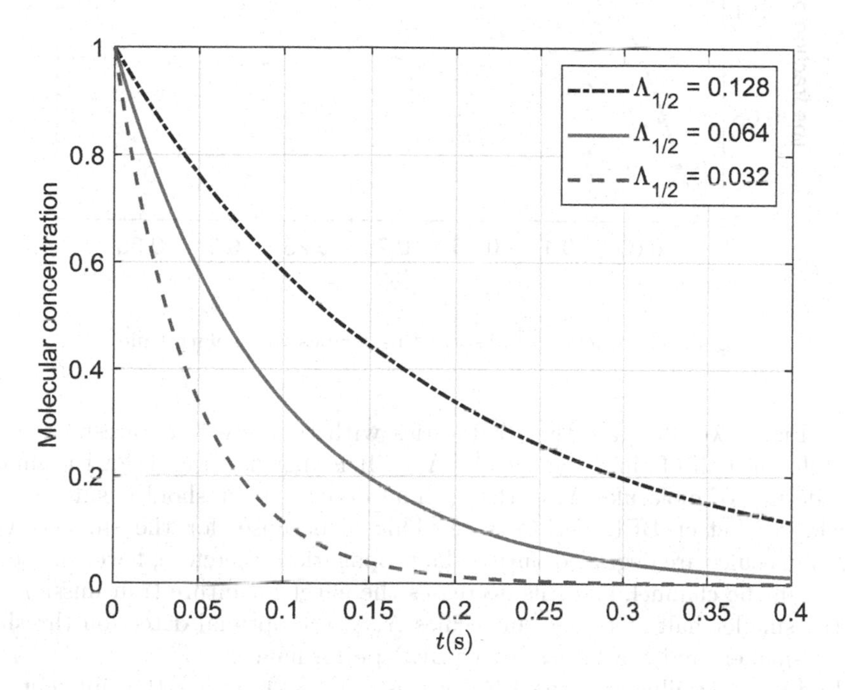

Fig. 2. The molecular concentration varies with time t. The initial molecular concentration C_0=1.

Figure 3 illustrates the fraction of molecules absorbed by the receiver varies with time under the different half lifetime of molecules $\Lambda_{1/2}$. It can be clearly seen from Fig. 3, for the larger half lifetime of molecules $\Lambda_{1/2}$, the higher fraction of molecules absorbed by the receiver due to the larger half lifetime of molecules $\Lambda_{1/2}$, the longer lifetime of the molecules, then, making more molecules are absorbed by the receiver.

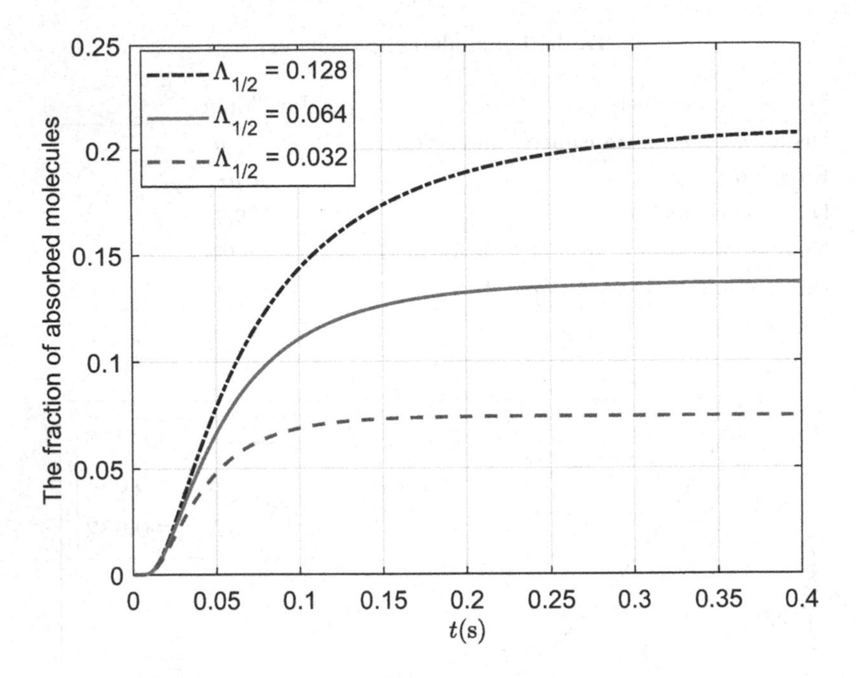

Fig. 3. The fraction of absorbed molecules varies with time t.

In Fig. 4, we illustrate the BER varies with the detection threshold under the different half lifetime of molecules $\Lambda_{1/2}$. It is shown in Fig. 4, for the smaller half lifetime of molecules $\Lambda_{1/2}$, the optimal detection threshold is smaller, and it achieves better BER performance. This is because, for the smaller $\Lambda_{1/2}$, more molecules are degraded during the propagation, therefore, fewer molecules remain in the channel, and this decreases the effect for future transmission. So, for the smaller half lifetime of molecules $\Lambda_{1/2}$, the optimal detection threshold is also smaller, and it achieves better BER performance.

In Fig. 5, we illustrate the BER varies with SNR under the different half lifetime of molecules $\Lambda_{1/2}$ based on the ML detection. As shown in Fig. 5, with the increase of SNR, the BER decreases, and the BER is also affected by the $\Lambda_{1/2}$. For the smaller $\Lambda_{1/2}$, it achieves better BER performance, due to the smaller $\Lambda_{1/2}$, the molecules degrade during the propagation and there are fewer molecules remaining in the channel namely lower ISI. It also verified the effectiveness of ML detection in the DBMC system with molecular degradation during the propagation.

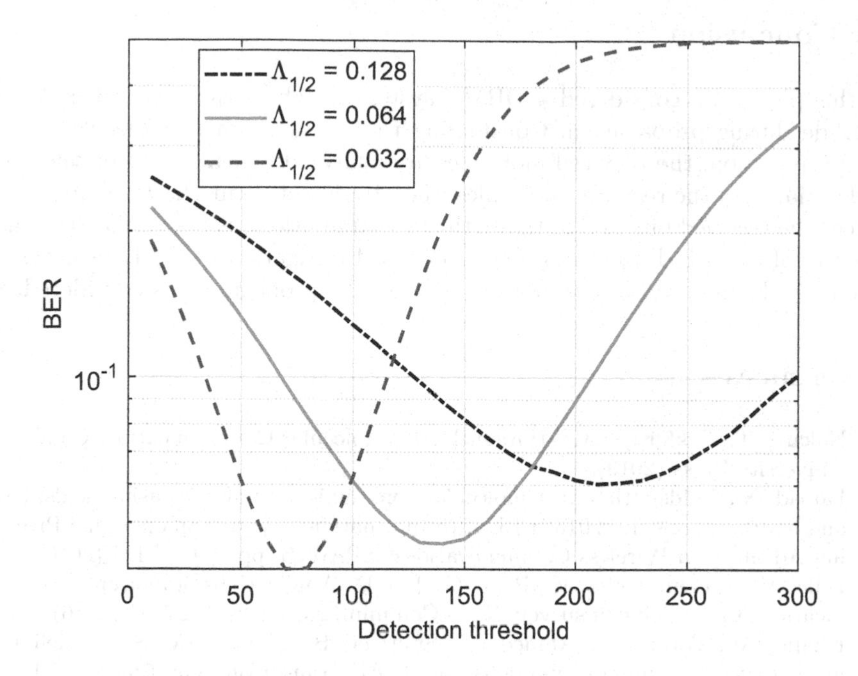

Fig. 4. The BER varies with the detection threshold.

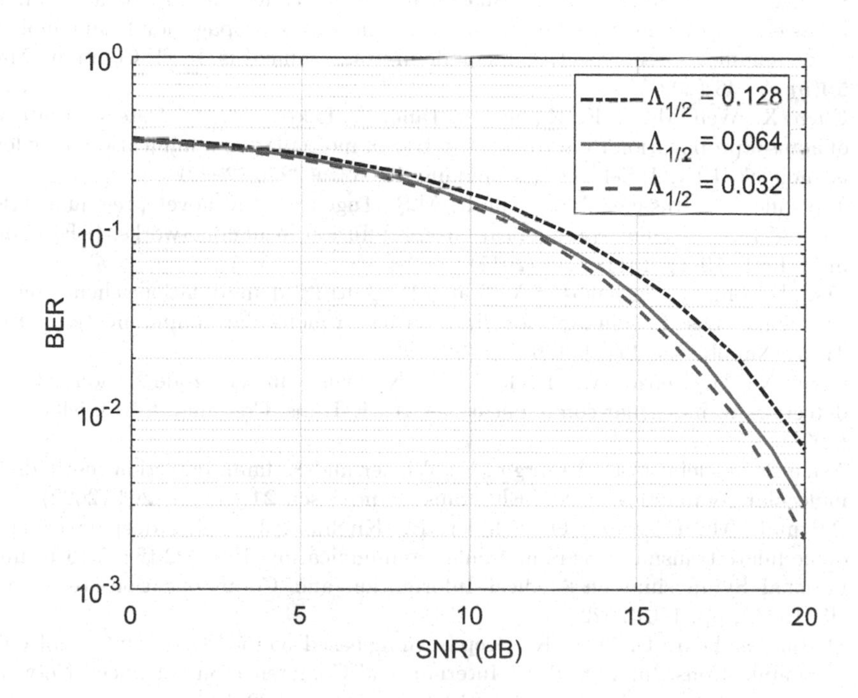

Fig. 5. The BER varies with SNR in the considered DBMC.

5 Conclusion

In this paper, we considered a DBMC system in which the released molecules degrade during propagation. Considering the degradation of molecules during the propagation, the received molecules are analyzed. Then, based on the mean and variance of the received molecules, the ML detection method is employed to detect the received bits and mitigate the ISI. Simulation results verified the effectiveness of the ML detection method in the ISI mitigation in the DBMC system when the degradation of the molecule during the propagation is considered.

References

1. Nakano, T., Eckford, A.W., Haraguchi, T.: Molecular Communication. Cambridge University Press (2013)
2. Farsad, N., Goldsmith, A.: A molecular communication system using acids, bases and hydrogen ions. In: 2016 IEEE 17th International Workshop on Signal Processing Advances in Wireless Communications (SPAWC), pp. 1–6. IEEE (2016)
3. Felicetti, L., Femminella, M., Reali, G., Liò, P.: Applications of molecular communications to medicine: a survey. Nano Commun. Networks **7**, 27–45 (2016)
4. Khalid, M., Amin, O., Ahmed, S., Shihada, B., Alouini, M.-S.: Modeling of viral aerosol transmission and detection. IEEE Trans. Commun. **68**(8), 4859–4873 (2020)
5. Schurwanz, M., Hoeher, P.A., Bhattacharjee, S., Damrath, M., Stratmann, L., Dressler, F.: Infectious disease transmission via aerosol propagation from a molecular communication perspective: Shannon meets Coronavirus. IEEE Commun. Mag. **59**(5), 40–46 (2021)
6. Chen, X., Wen, M., Ji, F., Huang, Y., Tang, Y., Eckford, A.W.: Detection interval of aerosol propagation from the perspective of molecular communication: how long is enough? IEEE J. Sel. Areas Commun. **40**, 3255–3270 (2022)
7. Tepekule, B., Pusane, A.E., Kuran, M.Ş, Tugcu, T.: A novel pre-equalization method for molecular communication via diffusion in nanonetworks. IEEE Commun. Lett. **19**(8), 1311–1314 (2015)
8. Zhai, H., Liu, Q., Vasilakos, A.V., Yang, K.: Anti-ISI demodulation scheme and its experiment-based evaluation for diffusion-based molecular communication. IEEE Trans. Nanobiosci. **17**(2), 126–133 (2018)
9. Qian, X., Di Renzo, M., Eckford, A.: K-means clustering-aided non-coherent detection for molecular communications. IEEE Trans. Commun. **69**(8), 5456–5470 (2021)
10. Aslan, E., Çelebi, M.E., Pekergin, F.: Wiener and Kalman detection methods for molecular communications. IEEE Trans. Nanobiosci. **21**(2), 256–264 (2022)
11. Bartunik, M., Keszocze, O., Schiller, B., Kirchner, J.: Using deep learning to demodulate transmissions in molecular communication. In: 2022 IEEE 16th International Symposium on Medical Information and Communication Technology (ISMICT), pp. 1–6 (2022)
12. Sharma, S., Dixit, D., Deka, K.: Deep learning based symbol detection for molecular communications. In: 2020 IEEE International Conference on Advanced Networks and Telecommunications Systems (ANTS), pp. 1–6 (2020)

13. Nakano, T., Okaie, Y., Liu, J.-Q.: Channel model and capacity analysis of molecular communication with Brownian motion. IEEE Commun. Lett. **16**(6), 797–800 (2012)
14. Liu, Q., Yang, K.: Channel capacity analysis of a diffusion-based molecular communication system with ligand receptors. Int. J. Commun. Syst. **28**(8), 1508–1520 (2015)
15. Heren, A.C., Yilmaz, H.B., Chae, C.-B., Tugcu, T.: Effect of degradation in molecular communication: impairment or enhancement? IEEE Trans. Mol. Biol. Multi-Scale Commun. **1**(2), 217–229 (2015)
16. Ahmadzadeh, A., Arjmandi, H., Burkovski, A., Schober, R.: Reactive receiver modeling for diffusive molecular communication systems with molecule degradation. In: 2016 IEEE International Conference on Communications (ICC), pp. 1–7. IEEE (2016)
17. Cho, Y.J., Yilmaz, H.B., Guo, W., Chae, C.-B.: Effective inter-symbol interference mitigation with a limited amount of enzymes in molecular communications. Trans. Emerg. Telecommun. Technol. **28**(7), e3106 (2017)
18. Vakilipoor, F., Ratti, F., Awan, H., Magarini, M.: Low complexity receiver design for time-varying poisson molecular communication channels with memory. Digit. Sig. Proc. **124**, 103187 (2022)

Signal Transmission Through Human Body via Human Oxygen Saturation Detection

Chengyi Zhang[1], Hao Yan[2], Qiang Liu[3], Kun Yang[3], and Lin Lin[1](✉)

[1] Tongji University, Shanghai, China
fxlinlin@tongji.edu.cn
[2] Shanghai Jiao Tong University, Shanghai, China
[3] University of Electronic Science and Technology of China, Anhui, China

Abstract. For a long time, people have carried out various studies on molecular communication and nano information network in order to realize biomedical applications inside human body. However, how to realize the communication between these applications and the outside body has become a new problem. In general, different components in the blood have different absorption rates of the different light. Based on this, we propose a new through-body communication method. The nanomachine in the blood vessel transmits signal by releasing certain substances which can influence blood oxygen saturation. The change of blood oxygen saturation can be detected by a outside body device measuring the attenuation of different light through blood. The framework of the entire communication system is proposed and mathematically modeled. Its error performance is discussed and evaluated. This research will contribute to the realization of the connection of communication systems inside and outside the human body.

Keywords: molecular communication · nanomachine · oxygen saturation · light absorption

1 Introduction

In recent years, molecular communication has become a research hotspot because nanomachine is expected to use in actual test. It is possible for us to use nanomachine to form the internet of nano things and complete the communication inside and outside the human body [1]. So far, the problem of designing a suitable interface between the nanoscale environment and the external macroscopic world remains an open research issue.

This work was supported in part by the National Natural Science Foundation, China, under Grant 61971314, 62071297; in part by the Fundamental Research Funds for the Central Universities under Grant 22120220629; in part by the Science and Technology Commission of Shanghai Municipality under Grant 19510744900, 19ZR1426500; and in part by the Sino-German Center of Intelligent Systems, Tongji University. Corresponding author: Lin Lin.

Y. Chen et al. (Eds.): BICT 2023, LNICST 512, pp. 190–199, 2023.
https://doi.org/10.1007/978-3-031-43135-7_18

There are many studies on the theoretical research of communication systems inside and outside the human body such as [2,3]. In [2], the authors set up nanomachines in the human body and send signals by stimulating nerve fibers through electrodes. This signal propagates through the nerves and produces a surface electromyography signal, which serves as the information received by the body surface receiver. However, the signal transmission through the nervous system is susceptible to the interference of the action caused by the subjective consciousness of the human body on the neural signal reception. In literature [3], the authors used blood vessels as communication channels and used smart probes fixed in them as a tool for information interaction inside and outside the human body. The probes are expected to release a substance that generates an allergic reaction on the skin surface, or is detectable in the infrared bandwidth or by ultrasound. However, the authors did not elaborate on how to implement these ideas.

We notice that the oximeter is often used outside human body to detect the blood oxygen saturation in blood vessels. We can utilize this technique to realize the communication system through human body. Blood oxygen saturation is one of the important basic data in clinical medicine, which can be inferred by measuring the attenuation of different light through blood. The signal transmission through the body can be achieved by altering blood oxygen saturation and detecting its change outside human body by optical method.

The main contributions of this paper are:

1) We propose a new through-body communication method that people can make use of blood oxygen saturation detection as a medium for information interaction inside and outside the human body. The framework of the entire communication system is proposed.
2) Our proposed system is mathematically modeled and the error performance is evaluated.

The rest of this paper is organized as follows. Section 2 introduces preliminary knowledge of optical technology and oxygen saturation, and the design of through-body communication system. Section 3 presents the mathematical model of our system. Section 4 presents the simulation results. Section 5 concludes the paper.

2 Design of Through-Body Communication System

In this section, we will make a further explanation about blood oxygen saturation and the principle of optical detection in this process. After that, we will introduce the design of our communication system with blood oxygen saturation detection as a medium for information interaction.

2.1 Blood Oxygen Saturation

The oxygen consumed by the human body mainly comes from the oxygen carried by hemoglobin. There are four kinds of hemoglobin in normal blood: oxygenated hemoglobin (HbO_2), deoxyhemoglobin (Hb), carboxyhemoglobin (COHb) and Methemoglobin (MetHb). In fact, both MetHb and COHb absorb red and infrared light. This will result in incorrect readings. MetHb is less than 2% and COHb is less than 3% of the total amount of hemoglobin in normal human body. Among them, deoxyhemoglobin is reversibly combined with oxygen, while carboxyhemoglobin and methemoglobin are not combined with oxygen. The blood oxygen saturation (S_{O_2}) is used to describe the change of oxygen content in blood. It refers to the percentage of bound oxygen volume in total blood volume. The function of hemoglobin is to carry oxygen to all parts of the body. The oxygen content of hemoglobin at any time is called blood oxygen saturation. It can be expressed as

$$S_{O_2} = C_{HbO_2}/(C_{HbO_2} + C_{Hb}).$$

As for the oxyhemoglobin dissociation curve, also called the oxygen dissociation curve (ODC), is a curve that plots the proportion of hemoglobin in its saturated (oxygen-laden) form on the vertical axis against the prevailing oxygen tension on the horizontal axis. This curve is an important tool for understanding how our blood carries and releases oxygen. Specifically, the oxyhemoglobin dissociation curve relates oxygen saturation (S_{O_2}) and partial pressure of oxygen in the blood (P_{O_2}) [4].

2.2 Principle of Optical Detection

As we mentioned before, hemoglobin has both oxygen carrying state and no-load state. Hemoglobin in carrying oxygen state is called oxyhemoglobin, and hemoglobin in no-load state is called deoxyhemoglobin. Oxyhemoglobin and deoxyhemoglobin have different absorption characteristics in the spectrum range of visible light and near infrared. Deoxyhemoglobin absorbs more red frequency light and less infrared frequency light. Oxyhemoglobin absorbs less red frequency light and more infrared frequency light. The principle of a fingertip pulse oximeter is based on this fact. When red light and infrared light irradiate the finger alternately, the photodiode in the fingertip pulse oximeter will produce a weak photocurrent that changes with the pulse. After converting, filtering and amplifying the photocurrent, the pulse waveform is obtained. The pulse frequency is calculated from the peak spacing, and the blood oxygen saturation is calculated from the photocurrent ratio of red light and infrared light.

According to literature [5], when light of a specific wavelength is incident on the fingertip, the transmitted light intensity can be divided into two parts: the pulsatile component and the non-pulsatile component in the fingertip tissue. When the arterial blood vessels in the light-transmitting area pulsate, the amount of light absorbed by the arterial blood will change accordingly, which is called current (AC) component. The absorption of light by other tissues such as skin, muscle, bone and venous blood is constant and is called direct current (DC) component. If the attenuation due to factors such as scattering and reflection is ignored and according to the Lambert-Beer law, when the wavelength is λ and the monochromatic light with the light intensity of I_0 is vertically incident, the transmitted light intensity can be written as

$$I = I_0 e^{-\varepsilon_0 C_0 L} e^{-\varepsilon_{HbO_2} C_{HbO_2} L} e^{-\varepsilon_{Hb} C_{Hb} L}, \tag{1}$$

where ε_0, C_0, L are the absorption coefficient, the concentration of light absorbing substances and the optical path length of non-arterial components in the tissue and venous blood. C_{HbO_2}, ε_{HbO_2} are the absorption concentration and coefficient of HbO_2 in arterial blood. C_{Hb} and ε_{Hb} are the absorption concentration and coefficient of Hb in arterial blood.

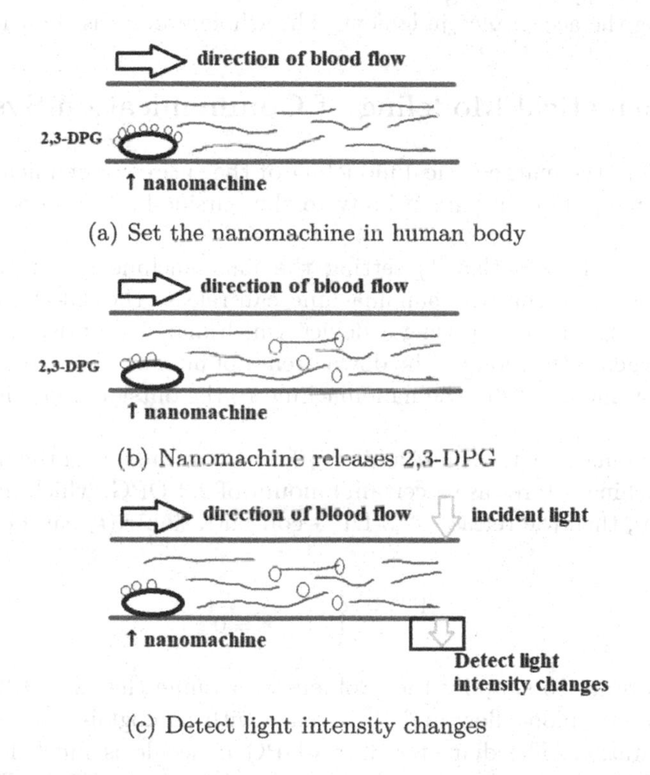

Fig. 1. The data transmission via human oxygen saturation detection.

2.3 System Design

In order to achieve a feasible change of blood oxygen saturation in blood vessel and make decisions based on the received light intensity at the receiving end, we need to release something to change the affinity of hemoglobin for oxygen. According to the literature [6], CO_2 concentration, pH value, temperature and

2,3 diphosphate glyceride (2,3-DPG) all affect the affinity of hemoglobin for oxygen, thus causing changes in blood oxygen saturation. For safety and practical considerations, the pH value and temperature in human blood vessels cannot be easily changed, and CO_2 is unable to be carried by nanomachine as a gas molecule, so we consider that 2,3-DPG is used to change blood oxygen saturation. Under the condition of other situations are the same, the oxyhemoglobin dissociation curve shifts to the right with the increase of 2,3-DPG [6], which also represents the decrease in blood oxygen saturation. We assume that the blood oxygen saturation variation caused by 2,3-DPG variation is much faster than that that of regular natural saturation changes. The basic idea of our system design is that nanomachine releases 2,3-DPG to change the blood oxygen concentration, thereby producing a change in the light intensity at the receiving end and realizing the acceptance judgment. The whole process is shown in Fig. 1.

3 Mathematical Modeling of Communication System

In this section, the mathematical modeling of the entire communication process from the nanomachine in human body to the outside-body processing unit are presented.

The general idea is that 1) setting the nanomachine on the upstream of human fingertips so that the nanomachine can release the 2,3-DPG when they need to send signal, and 2) using a device which make an optical measurement of blood oxygen saturation on the downstream of human fingertips. In this way, the data transmission from the nanomachine to the outside body device can be realized.

If the nanomachine transmits a bit $x\epsilon\{0,1\}$ to the outside of the human body, the nanomachine will release a certain amount of 2,3-DPG, which may be set as M molecules, then the released 2,3-DPG concentration $A(t)$ can be defined as

$$A(t) = \begin{cases} M, & x = 1 \\ 0, & x = 0 \end{cases} \tag{2}$$

Next, in order to simplify the problem,we assume that 2,3-DPG molecules first move under blood flow and then react with hemoglobin to reduce blood oxygen saturation. The diameter of 2,3-DPG molecule is much less than the diameter of the transverse palmar branch since the volume of 2,3-DPG is around 10^{-19} mm^3 and the diameter of the transverse palmar branch is around 0.4 mm [7]. In the case of limited transmission distance, we can regard the process of 2,3-DPG molecular transmission in fingertip vessels as a borderless system. We adopt a model in [8] for 3-D advection-diffusion where the 2,3-DPG concentration at time t is presented as

$$C(t) = \frac{M}{(4\pi Dt)^{3/2}} e^{-\frac{(d-vt)^2}{4Dt}}, \tag{3}$$

where v is the speed of blood flow, and d is the distance between nanomachine and the receiver. D is the diffusion coefficient for 2,3-DPG molecules in blood vessel.

The sensing of molecular concentration occurs within a spherical receptive space with volume V. Additive counting noise is generated due to the random motion of the information molecules. From [9], the total noisy molecule concentration Z(t) is given by

$$Z(t) = C(t) + n(t), \tag{4}$$

where $n(t)$ is the non-stationary and signal dependent additive noise. When 2,3-DPG molecules arrive at the receiver area, the blood oxygen saturation S_s can be expressed as

$$S_s = (C_{HbO_2} - f(M,t))/(C_{HbO_2} + C_{Hb}). \tag{5}$$

where $f(M,t)$ is the function which represents the decrease part in HbO_2 concentration after the release of 2,3-DPG.

Reference [6] gives a clear derivation of the relationship between the 2,3-DPG and S_{O_2}, which is too complex. We obtain another way to calculate it approximately. We adopt a model in [10] to calculate oxygen saturation. Its equation for the ODC describes the oxygen saturation S_{O_2} as a function of oxygen partial pressure P_{O_2} relative to the half-saturation level P_{50}

$$S_{O_2} = (P_{O_2}/P_{50})^n/[1 + (P_{O_2}/P_{50})^n], \tag{6}$$

where n is the Hill exponent. The value n = 2.7 was found to fit well to the data for normal human blood in the saturation range of 20–98%. According to [6], when pH = 7.24, P_{CO_2}= 40 mmHg, T=37°C, the relationship between P_{50} and C can be calculated as

$$P_{50} = 26.8 + 795.63(C - 0.00465) - 19660.89(C - 0.00465)^2. \tag{7}$$

From literature [11], P_{O_2} in arterial blood is around 90 mmHg. Thus, we can calculate the relationship between the 2,3-DPG and S_{O_2} to choose the best 2,3-DPG concentration based on (6) and (7).

The output decoding can also be performed according to the general method of oximeter measurement. Two beams of light with different wavelengths (red light and infrared light) are used as the incident light in the measurement of blood oxygen saturation. When the wavelength of the infrared light is taken near 805nm, the blood oxygen saturation can be expressed as

$$S_{O_2} = A * \frac{I_{AC}^{\lambda_1}/I_{DC}^{\lambda_1}}{I_{AC}^{\lambda_2}/I_{DC}^{\lambda_2}} - B, \tag{8}$$

where A, B are expressions about the absorption coefficient, which can generally be regarded as constants. $I_{AC}^{\lambda_1}$, $I_{AC}^{\lambda_2}$ are respectively the pulsatile component of transmitted light intensity when the light with a wavelength of λ_1 or λ_2 vertically enters the arterial blood of the fingertip of the human body. $I_{DC}^{\lambda_1}$,

$I_{DC}^{\lambda_2}$ are respectively the non-pulsatile component of transmitted light intensity when the light with a wavelength of λ_1 or λ_2 vertically enters the venous blood of the fingertip of the human body. In this way, we can detect the oxygen saturation at the receiver as S_r. The decision equation of the receiver is

$$\hat{x} = \begin{cases} bit"1", & S_r \leq S_{th} \\ bit"0", & S_r > S_{th}, \end{cases} \quad (9)$$

where S_{th} is the receiver's decision threshold and $\hat{x}$ is recovered signal.

4 Stimulation Results

In this section, the error rate of the through-body communication system is evaluated by MATLAB. The distance between nanomachine and receiver is chosen from 8 mm to 12 mm since the distance between the middle transverse palmar branch and the distal transverse palmar branch is about 10.33 mm [7]. Considering that the reasonable range of the blood flow velocity is from 30 mm/s to 126 mm/s [12], we set the velocity from 30 mm/s to 50 mm/s. We assume that the diffusion coefficient D is 200 mm^2/s in blood vessel. We define M in (3) is 50000. From (4) and (5), we can know the received number of 2,3-DPG is directly related to blood oxygen saturation. In order to simplify the simulation calculation process, we define the threshold of molecule number M_{th} is 8. We send 100,000 bits of data in a simulation and the number of molecules received is counted and judged by a spherical receiver with a volume of 1 mm^3. What's more, for the probabilities of sending bit "1" and bit "0" at the nanomachine, we choose 0.5 for both of them.

Considering that there may be 2,3-DPG that cannot be cleared in time and remain in the blood vessel, we add the inter-symbol interference (ISI) influence and additive noise. We assume that ISI effect can last for eight time slots. Every time slots lasts $t = d/v$ since the concentration of 2,3-DPG becomes maximum at this time. Besides, we add gaussian white noise with a signal-to-noise ratio of 20dB in the channel. Both the signal of current time slot and the legacy signals of the previous eight time slots are disturbed by additive noise. Table 1 shows some important parameters used in the simulation.

Table 1. Stimulation parameters

Parameter	Symbol	Value
Distance between nanomachine and receiver	d	8 mm to 12 mm
Blood flow velocity	v	30 mm/s to 50 mm/s
Diffusion coefficient	D	200 mm^2/s
Number of 2,3-DPG molecules	M	50000
Volume of the receiver	V	1 mm^3

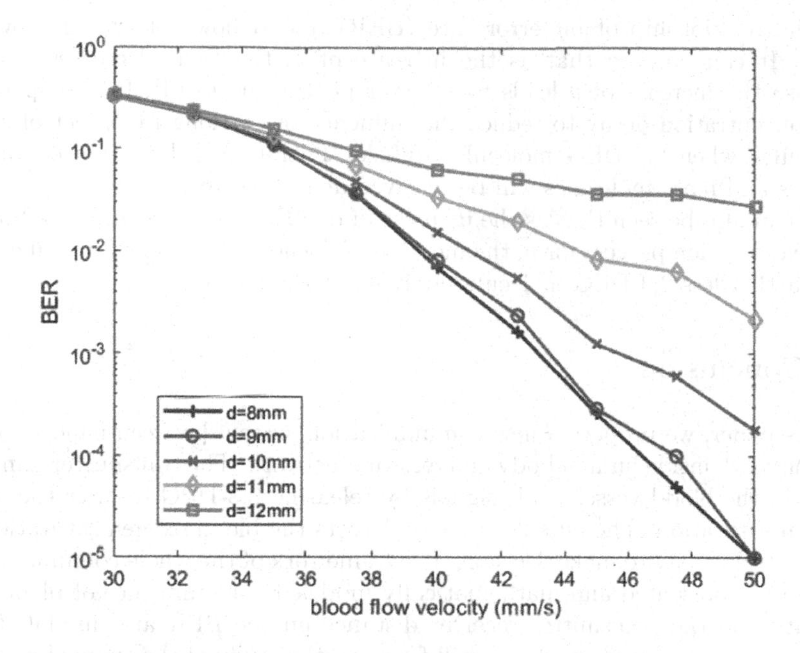

Fig. 2. The relationship of BER and blood flow velocity for different distances between nanomachine and receiver.

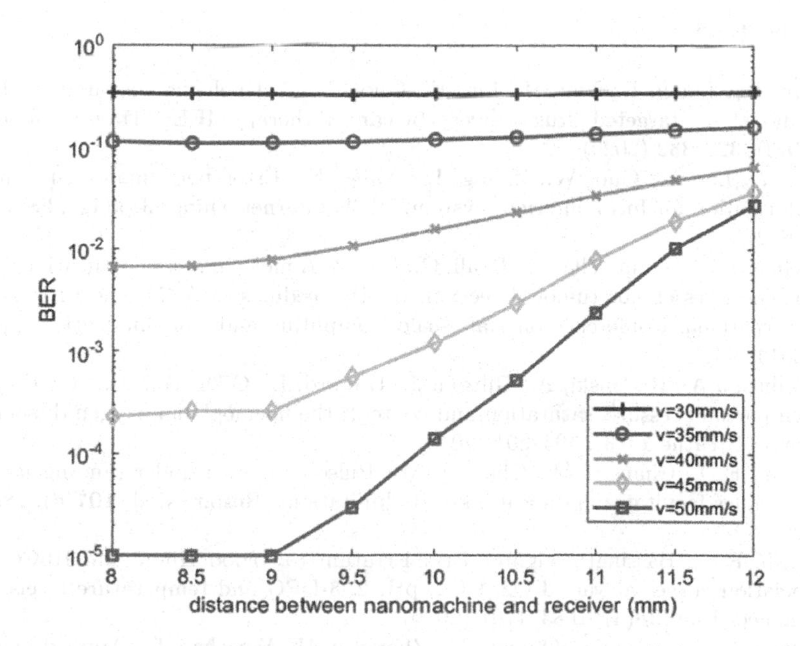

Fig. 3. The relationship of the BER and the distance between transmitter nanomachine and the receiver for different blood flow velocities.

The relationship of bit error rate (BER), blood flow velocity is shown in Fig. 2. It can be seen that as the increase of v, the BER decreases. This is because the increase of v leads to a larger peak value in (3). It also speed up the concentration decay to reduce the influence of the long tail effect of signal molecules when 2,3-DPG molecules diffuse. Therefore, ISI is reduced in this process and more molecules will be received at the receiver.

It can also be seen that as the increase of d, BER decreases in Fig. 3. Similar to the discussion process for v, the increase of d leads to smaller peak value $C(t)$, and further less 2,3-DPG molecules arriving at the receiver.

5 Conclusion

In this paper, we propose a new communication system between nano information network inside human body and external network. The transmitter nanomachine in the blood vessel sends signals by releasing 2,3-DPG to alter the blood oxygen saturation. The outside receiver detects the blood oxygen saturation by optical technology to make decision. The framework of the whole communication system are presented and mathematically modeled. The influence of blood flow velocity and the transmitter receiver distance on the BER are simulated and discussed. In our future work, we will focus on the study of the channel capacity of our proposed system.

References

1. Lin, L., Huang, F., Yan, H., Liu, F., Guo, W.: Ant-behavior inspired intelligent nanonet for targeted drug delivery in cancer therapy. IEEE Trans. Nanobiosci. **19**(3), 323–332 (2020)
2. Li, Y., Lin, L., Guo, W., Zhang, D., Yang, K.: Error performance and mutual information for IoNT interface system. IEEE Internet Things J. **9**(12), 9831–9842 (2022)
3. Felicetti, L., Femminella, M., Reali, G., Liò, P.: A molecular communication system in blood vessels for tumor detection. In: Proceedings of ACM The First Annual International Conference on Nanoscale Computing and Communication, pp. 1–9 (2014)
4. Collins, J.A., Rudenski, A., Gibson, J., Howard, L., O'Driscoll, R.: Relating oxygen partial pressure, saturation and content: the haemoglobin-oxygen dissociation curve. Breathe **11**(3), 194–201 (2015)
5. Chan, E.D., Chan, M.M., Chan, M.M.: Pulse oximetry: understanding its basic principles facilitates appreciation of its limitations. Respir. Med. **107**(6), 789–799 (2013)
6. Dash, R.K., Bassingthwaighte, J.B.: Erratum to: blood HbO2 and HbCO2 dissociation curves at varied O2, CO2, pH, 2, 3-DPG and temperature levels. Ann. Biomed. Eng. **38**(4), 1683–1701 (2010)
7. Lihua, C., Hongtai, H., Qunwu, H., Zhengrui, H., Wenchao, L.: Applied anatomy of proper digital palmar artery. Chin. J. Anat. **29**(4), 503–505 (2006)

8. Farsad, N., Yilmaz, H.B., Eckford, A., Chae, C.B., Guo, W.: A comprehensive survey of recent advancements in molecular communication. IEEE Communications Surveys & Tutorials **18**(3), 1887–1919 (2016)
9. Lin, L., Wu, Q., Liu, F., Yan, H.: Mutual information and maximum achievable rate for mobile molecular communication systems. IEEE Trans. Nanobiosci. **17**(4), 507–517 (2018)
10. Hill, A.V.: The possible effects of the aggregation of the molecules of hemoglobin on its dissociation curves. J. Physiol. **40**, iv–vii (1910)
11. Ortiz-Prado, E., Dunn, J.F., Vasconez, J., Castillo, D., Viscor, G.: Partial pressure of oxygen in the human body: a general review. Am. J. Blood Res. **9**(1), 1 (2019)
12. Yusheng, F., Wang, Q., Yi, J., Song, D., Xiang, X.: A numerical model of blood flow velocity measurement based on finger ring. J. Healthc. Eng. **2018**, 3916481 (2018)

Simple ISI-Avoiding and Rate-Increasing Modulation for Diffusion-Base Molecular Communications

GuoYing Lin[1](✉), Kun Yang[2], and Qiang Liu[3]

[1] Yangtze Delta Region Institute (Quzhou), University of Electronic Science and Technology of China, Quzhou 324000, Zhejiang, China
1298130342@qq.com
[2] University of Essex, Colchester CO4 3SQ, UK
[3] University of Electronic Science and Technology of China, Chengdu, China

Abstract. Molecular communication (MC) is a significant technology in the field of nanobiology, which uses molecules as message carriers to transmit information. Diffusion channel model is the most commonly channel model base on Brownian motion in molecular communication. In single-input single-output (SISO) molecular communication model, inter-symbol interference (ISI) exists due to the long tail effect. In this study, inspired by the D-MoSK modulation scheme, where different types molecules used for encoding, A new simple modulation is proposed, which can not only reduce the ISI interference effectively, but also improve the transmission rate to a certain extent. Numerical results show that compared with the current modulation scheme, the proposed scheme makes the system achieve better BER performance and the transmission rate is also improved.

Keywords: Molecular communication · Molecule shift keying · Transmission rate · Modulation · Diffusion channel

1 Introduction

Human cells can sense each other and store biological information. This magical biological structure promotes the emergence of nanomachines, which can realize simple computing and information processing functions [1]. Communication equipment in the field of nano-network is completely different from traditional communication. Communication under nano-network requires lower energy consumption, smaller size and biological compatibility [2, 3]. Molecular communication builds a bridge between engineered nanotechnology and natural Bionanotechnology, which is a new communication rule in the microenvironment [4]. The way of motion of a messenger molecule in communication of a diffusing molecule is passive and scarcely consumes any energy, which is also the most common way in molecular communication [1, 5, 17]. The most basic and simple way of communicating information to the biological environment is through free diffusion of molecules [5, 6]. Although molecular communication has encountered many obstacles in

Y. Chen et al. (Eds.): BICT 2023, LNICST 512, pp. 200–209, 2023.
https://doi.org/10.1007/978-3-031-43135-7_19

practical application, it has been developing forward. Inspired by traditional electromagnetic communication, molecular communication has many ways to encode information into messenger molecules. Although molecular communication is a new communication mode in micro environment, practical methods in traditional communication can also be used by molecular communication. For example, most modulation methods in molecular communication come from traditional electromagnetic communication. Concentration shift keying (CSK) is similar to amplitude shift keying in traditional electromagnetic communication. It uses the concentration of the received messenger molecules as the amplitude of the signal. If the number of messenger molecules reaching the receiver exceeds the threshold τ within a time slot, the receiver will decode to "1", otherwise to "0" [7, 8]. Molecules shift keying (MoSK) is similar to frequency shift keying, which uses different kinds of messenger molecules to encode, and different kinds of messenger molecules are equivalent to spectrum resources [7, 9].

In this paper, we propose a simple modulation method that uses different types molecules to encode and uses a simple mapping to improve transmission rates. Mathematical calculation and simulation experiments are based on diffusion molecular communication in the hope of inspiring related research and proposing better methods. In Sect. 2, we introduce the simple SISO model of molecular communication using multiple messenger molecules in diffused channels, and the basic formula of the diffusion motion of messenger molecules is derived. In Sect. 3, a simple modulation method which can avoid ISI interference and improve transmission rate is proposed. In addition, the ISI formula and BER performance of the system under the diffusion model of the second part are derived. The numerical simulation results are in the Sect. 4 and the conclusions are in the Sect. 5.

2 System Model

A simple SISO molecular communication model is shown in Fig. 1. The considered model consists of a set of transmitters (Tx) and receptors (Rx), and the distance between them is denoted by d. The receptor is assumed to be circular with radius R. The medium between the transmitter and the receiver is assumed to be a diffusion channel. Black dots of different shapes in Fig. 1 represent different types of messenger molecules, which are sent from the transmitter to the channel and reach the receiver through Brownian motion.

In the diffusion channel of molecular communication, the messenger molecules generally do random diffusion motion in the fluid medium, and their arrival at the receptor is a random probability event. To facilitate mathematical modeling and research analysis, assuming that the position of the transmitter is at the coordinate axis origin and the messenger molecules are released to the channel at $t = 0$, then the probability density function of the signal molecule can be written as [10]

$$p_A(d, t_s) = \frac{R}{R+d} erfc\left(-\frac{d}{\sqrt{4Dt_s}}\right) \quad (1)$$

where $erfc(x)$ is the complementary error function, R represents the radius of the receiver and d represents the distance between the transmitter and the receiver.

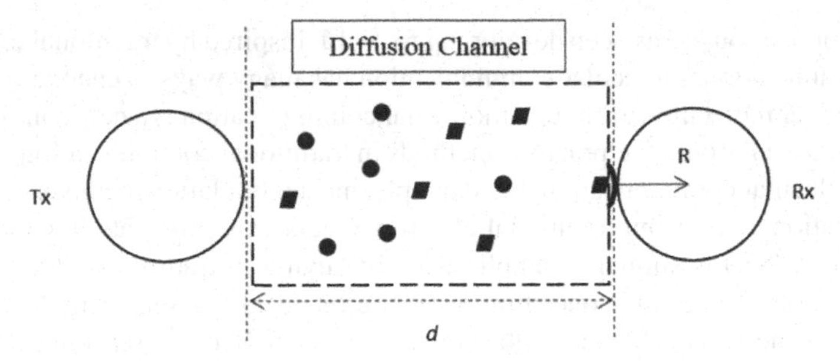

Fig. 1. SISO molecular communication system

A large number of studies and analyses show that the number of messenger molecules N_A received by the receiver in the time period t_s is related to the number of messenger molecules n transmitted by the transmitter, and it follows binomial distribution as [11, 12]

$$N_A \sim B(n, p_A(d, t_S)) \tag{2}$$

A binomial distribution $B(n, p(d, t_s))$ can be approximated with a normal distribution $N \sim N(np, np(1-p))$ when p is not close to one or zero and np is large enough, then (3) can be approximated as

$$N_A \sim N(np_A(d, t_S), np_A(d, t_S)(1 - p_A(d, t_S))) \tag{3}$$

3 Proposed Modulation Method

The cooperation between biological cells cannot be achieved without information and material exchange, which mainly relies on a variety of different protein carriers to transport different ions. The natural idea was to use a variety of different messenger molecules for molecular communication. Regardless of the existence of nanomachin es capable of detecting multiple messenger molecules in reality, using multiple messenger molecules for modulation will inevitably simplify the structure of the transmitters and receptors.

The model of molecular communication we study is that the transmitter encodes the message only to the type of molecule, and the receiver receives the messenger molecule and decodes it into the corresponding message. In addition, the transmitter can send only one type of signal molecule or no molecule at all within a time slot t. Then the sequence of molecules sent by the transmitter over a continuous period of time can be represented as a string of symbols, they are either A, B… or - (no transmission).

The ISI in molecular communication system is mainly caused by the long tail effect. The Brownian motion of the signal molecule in the diffusion channel results in a long tail for the time probability density of the signal molecule reaching the receptor. Practically we ignore the long tail effect that leads to long memories and we replace it with finite memory. This means that the receptors in the current time slot will only be affected by the messenger molecules remaining in the previous limited time slots. The memory length under our diffusion channel is assumed to be *k* time slot. In order to avoid ISI, same type messenger molecules should not be emitted within *k* time slot, which is the ISI-avoiding scheme [14].

For simplicity, our modulation method uses two different types of messenger molecules for analysis and simulation. One of them is A molecule and the other is B. The symbol '-' means that no molecule is emitted. Under the ISI avoidance scheme, *k* should be 2, that is, two adjacent symbols in the transmission sequence cannot be the same. For example, 'A, A, B' is invalid, but 'A, B, A' or '–, –, A' is allowed. Then the transmission sequence of the transmitter in a period of time may be 'A, B, A, –, –, A,...'. When we divide every two symbols into a combination, we can get a symbol set with an element length of 2:{A –, – A, B –, – B, AB, BA, –}. There are seven elements in the set, and each element represents the combination of molecular types that the transmitter may send. It is worth noting that if an ordinary modulation method is used, that is, one type of molecule carries 1 bit of information, then this transmission sequence can only transmit 14 bit of information at most. However, it is assumed that each element in the symbol set appears with equal probability, Then use Huffman coding to map all elements in the set to a longer bit sequence, so as to improve the transmission rate. The result is that the symbol set is mapped to the bit sequence set {110, 011,100, 010, 111,101,00}, and the transmission rate is increased by more than 40%. Fig. 2 shows the state diagram of the proposed modulation method, in which each edge is marked with the transmitted symbol combination and the corresponding bit sequence combination. More precisely, the left side of '/' represents the transmitted symbol, and the right side represents the corresponding bit sequence.

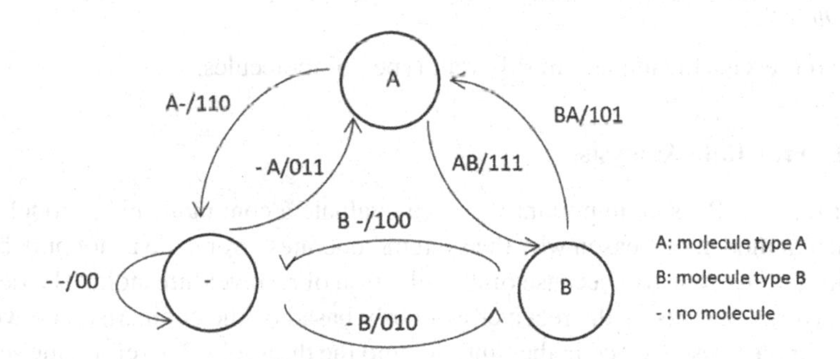

Fig. 2. State diagram of the modulation method.

3.1 Inter-Symbol Interference

Theoretically, transmitting information according to our modulation method can effectively avoid ISI, but in order to make the experimental model and simulation data more authentic, we still consider ISI [13].

The random Brownian motion of molecules is the main cause of ISI. Due to the randomness of messenger molecules, it is bound to lead to the residual molecules in the channel failing to reach the receiver within the corresponding time slot, thus affecting the subsequent time slot and causing ISI.

Suppose that $N_{ISI,its}$ represents the number of molecules that were emitted i time slots before and reach the receptor at current time slot. According to [15, 16], $N_{ISI,its}$ is a random variable, they consist of the subtraction of two normal distributions:

$$\begin{aligned} N_{ISI,it_s} &\sim \frac{1}{2}N(np(d,(i+1)t_S), np(d,(i+1)t_S)(1-p(d,(i+1)t_S))) \\ &-\frac{1}{2}N(np(d,it_S), np(d,it_S)(1-p(d,it_S))) \end{aligned} \tag{5}$$

where the factor 1/2 is due to equal probability of transmission of bits 0 and 1. The first term indicates the total number of molecules that are emitted at that time slot and absorbed by the receiver within all subsequent $i + 1$ time slots and the second term indicates those molecules that were absorbed within the subsequent i time slots. Then the number of ISI molecules received by the current time slot can be expressed as:

$$N_{C,ISI} = \sum_{i=2_{i\in E}}^{\infty} N_{ISI,\,it_s} \tag{6}$$

It can be further deduced that under our modulation method, the number of molecules of ISI received in the current time slot is:

$$\begin{aligned} N^{m}_{\substack{ISI,it_s \\ i\geq 2}} &\sim \frac{1}{m+1}N(np(d,(i+1)t_S), np(d,(i+1)t_S)(1-p(d,(i+1)t_S))) \\ &-\frac{1}{m+1}N(np(d,(i)t_S), np(d,(i)t_S)(1-p(d,(i)t_S))) \end{aligned} \tag{7}$$

where m represents the number of different types of molecules.

3.2 Bit Error Rate Analysis

Bit error rate (BER) is an important factor to evaluate a communication model. In this experimental model, the reason why the system generates error code is not only because of ISI interference, but also because of the selection of receiver threshold. The detection of messenger molecules by the receiver is mainly based on the comparison between the number of molecules received in the time slot and the threshold. Therefore, the selection of receiver threshold is critical to reduce the system bit error rate. There is a detection method called MAP detection, which can minimize the impact of receiver threshold on bit error. Let Z denote the number of molecules observed. Then, the two detection hypotheses are:

$$\begin{aligned} H_0 &: Z = N_0 \sim (\mu_0, \sigma_0^2) \\ H_1 &: Z = N_1 \sim (\mu_1, \sigma_1^2) \end{aligned} \tag{8}$$

The MAP detection is to obtain the point estimation of the quantity that is difficult to observe based on empirical data. Similar to the MLE, but with the greatest difference, the MAP incorporates the prior distribution of the quantity to be estimated. Therefore, the MAP can be regarded as the regularized MLE.

Applying MAP, the formula can be derived as follow:

$$\begin{aligned}\frac{P(H_0|Z)}{P(H_1|Z)} &= \frac{P(H_0)P(Z|P_0)}{P(H_1)P(Z|P_1)} \\ &= \frac{\sigma_1^2}{\sigma_0^2}\exp\{\frac{(Z-\mu_1)^2}{2\sigma_1^2} - \frac{(Z-\mu_0)^2}{2\sigma_0^2}\}\end{aligned} \tag{9}$$

By taking logarithm and setting to zero, the optimal decision threshold becomes:

$$\tau = \frac{-B + \sqrt{B^2 - 4AC}}{2A} \tag{10}$$

where:

$$\begin{aligned} A &= -\frac{1}{2}\left(\frac{1}{\sigma_1^2} - \frac{1}{\sigma_0^2}\right) \quad B = \frac{\mu_1}{\sigma_1^2} - \frac{\mu_0}{\sigma_0^2} \\ C &= \ln\left(\frac{\sigma_0}{\sigma_1}\right) - \frac{1}{2}\left(\frac{\mu_1^2}{\sigma_1^2} - \frac{\mu_0^2}{\sigma_0^2}\right)\end{aligned} \tag{11}$$

The BER for the information transmitted can be written as

$$\begin{aligned} P_e &= \frac{1}{2}(p(N_0 > \tau) + p(N_1 < \tau)) \\ &= \frac{1}{2}\left(Q\left(\frac{\tau - \mu_0}{\sigma_0}\right) + 1 - Q\left(\frac{\tau - \mu_1}{\sigma_1}\right)\right)\end{aligned} \tag{12}$$

4 Numerical Results

In this part, we will show the BER performance of the proposed modulation method and compare it with CSK modulation method under the same conditions. According to experience and selection of simulation model, we set some default simulation parameters as shown in Table 1: the distance between transmitter and receptor $d = 25\ \mu m$; the radius of the receptor $R = 10\ \mu m$; the Diffusion coefficient $D = 100\ \mu m^2/s$; the symbol duration $t_s = 30$ s; and the related time slot $k = 8$.

In Fig. 3, the comparison between the simulation results of theoretical derivation and numerical simulation results is shown. The theoretical simulation results are derived from Formula (5)–(7), while the numerical simulation results are obtained by.

transmitting messenger molecules. The ordinate represents the BER performance, and the abscissa represents the number of messenger molecules released, that is, energy or power. The simulation results show that the numerical simulation results are very close to the theoretical analysis results on the whole, and with the increase of simulation data from 1000 bit to 100000 bit, the numerical simulation curve is more smooth, which proves the authenticity of the model.

Figure 4 shows the BER performance comparison between the proposed modulation technology and other modulation technologies. The simulation parameters use the default

Table 1. Simulation parameters.

Parameter	Value
Distance between transmitter and receptor (d)	25 μm
Radius of the receptor (R)	10 μm
Diffusion coefficient (D)	100 μm^2/s
Symbol duration (t_s)	30 s
Related time slot (k)	8

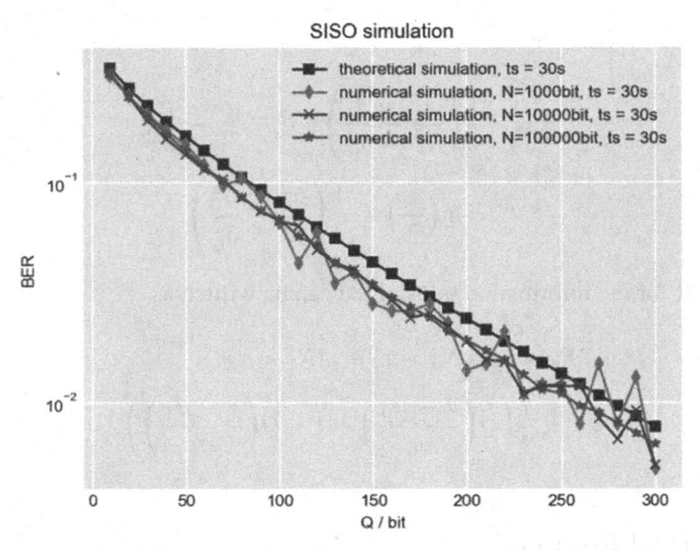

Fig. 3. Theoretical simulation and numerical simulation

parameters. Compared with CSK and MoSK modulation, the proposed modulations obviously have better BER performance. With the increase of Q, the gap will become more and more obvious. This also proves that the proposed modulation method can effectively reduce the negative impact of ISI and improve the system performance.

The Fig. 5 shows the distance between transmitter and receiver in relation to BER performance. The six curves respectively represent BER performance curves when $d =$ 10 μm, $d = 15$ μm, $d = 25$ μm, $d = 35$ μm, $d = 45$ μm, $d = 100$ μm. It is not difficult to find that with the increase of d, the BER performance of the system is getting worse. This is because the increase of the receiving and transmitting distance decreases the probability of the large receiver of the messenger molecule, which leads to the increase of the bit error rate and the decline of BER performance. However, when d increases to a certain extent, the BER performance degradation becomes limited. At this time, it can also be understood that too much d has less impact on the system performance.

Similarly, Fig. 6 shows the impact of different symbol duration on system BER performance. The six curves respectively represent BER performance curves when t_s

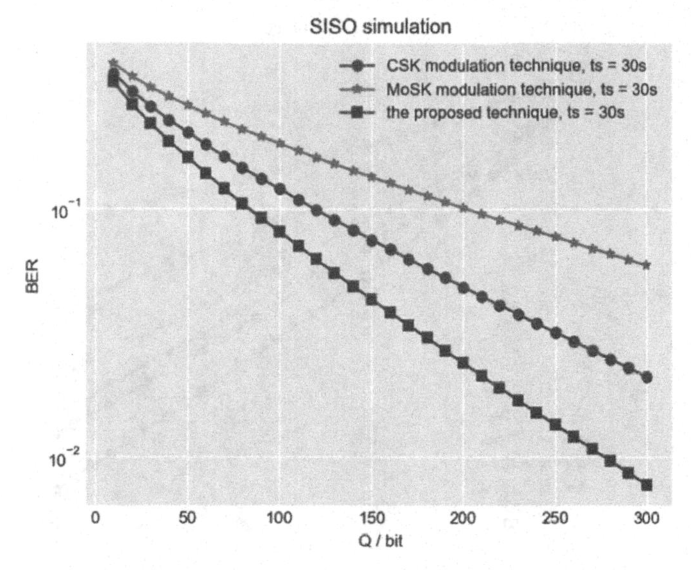

Fig. 4. BER performance of different modulation technique

$= 5$ s, $t_s = 10$ s, $t_s = 20$ s, $t_s = 30$ s, $t_s = 40$ s, $ts = 100$ s. In addition to the change of symbol duration t_s, other simulation parameters are set by military default. We can observe that the BER performance of the system is getting better and better with the continuous increase of t_s. This is because the messenger molecule has enough time to move until the receiver is absorbed, which improves the overall probability to a large extent and reduces the bit error rate. However, when t_s increases to a certain extent, perhaps when $t_s = 30$–40 s, the BER performance of the system does not change much. At this time, the impact of t_s on the system performance is very limit.

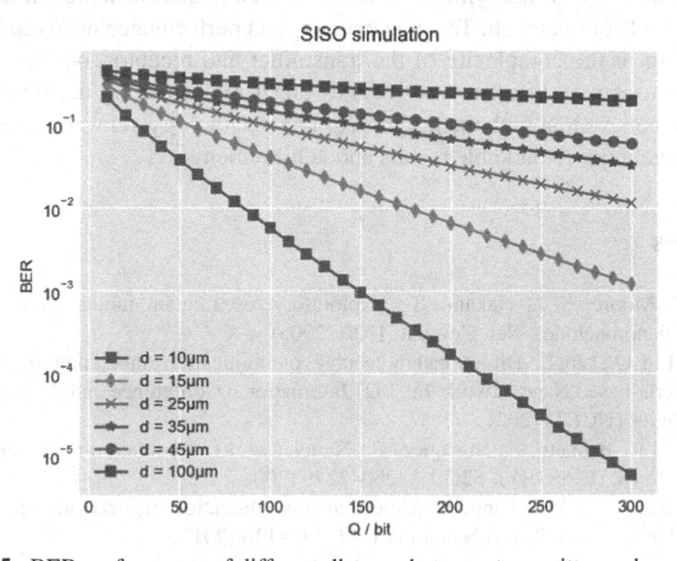

Fig. 5. BER performance of different distance between transmitter and receptor

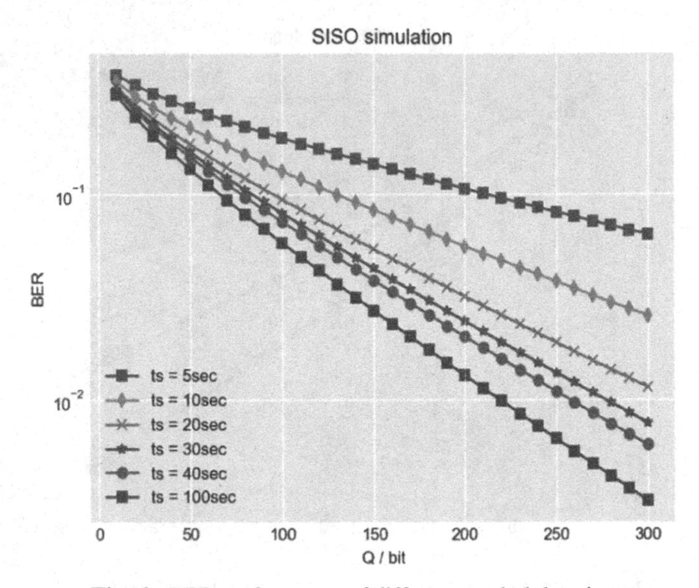

Fig. 6. BER performance of different symbol duration

5 Conclusion

In this paper, we propose a simple modulation scheme. Through the comparison of simulation data, we can draw a conclusion that the proposed modulation scheme can effectively reduce the impact of ISI, and has better BER performance than other modulation schemes. In addition, the proposed modulation method can enable the messenger molecules to carry more information and improve the efficiency of information transmission. According to the calculation and analysis above, this simple modulation method can improve the bps (bits per symbol) by about 40% compared with the ordinary method, and reduce the ISI effect well. To some extent, good performance needs to be achieved at the expense of the complexity of the transmitter and receptor, but this modulation scheme only uses two different types of messenger molecules. In the future work and research, we believe that with the progress of nanotechnology, relevant research will be able to achieve more remarkable results and achievements.

References

1. Suda, T., Moore, M.J., Nakano, T.: Exploratory research on molecular communication between nanomachines. Nat. Comput. 1–30 (2005)
2. Yue, G., Liu, Q., Lin, L.: Directional molecular communication among nanomachines in massive bacteria-based Nanonetwork. In: 2021 International Conference on UK-China Emerging Technologies (UCET) (2021)
3. Akyildiz, I.F., Brunetti, F., Blazquez, C.: Nanonetworks: a new communication paradigm. Comput. Netw. (Elsevier) J. **52**(12), 2260–2279 (2008)
4. Nakano, T., Moore, M.J., Fang, W.: Molecular communication and networking: Opportunities and challenges. IEEE Trans. Nanobiosci. **11**, 135–148 (2012)

5. Yeh, P.C., Chen, K.C., Lee, Y.C., et al.: A new frontier of wireless communication theory: diffusion-based molecular communications. IEEE Wirel. Commun. **19**(5), 28–35 (2012)
6. Nakano, T., Suda, T., Okaie, Y., et al.: Molecular communication among biological nanomachines: a layered architecture and research issues. IEEE Trans. Nanobiosci. **13**(3), 169–197 (2014)
7. Kuran, M.S., Yilmaz, H.B., Tugcu, T., Akyildiz, I.F.: Modulation techniques for communication via diffusion in nanonetworks. In: Proceedings IEEE International Conference Communication Kyoto, Japan, pp. 1–5 (2011)
8. Arjmandi, H., Gohari, A., Kenari, M.N., Bateni, F.: Diffusion-based nanonetworking: a new modulation technique and performance analysis. IEEE Commun. Lett. **17**, 645–648 (2013)
9. Atakan, B., Galmes, S., Akan, O.B.: Nanoscale communication with molecular arrays in nanonetworks. IEEE Trans. Nanobiosci. **11**(2), 149–160 (2012)
10. Pierobon, M., Akyildiz, I.F.: Intersymbol and co-channel interference in diffusion-based molecular communication. In: IEEE International Conference on Communications, pp. 6126–6131 (2012)
11. Saeed, M., Bahrami, H.R.: Performance of MIMO molecular communications in diffusion-based channels. Int. J. Commun. Syst. (2017)
12. Moore, M.J., Suda, T., Oiwa, K.: Molecular communication: modeling noise effects on information rate. IEEE Trans. Nanobiosci. **8**(2), 169–180 (2009)
13. Jiang, C., Chen, Y., Liu, K.J.R.: Inter-user interference in molecular communication networks. In: ICASSP, pp. 5725–5729 (2014)
14. Arjmandi, H., Movahednasab, M., Gohari, A., Mirmohseni, M., Nasiri-Kenari, M., Fekri, F.: ISI-avoiding modulation for diffusion-based molecular communication. IEEE Trans. Mol. Biolog. Multi-Scale Commun. **3**(1), 48–59 (2017). https://doi.org/10.1109/TMBMC.2016.2640311
15. Noel, A., Cheung, K.C., Schober, R.: Improving receiver performance of diffusive molecular communication with enzymes. IEEE Trans. **13**(1), 31–43 (2014)
16. Liu, Q., Lu, Z., Yang, K.: Modeling and dual threshold algorithm for diffusion-based molecular MIMO communications. IEEE Trans. Nanobiosci. **20**(4), 416–425 (2021). https://doi.org/10.1109/TNB.2021.3077297
17. Liu, Q., Yang, K., Xie, J., Sun, Y.: DNA-based molecular computing, storage, and communications. IEEE Internet Things J. **9**(2), 897–915 (2022). https://doi.org/10.1109/JIOT.2021.30836

Range Expansion in Neuro-Spike Synaptic Communication: Error Performance Analysis

Abhinav[1], Lokendra Chouhan[2](✉), and Prabhat K. Sharma[1]

[1] Visvesvaraya National Institute of Technology, Nagpur, India
[2] Indian Institute of Information Technology, Chittoor, India
lokendrachouhan22@gmail.com

Abstract. In this paper, a neuro-spike synaptic cooperative communication channel model is exploited. In the considered model, a neuro-spike relay (NSR) is placed in the synaptic gap of two neurons to extend the range of communication. For the analysis, a time-slotted channel is exploited, where we transmit a binary bit in each time-slot for the transmission of information from the pre-synaptic neuron called neuro-spike source (NSS) to the post-synaptic neuron called neuro-spike destination (NSD). Further, the considered model is analyzed in terms of the probability of detection and probability of false alarm. Moreover, the effect of ISI due to the transmission of molecules from the previous time-slots, and noise arises from unintended neurons are also considered in the analysis. Furthermore, the closed-form expression for the end-to-end probability of error is also computed for the cooperative link. Above all, the analytical expressions are validated using Monte-Carlo simulations.

Keywords: Neuro-spike communication · synaptic gap · neuro-spike relay · Poisson distribution

1 Introduction

A molecular communication (MC) is one of the trending communication technologies in recent days, in which the characteristics of molecules can be used to transmit the information from one end to other [1]. Few examples of an MC systems are cell-cell signaling [2], bacterial communication [3], and synaptic or neuro-spike communication [4]. The neuro-spike communication is one of the widely investigated forms in molecular communication [5–7]. In this, both electrochemical impulse and neurotransmitter are used for the transmission of molecules from neuro-spike source (NSS) to neuro-spike destination (NSD) neurons. Inside the neuron, the information is transmitted in the form of electrochemical impulse which is called as axon potential and is transmitted to the end of a neuron.

Y. Chen et al. (Eds.): BICT 2023, LNICST 512, pp. 210–223, 2023.
https://doi.org/10.1007/978-3-031-43135-7_20

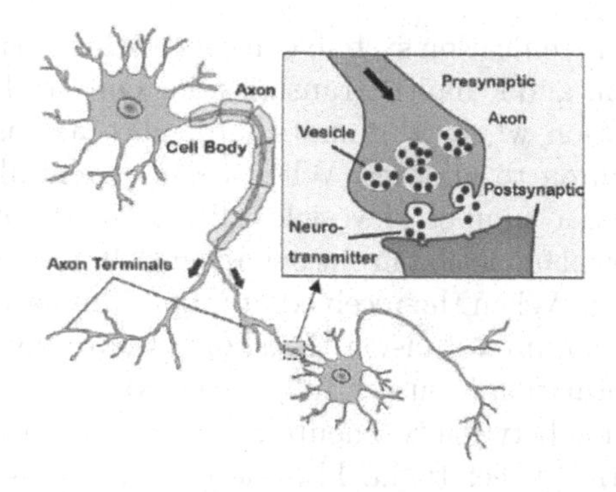

Fig. 1. Basic structure of neuron [8].

Neurons communicate with one another at junctions called *synapses*. At a synapse, one neuron sends a message to a target neuron-another cell. Mostly, synapses are chemical, i.e., these synapses communicate using chemical messengers. In a few cases, synapses are electrical, i.e., in these synapses, ions flow directly between the cells. At a chemical synapse, an action potential triggers the pre-synaptic neuron to release neurotransmitters. These molecules bind to receptors on the postsynaptic cell and make it more or less likely to fire an action potential. The pre-synaptic and post-synaptic neurons have a gap in between which is known as the synaptic cleft. The synaptic communication has nerve cell as a major part and it basically consist of three parts:

- Dendrites:- These are the projections of neuron that receive signal from other neuron. It acts as sensor or receptor of a basic communication system. When a random stimulus is applied to neurons its dendrites fires and generates an electrical signal which propagates through the axon to its tail. Dendrites branches as they move towards their tip just behave like tree branches.
- Axon:- Each neuron in our body has a cable like structure which is very thin. It is several times thinner than human hair. It acts as a carrier for the electrical pulse generated at the dendrites. Depending upon the types of neuron axon length varies.
- Axon terminal:- At this end, packets of neurotransmitter are released in the synaptic cleft. The released neurotransmitter diffuses in the synaptic cleft follows brownian motion. Brownian motion is the random motion of particles. The basic structure of neuron is demonstrated in Fig. 1.

It is very well known fact that neuro-degenerative diseases are incurable in many cases [9]. For instance, connections of neurons are lost which may cause death of a person. Further, in the recent experiments, it was seen that loss of neuron connectivity increases the average gap of two neurons, which is the primary cause of Alzheimer's disease. Size of neuron is found to be very small and it ranges from 4 to 100 μm [10]. In neuro-spike communication, usually, out of one hundred billion nerve cells in the human brain, each one of them communicates with thousand others. The stronger connection between neurons can be made if a neuro-spike relay (NSR) is placed in between the NSS and NSD.

A neuro-spike-communication system consists of three-part, i.e., axonal pathway, spike generation, and synaptic transmission. Our study mainly focuses on synaptic communication where glutamate molecules act as a carrier of information from one neuron to another. When axon potential reaches the tail of transmitting neurons, a number of vesicles diffuse in the synaptic gap. These molecules follow Brownian motion in the synaptic cleft and reach the dendrites of the destination neuron. When the received number of molecules at post-synaptic neurons exceeds the required decision threshold, the receiver confirms that the particular bit of information is successfully received.

The communication between two neurons gets weaker because of the increase in average path length between them. This increase in average path length occurs due to death or progressive degradation of nerve cells. Several works have studied synaptic communication in different aspects. For instance, in [11] authors have discussed capacity analysis of neuro-spike communication using temporal modulation. Further, authors have discussed the role of ISI in synaptic communication in [12]. The interfacing of nano-machine and neuron communication has been provided in [13]. Moreover, in [14], the joint optimization of input spike rate and decision threshold at the receiver to maximize achievable bit rate has been exploited. Authors have discussed the diffusion-based model for synaptic communication in [15].

The error probability for a cooperative communication system consisting of K receivers in a diffusive medium was provided in [16]. In [17], authors have discussed ML detection for a cooperative communication system in the diffusive channel. In [18], authors have discussed the optimal positioning of a relay machine in the cooperative diffusive molecular communication channel. In [19], authors have discussed characterizing the three-dimensional diffusive molecular communication channel with an absorbing receiver. In [20], authors have discussed optimal receiver design for diffusive molecular communication channels with flow and additive noise. In [21], authors have analyzed a communication network consisting of a single transmitter and receiver incorporated by multiple relays in between them. In [22], authors have considered a passive receiver that receives molecules without changing its characteristics, and further, they can diffuse in the environment. In [23,24], authors have assumed that molecular concentration remains uniform throughout the channel and source and destination nano-machine can count the number of molecules within the spherical receiver boundary. Note that, the concentration of neuro-transmitter attenuates with the distance [25]. Thus, neuro-spike direct communication between source and destination may leads to the information loss due to the degraded neuro-transmitters. One way to reduce the imapct of degradation on information loss is by adding the additional nodes (known as neuro spike relay) between NSS and NSD which can increase the communication channel length.

To the best of the author's knowledge, no authors have considered the error performance of cooperative neuro-spike communication. Based on the above research gap and motivations, in this paper we consider range expansion using

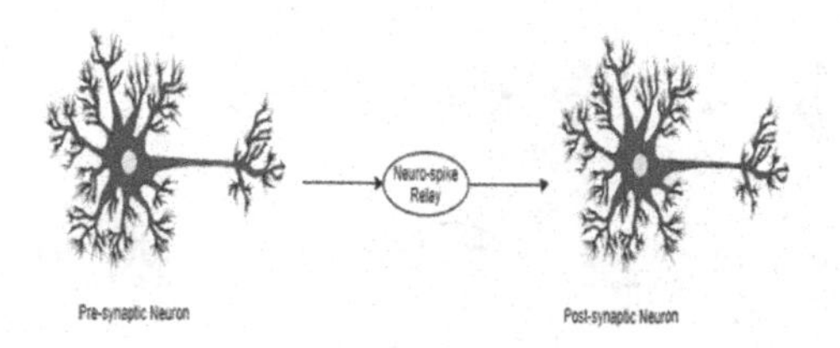

Fig. 2. Two neurons assisted by a NSR.

the relay-assisted method in neuro-spike synaptic communication. In this context, the following are our key contributions through this work:

- First time in the literature, for the range expansion the relay-assisted neuro-spike Synaptic communication is considered.
- We consider the effect of re-uptake probability and spill-over effect for calculation of results.
- The impacts of inter-symbol-interference (ISI) and unintended molecules from other sources are also considered.
- Further, the considered system is presented in the form of equivalent binary channel and is analyzed in terms of probability of detection and probability of false alarm.
- Moreover, the expressions for average probability of error considering ISI is derived. We evaluate the error performance of the system considering various environment parameters, e.g., diffusion coefficient of neuro transmitter.

Above all, the analytical expressions are verified using Monte-Carlo simulation.

The rest of the paper is organized as follows. In Sect. 2, we present a system model of the cooperative neuro-spike communication system. Also in this section, we provide the preliminaries required for neuro-spike communication. The formulation of analytical frameworks for direct and NSR-assisted communication links is provided in Sect. 3. Numerical results are examined in Sect. 4. At the end, we conclude our paper in Sect. 5.

2 System Model

As shown in Fig. 2, we consider a relay-assisted neuro-spike communication system, in which a pre-synaptic and a post-synaptic neurons are placed at a particular distance as a source and destination, respectively. To enhance the range of communication, a neuro-spike relay is also placed in between the pre-synaptic and post-synaptic neurons. Let us consider a binary information transmission between pre-synaptic and post-synaptic neurons. And to do so, a time-slotted channel is assumed, where pre-synaptic neuron transmits K binary information bits towards the post-synaptic neuron. Let us consider the binary information sequence of length K bits which is to be transmitted, is presented by $\{x[1], x[2], \ldots, x[i], \ldots, x[K]\}$, where $x[i] \in \{0, 1\}$, $\forall i$. Thus, for K bits, total 2^K

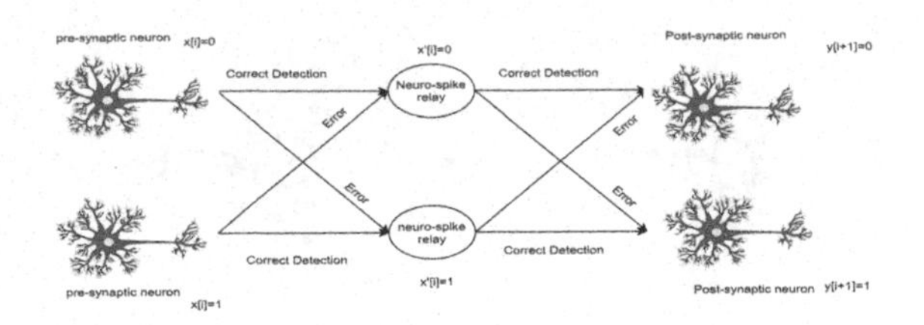

Fig. 3. System model for two neurons cooperated by a NSR.

symbols can be encoded and transmitted from the source. We assume that the priory probability of a particular bit being transmitted as 0 and 1 be α_0 and α_1, respectively. For relaying, a decode and forward strategy is used, in which pre-synaptic neuron transmits a particular bit in the ith time slot, later on, that bit is received and decoded first at the NSR, subsequently, the decoded bit is forwarded to the destination in the same time slot. At the end, the information bit is received and decoded at the post-synaptic neuron, eventually, in $(i+1)^{th}$ time slot. Therefore, the transmission of complete K bits of information from source to destination neurons takes $(K+1)$ time slots.

We consider a binary concentration shift keying (BCSK) for the transmission of information sequence, i.e., the pre-synaptic neurons and NSR transmit $Q_T[i]$ and $Q_C[i]$ number of molecules, respectively, for the transmission of binary bit 1, and zero molecules are used for transmission of binary bit 0 at the beginning of ith time slot. Further, it is assumed that source neurons and neuro-spike relay emits the different types of molecules, such that there is no self-interference at the neuro-spike relay [26,27]. Due to the random movement of molecules from pre-synaptic to post-synaptic neurons, the molecules may reach to the post-synaptic neuron out of order, eventually, leads to the stochastic channel behaviour. One of the possible ways to characterizes the stochastic behaviour of channel between the pre-synaptic and post-synaptic neurons is by using transition probability matrix. The transition probability matrix for the links from source to NSR, NSR to destination, and source to destination can be described as below

$$\begin{bmatrix} P_{00}^{sr} & P_{01}^{sr} \\ P_{10}^{sr} & P_{11}^{sr} \end{bmatrix}, \begin{bmatrix} P_{00}^{rd} & P_{01}^{rd} \\ P_{10}^{rd} & P_{11}^{rd} \end{bmatrix}, \text{ and } \begin{bmatrix} P_{00}^{sd} & P_{01}^{sd} \\ P_{10}^{sd} & P_{11}^{sd} \end{bmatrix},$$

respectively, here P_{lm}^{sr}, P_{lm}^{rd} and P_{lm}^{sd} ($\{l, m\} \in \{0, 1\}$) are the transition probabilities corresponding to the transmitted symbol l and received symbol m from source to relay, relay to destination, and source to destination, respectively (Fig. 3).

2.1 Preliminaries

Neuro-spike communication consists of three processes, namely the Vesicle release, diffusion, and ligand-receptor binding. Vesicles are a group of neurotransmitters enclosed inside a closed volume of thin membrane and it is located just behind the pre-synaptic membrane of the pre-synaptic or transmitting neuron. When the axon potential reaches the neuron tail, the bundle of glutamate molecules released diffuses in the synaptic cleft. In the synaptic gap, information transmission can be modeled in the form of molecular communication for which the study can have two different approaches; first, microscopic approach [28], in which the focus is on the probability of arrival of a single molecule. And the second is the macroscopic approach, in which the molecular concentration is considered at the receiver corresponding to the impulse response of the channel [29]. In this paper, we use a macroscopic model in which transmitted molecules in the cleft follow a Brownian motion. The concentration of molecules at a different location in the three-dimensional space is defined by Fick's equation.

$$\frac{\partial C(x,y,z,t)}{\partial t} = D\nabla^2 C(x,y,z,t)\ ;\ t \geq 0\ , \tag{1}$$

where D is the diffusion coefficient, ∇^2 is the Laplacian operation, $(x,y,z) \in \mathbb{R}^2 \times [0,L]$, and L is the gap between the tail of transmitting neurons and head of receiving neurons. Let the concentration of glutamate molecules at a time t in the three-dimensional space is denoted by $C(x,y,z,t)$. At $t = 0$, concentration of glutamate molecules is describe by the equation:

$$C(x,y,z,0) = N_{glu}\,\delta(x,y,z)\ , \tag{2}$$

where N_{glu} is an initial number of glutamate molecules within the vesicle. It is also assumed that there is no flux of glutamate molecules between pre-synaptic and post-synaptic membranes. In recent studies, it is observed that all the molecules in the vesicles of a pre-synaptic neuron do not reach the dendrites of a post-synaptic neuron [30]. Some molecules diffuse long enough to actuate neurons placed outside the synaptic cleft, even sometimes, these molecules can actuate neighboring neurons. This activity of actuating non-intended neurons is called *spill-over* and having significant physiological effects on synaptic transmission. Considering *spill-over* effect and reuptake process[1], the solution of (1) in terms of Fourier series is obtained as [11]

$$\begin{aligned} C(x,y,z,t) &= \frac{N_{glu}}{4\pi L D t} \times \exp\left(-\frac{(x^2+y^2)}{4Dt}\right) \\ &\times \left[1+2\sum_{1}^{N}\left[(1-p_u)^n \cos\left(\frac{n\pi z}{L}\right)\exp\left(Dt\left(\frac{n\pi}{L}\right)^2\right)\right]\right], \end{aligned} \tag{3}$$

where n represents number of Fourier modes taken for the evaluation of (3).

Remark: When molecules are transmitted from the pre-synaptic neurons, a part of the transmitted glutamate molecule is absorbed by the pre-synaptic terminal for recycling. Therefore, when re-uptake effect is considered, the above solution of the differential equation can be modified. Here, the coefficients of the

[1] The reuptake process is defined as the re-absorption by a neuron of a neurotransmitter following the transmission of a nerve impulse across a synapse [25].

Fourier series get modified based on reuptake probability, i.e., when $p_u = 0$, then no molecules are absorebd by the pre-synaptic membrane, and if $p_u = 1$, then all the molecules which hit the pre-synaptic membrane are absorbed.

Now, we defined a substantial terminology in the context of neuro-spike communication known as *spike generation*. The *spike generation* occurs when the potential of the membrane of a destination neuron reaches the threshold it fires a spike in the form of an electrical signal and propagates via the axon of the post-synaptic neuron. The neurotransmitter that arrives at the post-synaptic neurons either gets reflected from the membrane or gets bonded with the receptor to form ligand-receptor complex [31]. The binding process is modeled in such a way that receptors regularly sample in a small volume, V_e. A receptor can either be in a bound or unbound state. If a neurotransmitter is in the unbound state and it samples at least once in small effective volume, V_e, then it goes to the bound state. Further note that binding of glutamate is also a reversible process, i.e., the ligand may dissociate with the receptor and may get rebounded again. Binding probability is calculated first by considering a single neurotransmitter in the synaptic cleft, i.e., $N_{glu} = 1$. Probability of finding that neurotransmitter in the cleft at any time t is obtained by integrating the concentration function $C(x, y, z, t)$ for spatial coordinates space (x, y, z) and is given by

$$P_e(t) = \iiint_{V_e} C(x, y, z, t)\, dx\, dy\, dz. \tag{4}$$

Let i be an intended information transmission time-slot, in which the pre-synaptic neuron transmits $N_{glu}[i]$ number of molecules towards the post-synaptic neuron. Thus, the total number of molecules at the post-synaptic neuron is given by

$$N^{sd}[i] = N^{sd}_{cr}[i] + N^{sd}_{int}[i] + N^{sd}_{n}[i], \tag{5}$$

where $N^{sd}[i]$ is the total number of molecules received in i^{th} time slot at the receiving neurons. $N^{sd}_{cr}[i]$ is the number of molecules received in the ith time slot at the receiving neuron due to transmission of molecules in the current time slot and it follows Binomial distribution given by $\mathcal{B}(N_{glu}[i]\, x[i]; P^{sd}_{cr})$, where P^{sd}_{cr} is the probability of arrival of molecules in the current time slot.

$N^{sd}_{int}[i]$ is the number of molecules received in the ith time slot due to transmission of molecules from the previous time slot, i.e., from 1st time slot to $(i-1)^{th}$ time slot and these time slots are called as interfering time slot. It follows binomial distribution given by $\mathcal{B}(N_{glu}[i-j]\, x[i-j]; P^{sd}_{int})$, where $j \in \{1, 2, \ldots, i\}$ and P^{sd}_{int} is the probability of arrival of molecules from the previous time slot i.e.$(i-1)^{th}$ slot. N^{sd}_{n} is the number of molecules at post-synaptic neuron due to the background noise which is known as synaptic noise. The main source of the synaptic noise is the background synaptic activity which is due to the multiple access of synapses from thousands of other synapses [32].

Since, for a large number of molecules and small arrival probability, the Binomial distribution can be well approximated as Poisson's distribution [33]. Thus, by applying Poisson approximation, $N^{sd}_{cr}[i]$, $N^{sd}_{int}[i]$ and $N^{sd}_{n}[i]$ can be approximated with average rate $\theta^{sd}_{cr}[i] = N_{glu}[i-j]\, x[i]\, P^{sd}_{cr}$, $\theta^{sd}_{int}[i] = N_{glu}[i]\, x[i-j]\, P^{sd}_{int}$ and $\theta^{sd}_{n}[i]$ respectively.

3 Detection Analysis and Error Probability Computation

Now we calculate the number of molecules obtained at NSR after transmission from source neurons and, eventually, the number of molecules received at destination neuron. We define two binary hypotheses for the transmission of binary symbols 0 and 1, i.e., null and alternate hypotheses, respectively.

3.1 NSR Assisted Communication Between Pre-synaptic and Post-synaptic Neurons

Let us consider an information bit is transmitted from a pre-synaptic neuron to NSR in i^{th} time slot, and is decoded and re-transmitted to the post-synaptic neuron in the same time slot, and later it arrived at the post-synaptic neuron in the $(i+1)^{th}$ time slot. The symbol detection problem for an NSR-assisted system can also be formulated as a binary hypothesis testing problem. The null and alternate hypothesis for the link between the pre-synaptic neuron and NSR can be written as

$$H_0^{sr}[i] : N_n^{sr}[i] + \sum_{k=1}^{i-1}(N_{int}^{sr}[k]), \qquad H_1^{sr}[i] : N_n^{sr}[i] + N_{cr}^{sr}[i]\sum_{k=1}^{i-1} N_{int}^{sr}[k], \tag{6}$$

where $H_0^{sr}[i]$ and $H_1^{sr}[i]$ are binary hypotheses corresponding to transmission of bits 0 and 1, respectively. Thus, the number of molecules $N^{sr}[i]$ under $H_0^{sr}[i]$ and $H_1^{sr}[i]$ follow Poisson distribution [33,34]

$$H_0^{sr}[i] : \mathcal{P}(\theta_0^{sr}[i]) ; \qquad H_0^{sr}[i] : \mathcal{P}(\theta_1^{sr}[i]), \tag{7}$$

where

$$\theta_0^{sr}[i] = \theta_{int}^{sr}[i] + \theta_n^{sr}[i] = N_{glu}[i]\sum_{j=1}^{i-1} x[i-j]P_{int}^{sr} + \theta_n^{sr}[i], \tag{8}$$

and

$$\theta_1^{sr}[i] = \theta_{int}^{sr}[i] + \theta_{ms}^{sr}[i] + \theta_{cr}^{sr}[i] = N_{glu}[i]\sum_{j=1}^{i-1} x[i-j]P_{int}^{sr} + \theta_{ms}^{sr}[i] + N_{glu}[i]x[i]P_{cr}^{sr}. \tag{9}$$

Similarly, number of molecules received from NSR to Post-synaptic neuron can be formulated in term of two binary hypotheses as

$$H_1^{rd} : N_1^{rd}[i+1] = N_n^{rd}[i+1] + N_{cr}^{rd}[i+1] + \sum_{k=2}^{i} N_{int}^{rd}[k], \tag{10}$$

$$H_0^{rd} : N_0^{rd}[i+1] = N_n^{rd}[i+1] + \sum_{k=2}^{i} N_{int}^{rd}[k]. \tag{11}$$

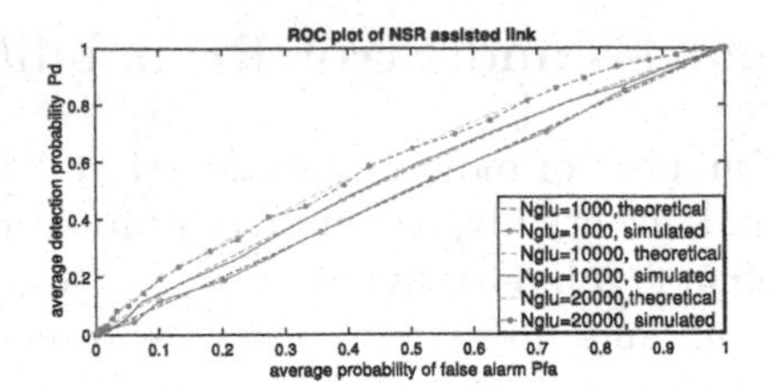

Fig. 4. ROC plot for NSR assisted link

The total number of molecules received at NSD follows the Poisson statistics and can formulated in terms of two hypotheses $H_0^{rd}[i+1]$ and $H_1^{rd}[i+1]$ as

$$H_0^{rd}[i+1] : \mathcal{P}(\theta_0^{rd}[i+1]) ; \qquad H_1^{rd}[i+1] : \mathcal{P}(\theta_1^{rd}[i+1]), \tag{12}$$

where the expressions for $\theta_0^{rd}[i+1]$ and $\theta_1^{rd}[i+1]$ can be obtained analogously from (8) and (9), respectively.

Now, depending upon the concentration of molecules received at the neuro-spike relay $N^{sr}[i]$ and at NSD $N^{rd}[i+1]$, the corresponding symbols are decoded. The symbol detection problems at relay and NSD are formulated as

$$N^{sr}[i] \underset{H_0^{sr}}{\overset{H_1^{sr}}{\gtrless}} \eta^{sr}[i]; \quad N^{rd}[i+1] \underset{H_0^{rd}}{\overset{H_1^{rd}}{\gtrless}} \eta^{rd}[i+1], \tag{13}$$

respectively. Here $\eta^{sr}[i]$ and $\eta^{rd}[i+1]$ are the optimum thresholds at relay and NSD, respectively. Let $X[i]$, $X'[i]$ and $Y[i+1]$ are the symbols at NSS, relay and NSD, respectively. In cooperative link, error occurs if any of the links is erroneous, i.e., $X[i] \neq X'[i]$and $X'[i] = Y[i+1]$ or $X[i] = X'[i]$ and $X'[i] \neq Y[i+1]$. Thus, the expression for probability of error in $(i+1)^{th}$ time slot can be written as:

$$P_e[i+1|X_1^{i-1}] = \alpha_1 P_e[i+1|X[i]=1, X_1^{i-1}] + \alpha_0 P_e[i+1|X[i]=0, X_1^{i-1}], \tag{14}$$

where $P_e[i+1|X[i]=1, X_1^{i-1}]$ is the probability of error when 1 was transmitted from the source neuron and $P_e[i+1|X[i]=0, X_1^{i-1}]$ is the probability of error at the receiver in $(i+1)^{th}$ time slot when 0 was transmitted from the transmitter for the given ISI sequence. Now the terms in (14) can be further expanded as

$$\begin{aligned}
&P_e[i+1|X[i]=1, X_1^{i-1}]) \\
&= \Pr(N^{sr}[i] < \eta^{sr}[i]|X[i]=1, X_1^{i-1}) \times \Pr(N^{rd}[i+1] < \eta[i+1]|X'[i]=0, X_1'^{i-1}) \\
&+ \Pr(N^{sr}[i] \geq \eta^{sr}[i]|X[i]=1, X_1^{i-1}) \times \Pr(N^{rd}[i+1] < \eta[i+1]|X'[i]=1, X_1'^{i-1}) \\
&= (1 - P_d^{sr}[i|X_1^{i-1}])(1 - P_f^{rd}[i+1|X_1'^{i-1}]) + (P_d^{sr}[i|X_1^{i-1}])(1 - P_d^{rd}[i+1|X_1'^{i-1}]),
\end{aligned} \tag{15}$$

and

$$\begin{aligned}
&P_e[i+1|X[i]=0,X_1^{i-1}] \\
&= \Pr(N^{sr}[i] \geq \eta^{sr}[i]|X[i]=0,X_1^{i-1})\Pr(N^{rd}[i+1] \geq \eta[i+1]|X'[i]=1,X_1'^{i-1}) \\
&+ \Pr(N^{sc}[i] < \eta^{sr}[i]|X[i]=0,X_1^{i-1})\Pr(N^{rd}[i+1] \geq \eta[i+1]|X'[i]=0,X_1'^{i-1}) \\
&= (P_f^{sr}[i|X_1^{i-1}])(P_d^{rd}[i+1|X_1'^{i-1}]) + (1-P_f^{sr}[i|X_1^{i-1}])(P_f^{rd}[i+1|X_1'^{i-1}]), \qquad (16)
\end{aligned}$$

where $\eta^{sr}[i]$ and $\eta^{rd}[i+1]$ are decision thresholds for NSR and post-synaptic neurons. The terms $P_d^{sr}[i]$ and $P_d^{rd}[i+1]$ denote the probabilities of detection at NSR and receiving neuron, respectively. Also, $P_f^{sr}[i]$ and $P_f^{rd}[i+1]$ denote the probabilities of false alarm at NSR and receiving neuron, respectively, which can be written as

$$P_f^{sr}[i|X_1^{i-1}] = \Pr\left(Y[i+1]=1|X[i]=0,X_1^{(i-1)}\right) = 1 - \sum_{m=1}^{\lfloor \eta^{sr}[i] \rfloor} \frac{\exp(-\theta_0^{sr}[i])(\theta_0^{sr}[i])^m}{m!}, \qquad (17)$$

and

$$P_d^{sr}[i|X_1^{i-1}] = \Pr\left(Y[i+1]=1|X[i]=1,X_1^{(i-1)}\right) = 1 - \sum_{m=1}^{\lfloor \eta^{sr}[i] \rfloor} \frac{\exp(-\theta_1^{sr}[i])(\theta_1^{sr}[i])^m}{m!}. \qquad (18)$$

Similarly, the probability of detection and false alarm for the second link, i.e., from NSR to destination neuron can be written as

$$P_f^{rd}[i+1|X_1^{i-1}] = 1 - \sum_{m=1}^{\lfloor \eta^{rd}[i+1] \rfloor} \frac{\exp(-\theta_0^{rd}[i])(\theta_0^{rd}[i])^m}{m!} \qquad (19)$$

and

$$P_d^{cd}[i|X_1^{i-1}] = 1 - \sum_{m=1}^{\lfloor \eta^{rd}[i+1] \rfloor} \frac{\exp(-\theta_1^{rd}[i])(\theta_1^{rd}[i])^m}{m!} \qquad (20)$$

The average probability of error $P_{e,avg}[i+1]$ is obtained by taking average over all the possible realizations of X_1^{i-1}, i.e.,

$$P_{e,avg}[i+1] = \sum_{X_1^{i-1} \in \mathcal{X}} \Pr(X_1^{i-1}) P_e(i+1 \mid X_1^{i-1}), \qquad (21)$$

where $\Pr(X_1^{i-1})$ is the probability of occurrence of one ISI symbol and can be evaluated as $\Pr(X_1^{i-1}) = \frac{1}{2^{i-1}}$.

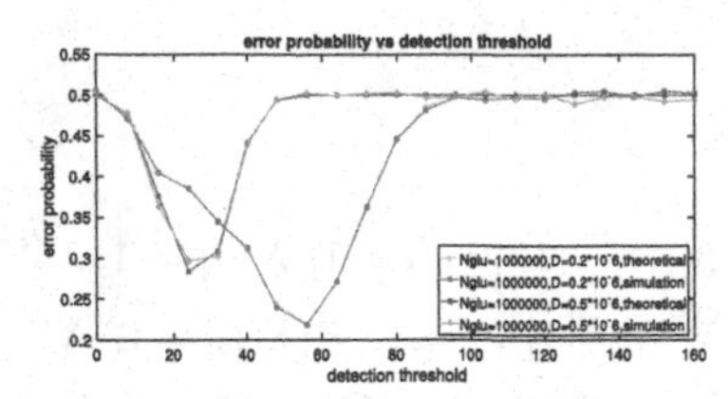

(a) Probability of error plot of NSR assisted link.

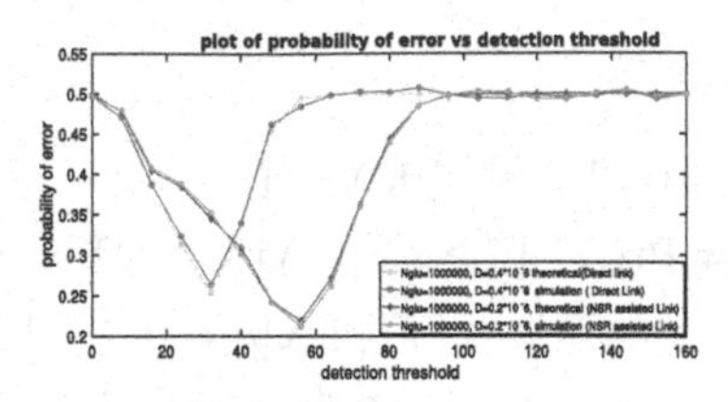

(b) Performance comparison between direct link and NSR assisted link in terms of probability of error

Fig. 5. Probability of error plots for NSR assisted link.

4 Numerical Results

In this section, the Monte-Carlo simulation is used to verify the analytical expressions. The system parameters for the simulation are as $K = 5$, $D_{glu} = 0.1 \times 10^{-9}\text{m}^2/\text{s}$, synaptic cleft width $L = 20\,nm$, effective volume $V_e = 0.5 \times 0.5 \times 0.5\,nm^3$, $N = 200$, and reuptake probability $P_u = 0.1$. All the system parameters remain unchanged throughout the simulations until otherwise stated.

The ROC plot for the NSR-assisted link is demonstrated in Fig. 4 considering a large number of transmitted molecules from the source neuron. The ROC is obtained by plotting the values of the probability of detection as a function of the probability of false alarm. First, it can be seen that the theoretical and analytical results for the probability of detection and false alarm are closely matched. Herein, it is observed that as the count of molecules sent from the pre-synaptic neuron increases, the probability of detection increases, and the false alarm probability decreases. Hence, the area under the ROC plot increases, which shows a performance improvement. It can also be observed from this figure that as the received number of molecules increases at the receiver, the area under the ROC plot increases which indicates better detection of the transmitted symbol. This is because detection probability increases with the increment in the value of received molecules. It can be seen that both the analytical and theoretical results match closely.

Figure 5(a) shows the plot of error probability versus detection threshold for NSR-assisted link. It is observed that the error probability achieves its minimum value at a particular value of the decision threshold which is an optimal threshold. It can also be marked that the probability of error increases with the increments in the value of the diffusion coefficient. Further, when the diffusion coefficient of the molecules increases, the total of molecules arriving at the receiver in the current time slot decreases, and the count of molecules at the destination in the interfering time slot increases. This results in an increase in error probability due to high ISI.

Figure 5(b) shows the comparison of direct and relay-assisted transmission systems. For fair comparison, we assume same distance between NSS and NSD in both the cases. Moreover, we ignore the number of resources taken in relay-assisted system. Herein, error probability as a function of the detection threshold is plotted for both transmission systems. For this comparison, all other parameters are kept unchanged, i.e., $Nglu = 300000$, $P_u = 0.1$, and the separation between pre-synaptic and post-synaptic neurons is 20 nm. It is observed that the probability of error is minimum for the relay-assisted link. Thus, we can conclude that by using NSR, the performance of the system can be improved.

5 Conclusion

In this paper, we considered a cooperative neuro-spike communication system where a neuro-spike relay was placed in the middle of the pre-synaptic and post-synaptic neurons. Pre-synaptic neuron act as a transmitter and post-synaptic neuron act as a receiver. The closed-form expressions for the probability of false alarm, probability of detection, and probability of error were derived for both direct and relay-assisted links. The obtained analytical expressions were also verified using Monte Carlo simulation. In the analysis, we also observed the variation of the ROC plot with a number of glutamate molecules. The variation of error probability with diffusion coefficient is also observed. It was concluded that with an increase in the diffusion coefficient of molecules, the performance of the system decreases. This is due to the increase in ISI and to overcome this issue, we should transmit a large number of molecules. The practical implementation of a cooperative neuro-spike communication system is still a challenging area. Future studies can focus on the treatment of serious health issues like loss of memory power, mood swings, self-neglect, etc.

References

1. Farsad, N., Yilmaz, H.B., Eckford, A., Chae, C.-B., Guo, W.: A comprehensive survey of recent advancements in molecular communication. IEEE Commun. Surveys Tuts. **18**(3), 1887–1919 (2016)
2. Scheiffele, P.: Cell-cell signaling during synapse formation in the CNS. Annu. Rev. Neurosci. **26**(1), 485–508 (2003)
3. Greenberg, E.P.: Bacterial communication: tiny teamwork. Nature **424**(6945), 134–134 (2003)
4. Liu, Q., He, P., Yang, K., Leng, S.: Inter-symbol interference analysis of synaptic channel in molecular communications. In: 2014 IEEE International Conference on Communications (ICC), pp. 4424–4429. IEEE (2014)
5. Balevi, E., Akan, O.B.: A physical channel model for nanoscale neuro-spike communications. IEEE Trans. Commun. **61**(3), 1178–1187 (2013)
6. Malak, D., Akan, O.B.: A communication theoretical analysis of synaptic multiple-access channel in hippocampal-cortical neurons. IEEE Trans. Commun. **61**(6), 2457–2467 (2013)
7. Malak, D., Akan, O.B.: Communication theoretical understanding of intra-body nervous nanonetworks. IEEE Commun. Mag. **52**(4), 129–135 (2014)
8. Ramezani, H., Khan, T., Akan, O.B.: Information theoretical analysis of synaptic communication for nanonetworks. In: IEEE INFOCOM 2018 - IEEE Conference on Computer Communications, pp. 2330–2338 (2018)
9. Emerit, J., Edeas, M., Bricaire, F.: Neurodegenerative diseases and oxidative stress. Biomed. Pharmacother. **58**(1), 39–46 (2004)
10. Hammond, C.: Chapter 1 - neurons. In: Hammond, C. (ed.), Cellular and Molecular Neurophysiology (Fourth Edition), fourth edition ed., pp. 3–23. Boston: Academic Press (2015). https://www.sciencedirect.com/science/article/pii/B9780123970329000017

11. Aghababaiyan, K., Shah-Mansouri, V., Maham, B.: Capacity and error probability analysis of neuro-spike communication exploiting temporal modulation. IEEE Trans. Commun. **68**(4), 2078–2089 (2020)
12. Liu, Q., He, P., Yang, K., Leng, S.: Inter-symbol interference analysis of synaptic channel in molecular communications. In: 2014 IEEE International Conference on Communications (ICC), pp. 4424–4429 (2014)
13. Mesiti, F., Balasingham, I.: Nanomachine-to-neuron communication interfaces for neuronal stimulation at nanoscale. IEEE J. Sel. Areas Commun. **31**(12), 695–704 (2013)
14. Aghababaiyan, K., Shah-Mansouri, V., Maham, B.: Joint optimization of input spike rate and receiver decision threshold to maximize achievable bit rate of neuro-spike communication channel. IEEE Trans. Nanobiosci. **18**(2), 117–127 (2019)
15. Khan, T., Bilgin, B.A., Akan, O.B.: Diffusion-based model for synaptic molecular communication channel. IEEE Trans. Nanobiosci. **16**(4), 299–308 (2017)
16. Fang, Y., Noel, A., Yang, N., Eckford, A.W., Kennedy, R.A.: Convex optimization of distributed cooperative detection in multi-receiver molecular communication. IEEE Trans. Mol. Biol. Multi-Scale Commun. **3**(3), 166–182 (2017)
17. Fang, Y., Noel, A., Yang, N., Eckford, A.W., Kennedy, R.A.: Symbol-by-symbol maximum likelihood detection for cooperative molecular communication. IEEE Trans. Commun. **67**(7), 4885–4899 (2019)
18. Tavakkoli, N., Azmi, P., Mokari, N.: Optimal positioning of relay node in cooperative molecular communication networks. IEEE Trans. Commun. **65**(12), 5293–5304 (2017)
19. Yilmaz, H.B., Heren, A.C., Tugcu, T., Chae, C.-B.: Three-dimensional channel characteristics for molecular communications with an absorbing receiver. IEEE Commun. Lett. **18**(6), 929–932 (2014)
20. Noel, A., Cheung, K.C., Schober, R.: Optimal receiver design for diffusive molecular communication with flow and additive noise. IEEE Trans. Nanobiosci. **13**(3), 350–362 (2014)
21. Ahmadzadeh, A., Noel, A., Schober, R.: Analysis and design of multi-hop diffusion-based molecular communication networks. IEEE Trans. Mol. Biol. Multi-Scale Commun. **1**(2), 144–157 (2015)
22. Noel, A., Deng, Y., Makrakis, D., Hafid, A.: Active versus passive: receiver model transforms for diffusive molecular communication. In: 2016 IEEE Global Communications Conference (GLOBECOM), pp. 1–6 (2016)
23. Noel, A., Cheung, K.C., Schober, R.: Diffusive molecular communication with disruptive flows. In: 2014 IEEE International Conference on Communications (ICC), pp. 3600–3606 (2014)
24. Ankit, Bhatnagar, M.R.: 3-D diffusive-drift molecular channel characterization for active and passive receivers. IEEE Trans. Mol. Biol. Multi-Scale Commun. **4**(2), 107–117 (2018)
25. Lesch, K.P., Bengel, D.: Neurotransmitter reuptake mechanisms. CNS Drugs **4**(4), 302–322 (1995)
26. Tavakkoli, N., Azmi, P., Mokari, N.: Performance evaluation and optimal detection of relay-assisted diffusion-based molecular communication with drift. IEEE Trans. Nanobiosci. **16**(1), 34–42 (2017)
27. Einolghozati, A., Sardari, M., Fekri, F.: Relaying in diffusion-based molecular communication. In: 2013 IEEE International Symposium on Information Theory, pp. 1844–1848 (2013)

28. Srinivas, K.V., Eckford, A.W., Adve, R.S.: Molecular communication in fluid media: the additive inverse gaussian noise channel. IEEE Trans. Inf. Theory **58**(7), 4678–4692 (2012)
29. Murin, Y., Farsad, N., Chowdhury, M., Goldsmith, A.: Communication over diffusion-based molecular timing channels. In: 2016 IEEE Global Communications Conference (GLOBECOM), pp. 1–6 (2016)
30. Johns, P.: Chapter 7 - synaptic transmission. In: Clinical Neuroscience, Johns, P. (ed.) Churchill Livingstone, pp. 81–89 (2014). https://www.sciencedirect.com/science/article/pii/B9780443103216000072
31. Levite, M.: Neurotransmitters activate T-cells and elicit crucial functions via neurotransmitter receptors. Curr. Opin. Pharmacol. **8**(4), 460–471 (2008)
32. Manwani, A.: Information-theoretic analysis of neuronal communication. California Institute of Technology (2000)
33. Soong, T.T.: Fundamentals of Probability and Statistics for Engineers. Wiley (2004)
34. Chouhan, L., Sharma, P.K., Varshney, N.: Optimal transmitted molecules and decision threshold for drift-induced diffusive molecular channel with mobile nanomachines. IEEE Trans. on NanoBiosci. **18**(4), 651–660 (2019)

Wearable Vibration Device to Assist with Ambulation for the Visually Impaired

Douglas E. Dow(✉), Jared J. Robbins, Kelley C. Roberts, Seth G. Bannish, and Bailey J. Cote

Program in Electrical Engineering, Wentworth Institute of Technology, Boston, Massachusetts 02115, USA
dowd@wit.edu

Abstract. People with visual impairment have increased difficulty in performing activities of daily living, such as walking without bumping into obstacles. Many assistive technologies are used to help with ambulation as one walks forward, such as a white walking cane or a service dog. These have proven to be of tremendous help, but the cane may miss suspended objects not touching the ground, and service dogs are not available to all who need them. Further assistive technologies continue to be developed and tested. In nature, those without visual acuity tend to obtain much information from their environment through the other senses, such as hearing or tactile touch. This study is exploring the mapping of obstacle detection to tactile vibration motors on the skin. Ultrasonic sensors were used to detect obstacles in the forward direction where the user would be walking, and calculate the distance. The distance was mapped to a vibration pattern, with the pattern being more intense for closer obstacles. A prototype was developed and had several tests run. Obstacle detection and distance were useful up to 3 m. The functional field of view was 10° to 30° from centerline, but became more narrow as the distance increased and for harder to detect obstacles. The distance was mapped to 3 different vibration patterns, and human subjects were able to distinguish the patterns in a consistent manner. The prototype shows promise, but more testing and development would be required toward widespread application.

Keywords: ultrasound · proximity sensor · arduino · microcontroller · tactile · haptic · vibration · wearable electronics

1 Introduction

The World Health Organization (WHO) reported in 2013 that an estimated 285 million people live with some form of visual impairment: 39 million were reported to be blind and 246 million had low vision [1]. Visual impairments contribute to challenges in activities of daily living (ADL), such as difficulty with navigation and mobilization. This difficulty increases risk of injury due to collision with undetected obstacles in their environment. For example, 88% of blind or visually impaired individuals reported at least one injury as a result of their condition, and 23% report the need for serious medical attention as a result of the injury [2].

Y. Chen et al. (Eds.): BICT 2023, LNICST 512, pp. 224–236, 2023.
https://doi.org/10.1007/978-3-031-43135-7_22

Standard techniques that are currently used for assisting visually impaired include the white cane and seeing-eye service dogs. Many individuals use the standard white cane to manually scan their surroundings in order to detect and avoid hazardous obstacles, especially low-lying obstacles near the ground. While the standard white cane is a low cost and widely used device, issues remain, such as the cane getting stuck in cracks or potholes and the inability to assist with the detection of hanging obstacles. Some people utilize a guide dog that was specifically trained to assist blind or visually impaired people. A guide dog may not be accessible to some due to the high cost and low availability [3].

Assistive technologies have been developed to provide support for blind or visually impaired individuals with certain tasks [3]. Assistive technologies for the visually impaired can be separated into three categories: vision enhancement (when a camera input is processed and then the results are visually displayed), vision replacement (which includes displaying information directly to the visual cortex of the brain or through an ocular nerve), or vision substitution (which constitutes non-visual display, such as tactile vibration or auditory). Vision enhancement has two disadvantages of not being able to assist individuals who are blind, and may distract the remaining visual capacity for those who are low vision. The vision replacement may be a robust solution in the future, but needs much more development. Visual substitution can build on techniques to map informational signals onto tactile vibrations that can be felt on the skin.

Mapping onto tactile vibrations has been reported for music [4], touch screens [5], virtual reality [6], and spatial tasks [7]. Several assistive technologies and systems have been reported. Audio Bracelet for Blind Interaction is an advanced vision substitution technology consisting of an auditory bracelet that uses auditory modality to convey spatial information of the user's surroundings [8]. Intelligent white canes have been developed that use color sensors, vibrations, and audio signals to alert the user during ambulation [9]. Although these are promising technologies, none have yet been able to fully support the needs of a blind or visually impaired individuals. Many users have found the auditory alerts to be distracting in a busy environment, and a bracelet worn on the wrist may not accurately pick up objects due to the constant mobility of the arms when walking.

There is still a need for the development of assistive technology that would assist the blind and visually impaired with ambulation, especially if it is more affordable and simpler to use [10]. Such a device should inform the user of surrounding obstacles that could be hazardous. A device with these capabilities might allow blind and visually impaired individuals to walk through their surroundings with a higher level of confidence, comfort, and safety. The purpose of this project was to develop and test prototype modules for a wearable, hands-free vision assistive technology system to help with navigation for blind or visually impaired individuals during ambulation.

2 Materials and Methods

2.1 Overview

The full system has three functional modules, each consisting of an ultrasonic sensor, microcontroller (MCU) and vibration motor. Figure 1 shows the block diagram of one functional module.

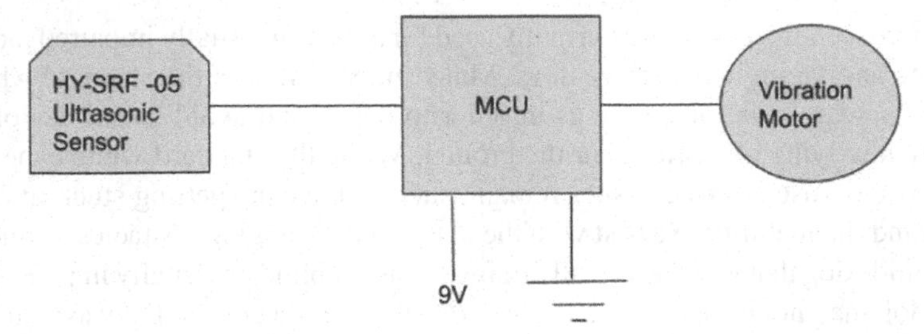

Fig. 1. Block diagram of the primary functional unit of the prototype. The ultrasonic sensor maps a spatial area, the microcontroller (MCU) converts the spatial information into a distance measurement, and then sends control signals to a vibration motor that provides haptic feedback to the user by vibrating on skin.

The primary functional unit for the prototype system had three modules as shown in Fig. 1: an ultrasonic sensor for obstacle detection, a MCU to map spatial information to vibration patterns, and a tactile vibrator. The full system of the prototype had three of these functional units of sensor-MCU-vibrator. Ultrasonic sensors were used for obstacle detection. Each ultrasonic sensor was connected to a microcontroller and the microcontroller drove a tactile vibration motor. The physical layout of the prototype is shown in Fig. 2. A control algorithm was developed for the MCU, within which the MCU received the time measurement of the reflected signal recorded by the ultrasonic sensor. The time measurement was converted to a distance value. The algorithm then determined a control pattern for output voltage pulses to be sent to a vibration motor, where the pattern of pulses was related to the distance. The system operated continuously while the device was powered on. A flow chart for the system is shown in Fig. 3.

2.2 Electrical

The ultrasonic sensor chosen for the prototype was the HY-SRF-05 Ultrasonic Sensor (CYD, Hong Kong, China). The sensor transmitted a 40 kHz ultrasonic pulse through the air, and any obstacles in its path would reflect the signal back to the receiver. The time was recorded from the echo of the wave as it reflected from an obstacle back to the receiver. The ultrasonic sensor module was chosen to use in the prototype due to low cost, availability and reliability. The full system of the prototype contained three functional units. Each functional unit consisted of an ultrasonic sensor, microcontroller, and vibration motor. The functional units were positioned so that their ultrasonic sensors pointed in the right, center, and left directions respectively to span the user's field of vision in the forward direction when walking forward.

The MCU module for the prototype was an Arduino Nano (Arduino, Monza, Italy, https://arduino.cc), which has an ATmega 328 8-bit microcontroller by Atmel (Microchip Technology, Westborough, MA). The Arduino integrated development environment (IDE) was used for code development and testing. The vibration motor selected for the prototype was the Tatoko DC Coreless Motor (Tatoko) due to its suitable size and vibration pattern. Each Arduino Nano was connected to an ultrasonic sensor and a Tatoko vibration motor (Fig. 1). The vibration motors were rated for 1.5–3 V and vibrated at

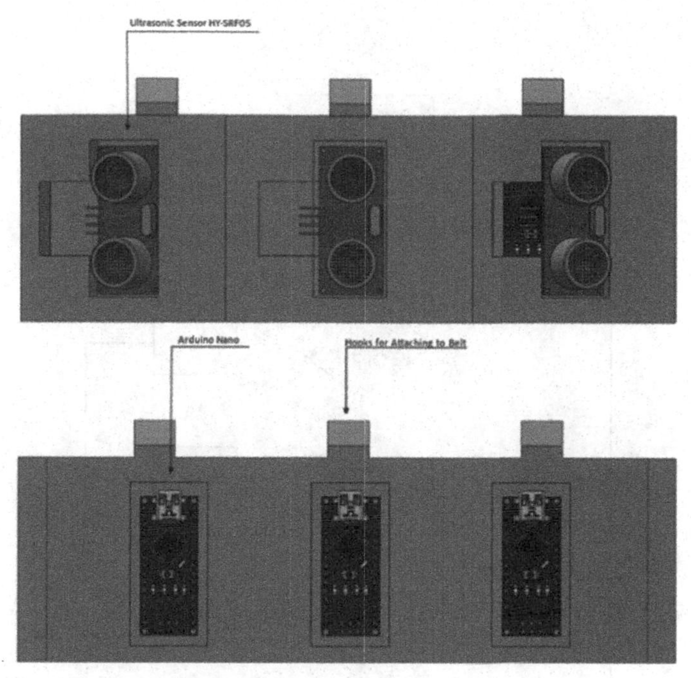

Fig. 2. The top diagram shows the front view of the model for the enclosure with the three ultrasonic sensors facing forward in the direction the user would be walking. The bottom diagram shows the inside view of the model with the MCU placed behind the corresponding ultrasonic sensor.

8000–16000 RPM. The Arduino output a voltage pulse pattern to the vibration motor which was related to the distance of an obstacle based on the readings from the ultrasonic sensor. The Arduino Nano was powered by a 9V battery. Power to the module could be turned on and off via a switch located on the battery enclosure.

2.3 Mechanical

Using Solidworks (Dassault Systems, Waltham, MA, USA), an enclosure was designed to mount the three functional units of an ultrasonic sensor, microcontroller and haptic vibrator. The ultrasonic sensor was mounted to the front of the enclosure which would face forward in the direction the user would be walking. A small cutout was provided to route the wires through the enclosure to connect the sensor with the MCU. The MCU was mounted inside the enclosure adjacent to its corresponding sensor. 3D models of the physical layout can be seen in Fig. 2.

The enclosure was mounted via hooks to an adjustable belt to be worn around the waist. The vibration motors were attached to the inside of the belt aligned with the corresponding ultrasonic sensor. Attaching the vibration motor inside the belt allowed for the user to feel the haptic vibrations on their skin.

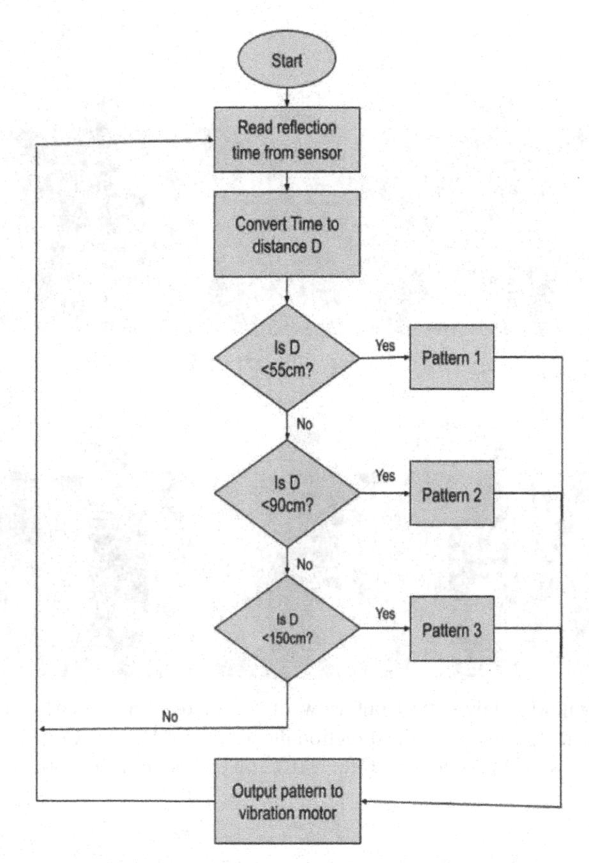

Fig. 3. Example of a flow chart of the algorithm in the MCU to read ultrasonic sensor for distance to the obstacle, and map that distance to a vibration pattern for tactile feedback.

2.4 Software

In order for the system to be able to process the information received from each ultrasonic sensor and convert into an output voltage pattern for the corresponding vibration motor, an algorithm was written and programmed in the Arduino version of C. A part of the algorithm is shown in Fig. 3 as a flow chart. The software read the voltage from the ultrasonic sensor on the analog to digital converter (ADC). The voltage corresponded to the time of the ultrasonic transmission and reflection. The read value for the time duration was converted to a distance. This distance would be from the sensor to the object that reflected the ultrasonic wave. At any moment of time, a sensor would only return one value for the primary object that reflected back the ultrasonic signal. So, the distance would be calculated for this object. The software mapped the distance to a pulse pattern for the vibration motor. Three patterns were generated with pulse width modulated (PWM) output voltage. Pattern 1 had intervals of 0.5 s, pattern 2 had intervals of 1.0 s, and pattern 3 had intervals of 3.0 s (Fig. 4). The voltage pulse pattern was then output to the vibration motor.

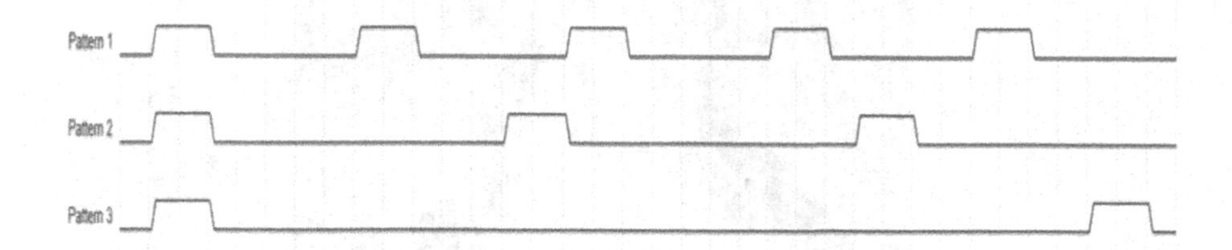

Fig. 4. Waveform to control the vibration motors. The positive pulse induces a burst of vibrations that last 0.2 s. The interval between pulses is different for the three patterns. The interval for Pattern 1 was 0.5 s, Pattern 2 was 1.0 s, and Pattern 3 was 3.0 s.

2.5 Prototype Development

A prototype was developed with three functional units of the ultrasonic sensor, MCU, and vibration motor placed in the enclosure and belt. The belt would go around the waist as they walked forward. The electronics were powered by batteries that were placed into the system with an on/off switch.

The ultrasonic sensors were mounted with screws to face forward. The ultrasonic sensors were oriented vertically because the vertical position allowed for better obstacle detection by seeming to limit interference compared with horizontal placement. The wires from the ultrasonic sensors were routed through openings in the enclosure and connected to the MCU. The MCU were mounted inside the enclosure using Velcro straps. The power cabling between the battery pack, MCU and vibration motors were routed through holes in the top cover of the enclosure. The enclosure was attached to the belt using hooks, which helped to maintain the enclosure and ultrasonic sensors in a more straight and forward direction.

The belt strap could be adjusted to be able to fit many different body types. A photograph of the final prototype being worn can be seen in Fig. 5 and a top view of the prototype can be seen in Fig. 6.

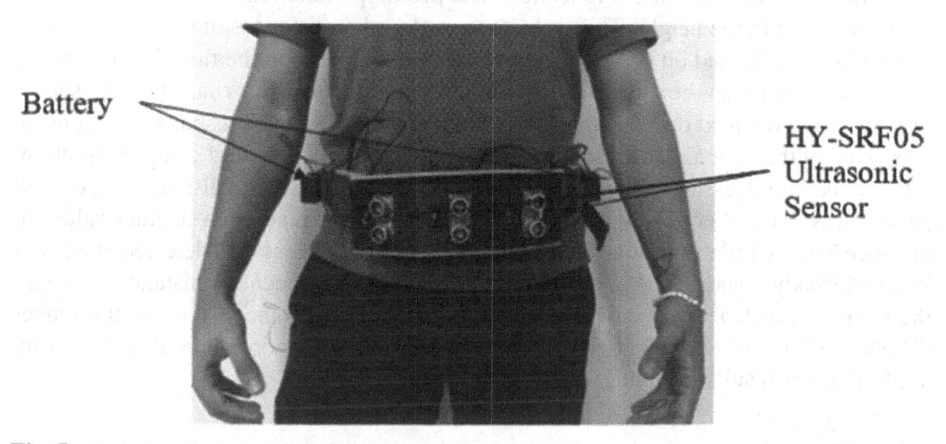

Fig. 5. Prototype being worn. This demonstrates the ultrasonic sensors being positioned to respond to obstacles in the user's forward field of view.

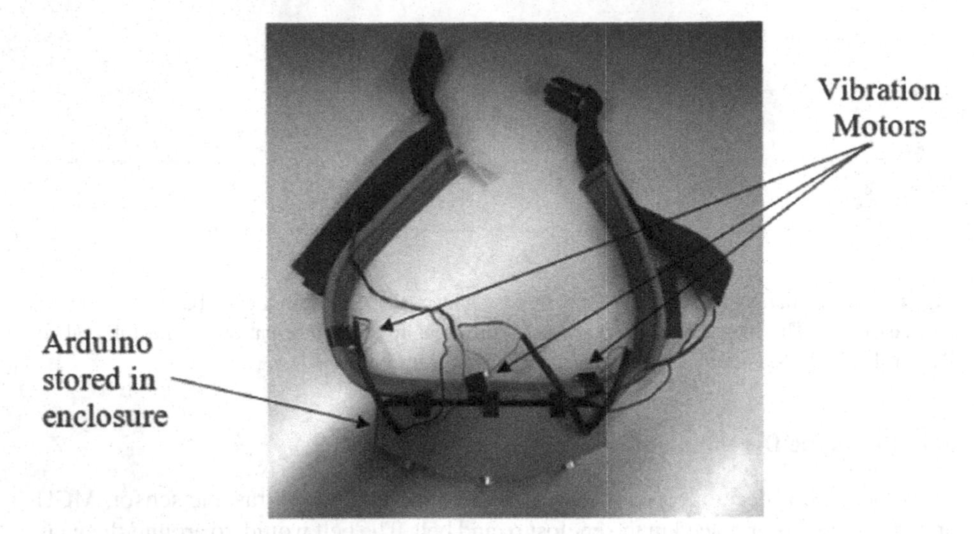

Fig. 6. Top view of prototype showing enclosure and vibration motors attached to the belt. An enclosure where electronics lay being covered is also shown.

3 Testing and Results

Several tests were conducted to assess the function of one functional unit, consisting of the ultrasonic sensor, MCU and vibration motor.

3.1 Object Detection Testing for Variability

A test was done to assess the performance of the sensing module for a static object. The ultrasonic sensor was placed so that the transmitter and receiver were in line with where the object was placed at different distances from the sensor. The object was a 1.8 m high metal pole with a 3.8 cm diameter. The sensor output a voltage value that corresponded to the time of transmission and reflection. The prototype functional unit was set up on a stand at the mid pole height. The obstacle was placed at distances of 0.6 m, 1.5 m and 3 m to the pole. Based on pre-measured tape marks marked on the floor in front of the pole. The sensor was set at each distance marker and the test was run. Then the sensor was moved to the next distance marker and test repeated. At each distance a screenshot was taken of the plot, and the distance values were reported as central value ± variation. After some initial testing of the ultrasonic sensor and conversion to distance, the overall pattern was that closer objects would be well detected and result in a steadier value for distance, having little variation, but farther objects would be less well detected and result in a less steady distance value, having more variation. At placement distance of 0.6 m, the measured distance was 0.61 m with no variability in distance value. At the longer distances of 1.5 m and 3.0 m, the measured distance had more variability of 0.3 m. Table 1 shows results from this testing of distances.

Table 1. Distance readings for placement of the prototype functional unit at specified distances to a metal pole (1.8 m high, 3.8 cm diameter). Distance reported as central value ± variation.

Placed Distance (m)	Measured Distance (m)
0.6	0.61 ± 0.00
1.5	1.50 ± 0.03
3.0	3.05 ± 0.03

3.2 Obstacle Detection for Range and Field of View

Another set of trials was performed to observe how well the ultrasonic sensor detects obstacles at the boundary of its field of view. The primary objective of the sensor system was to detect obstacles in front of a walking user. The ultrasonic sensors have a field of view where it can detect obstacles. If an obstacle was outside this field of view, it may not be detected. This boundary was important to assess how wide the coverage was in front of a walking user. Obstacles of different shape and material were used to determine whether those factors affect detection. Two obstacles used for this test as follows: 1) a 1.8 m high metal coat pole with a 3.8 cm diameter, and 2) a plastic, 1.2 m tall cylindrical fan enclosed in a plastic case with a 19 cm diameter. The prototype functional unit was placed about 1.2 m above the floor to simulate the height as if worn by a user. The obstacles were placed in front of the sensor at a particular distance and angle from the center line. A protractor was used to measure the angles from the center line that was straight out from the sensor. Angles were marked with tape on the floor at 0° (straight out from the sensor), 10°, 20° and 30°. The obstacles were moved to these points, and the distance readings were made.

Based on the resulting plots of distance over time, a classification was made for each trial of "Clear" or "Not Clear". The classification of Clear indicated the object was detected on the plot, with a persistent center line and modest variation. The classification of Not Clear indicated the object was not detected on the plot, lacking a persistent center line and having much variation. Figure 7 shows the classification of the obstacles at different distances and angles.

Figure 7 is a visualization of the results. The boxes display the distance of the object from the sensor and the angle, with 0° being straight ahead in the direction a user would walk. The color of each box indicates which objects (wider fan or narrower pole) were clearly detected. At 0.6 m, both object types were clearly detected from 0° to 20°, but at 30° only the wider fan was clearly detected. Maybe the wider object provided more surface area to reflect the ultrasonic signal. At 1.5 m both object types were clearly detected, but only for 0° to 10°. In contrast, for 20° to 30° only the wider fan was detected. At 2.4 m only the wider fan was clearly detected, and only for the angles of 0° to 20°. These results are summarized in Table 2.

3.3 Mapping of Distance and Vibrations

The MCU was programmed to detect the distance to an object, and map that distance to one of four vibration patterns as shown in Table 3. The duration of a single burst of

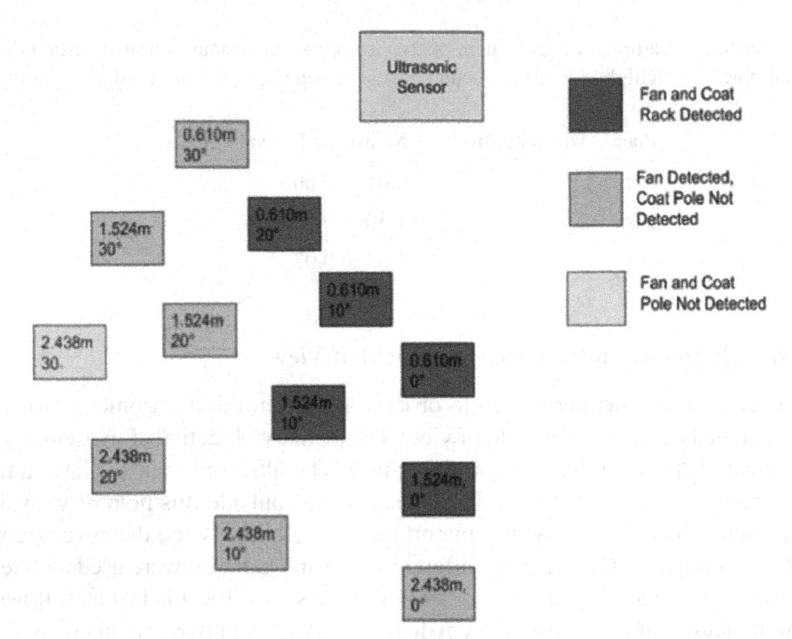

Fig. 7. Location of Obstacle Detection, the green box locations mark places where both the pole (narrow diameter) and fan (wide diameter) were consistently detected, the yellow box locations mark places where the pole was not consistently detected, but the fan still was, and the red box marks a spot where neither the pole nor the fan was consistently detected. (Color figure online)

Table 2. Results of object parameters and whether clearly detected by the prototype functional module utilizing the ultrasonic sensor. The objects were either Wide (19 m, diameter plastic cylindrical fan) or Narrow (1.8 cm, diameter metal pole).

Distance (m)	Angle (degree)	Clearly Detected
0.6	0°, 10°, 20° 30°	Wide or Narrow Wide
1.5	0°, 10° 20°, 30°	Wide or Narrow Wide
2.4	0°, 10°, 20° 30°	Wide Neither

vibration was 0.2 s. Then there would be a delay before the next burst of vibration would begin. This delay was the burst interval, which had values of 0.5 s, 1.0 s, and 3.0 s. Figure 4 shows the waveform of control signal to the vibration motor. When the voltage was high, the motor would vibrate.

The software on the MCU mapped the distance to a pattern of vibration that the user would feel. A distance of 3.0 m or more would not result in any pattern, being too far away to warn the user yet. A distance of less then 3.0 m would result in a vibration pattern. The smaller the distance the more frequent the vibration bursts. Each burst was

0.2 s, and the interval between bursts would vary from 3.0 s for farther objects and 0.5 s for closer objects. Table 3 shows the mapping of distance to vibration pattern.

Table 3. Mapping by the software running on the MCU for distance to vibration pattern. Each burst lasted 0.2 s. Distances beyond 3.0 m resulted in no vibration pattern.

Distance (d) to Detected Object (m)	Interval between Bursts (s)	Description
$d < 0.25$	0.5	Intense pattern (danger)
$0.25 \leq d < 0.50$	1.0	Moderate pattern (adjust path)
$0.50 \leq d < 3.0$	3.0	Warning (object ahead)
$d > 3.0$	none	Fine (continue walking forward)

A trial was set up to observe this mapping of distance to vibration pattern. The functional module of the prototype having the sensor, MCU and vibration motor was placed on a stationary stand. An obstacle (metal pole, 1.8 cm diameter) to be detected was placed straight in front of the sensor at the specified distance. The distance between the sensor and object progressively increased from 26 cm, and each increase in distance was 13 cm until about 3 m. Following obstacle placement, two observations were made and recorded: which vibration pattern was being performed by the vibration motor, and what voltage did the control signal have that was sent to the motor. Figure 8 shows the result of distance to vibration pattern sent to the vibration motors. The vibration pattern was indicated in the plot as the interval between the bursts of vibrations.

The observed generated vibration patterns (Fig. 8) closely followed what the software on the MCU was intending to map for each tested distance (Table 3). Exceptions were observed at the longer distance near 3.0 m. The measurement at 2.9 m did not have the correct interval of 3.0 s, instead no value as plotted as a 0. Possibly, the prototype failed to detect the obstacle at this longer distance and so did not map to an interval value. Moreover, when a vibration pattern was generated, the voltage value for each of the pulses to control the vibration motors had a stable value, and did not decline through the tested range.

3.4 Discrimination of Vibration Pattern

Once an object was detected by one of the ultrasonic modules, the corresponding haptic motor vibrated. The pattern of the vibration indicated the distance to the object. There were three different patterns to indicate a close, medium or far distance. Testing was done to ensure that the user could differentiate the three states of vibration. If the user could not tell the difference between the vibration pattern, then the prototype would be of limited use. The purpose of the vibration pattern was to communicate to the user that there is an obstacle ahead and about how far away it is. To test this, 10 volunteer subjects

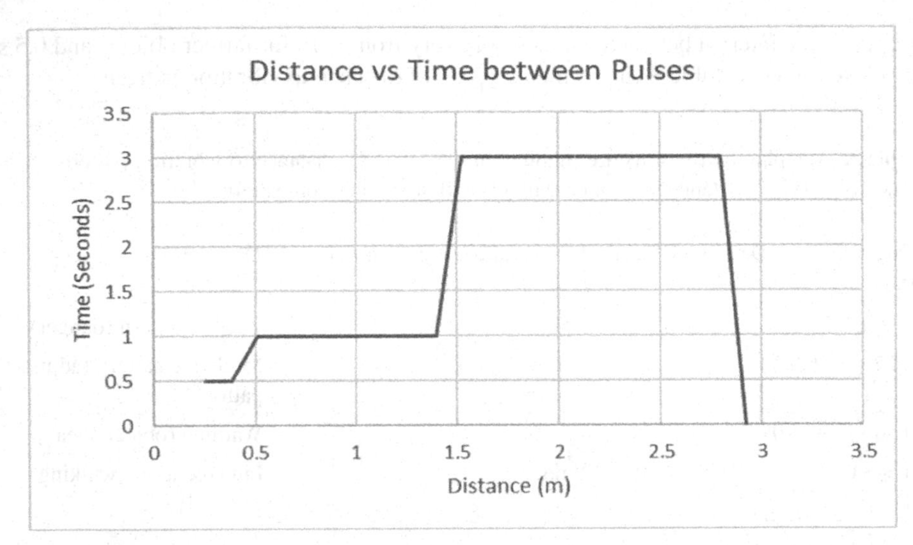

Fig. 8. Mapping of distance from the obstacle to the vibration pattern as indicated by the interval between the bursts of vibrations. The x-axis shows the distance between obstacle and the prototype functional module having the ultrasonic sensor. The y-axis shows the interval in seconds between the bursts of vibration. The interval was smaller as the distance decreased to indicate a higher level of urgency for the user to adjust their walking pattern to not bump into the obstacle. The voltage of the control signal sent to the vibration motors was consistent and did not decline over the tested intervals.

were recruited to assess whether they could distinguish between any two consecutive vibration patterns.

The subjects in this study were able-bodied, with no known sensory perception deficits. The ages of these volunteers ranged from 19 to 55, with 3 identifying as female and 7 as male. These pairs of vibration patterns were ordered randomly between the volunteers. The patterns of vibrations were labeled as Pattern 1, Pattern 2, and Pattern 3 as shown in Fig. 4. For the test, the prototype functional module having the vibration motors was placed adjacent to the abdomen over the volunteers' clothing. For each trial, the volunteers were alerted that a two-vibration pattern sequence was about to occur. Then, after both patterns were completed, the volunteer was asked whether the two patterns were the same or different. Each volunteer underwent 3 sets, with 6 trials per set. Each trial did the following pattern pairs in random order: 1–1, 1–2, 1–3, 1–1, 2–2, and 3–3. Thus, each subject had $3 \times 6 = 18$ trials, with the total numbers of trials for all 10 volunteers was $18 \times 10 = 180$ trials.

The subjects were correct for most of the trials on whether the two consecutive patterns were the same or different. Only 3 subjects had an error, and each had exactly 1 error out of their 18 trials. So, there were a total of 3 errors for 180 trials. Thus, the percent error was 1.7%. Overall, the subjects were able to distinguish between the 3 patterns generated by the vibration motors.

4 Discussion and Future Directions

The prototype of the system showed promise as a potential assistive technology for visual impairment being able to detect obstacles and provide haptic feedback to the user. Obstacles up to 1.5 m meter in front of the ultrasonic sensor module were consistently detected. The three vibration patterns could be distinguished by the users.

Considering the results for the obstacle detection for the application of a visually impaired user walking forward at a moderate pace, the obstacle detection would be helpful, but improvements would be desired. The more challenging narrow object (1.8 cm diameter metal pole) was clearly detected at 0.6 m and 1.5 m, but not consistently at 2.4 m. A warning range of 1.5 m would allow the user to stop or change direction, but a warning at an even greater distance (2.4 m) would allow more time to plan a more natural adjustment in walking speed or direction. Further testing would help clarify how consistent and potentially helpful the prototype system would be to a visually impaired person while walking.

The next step of development would be a system test of the whole prototype system being worn during ambulation through obstacles. Additionally, improving the vibration patterns and placement to provide improved feedback to the user. This would include adding more patterns to differentiate more obstacle distances.

References

1. Fernandes, H., Costa, P., Filipe, V., Paredes, H., Barroso, J.: A review of assistive spatial orientation and navigation technologies for the visually impaired. Univ. Access Inf. Soc. **18**(1), 155–168 (2019)
2. Buchs, G., Simon, N., Maidenbaum, S., Amedi, A.: Waist-up protection for blind individuals using the EyeCane as a primary and secondary mobility aid. Restor. Neurol. Neurosci. **35**(2), 225–235 (2017)
3. Froneman, T., van den Heever, D., Dellimore, K.: Development of a wearable support system to aid the visually impaired in independent mobilization and navigation. In: Presented at the 2017 39th Annual International Conference of the IEEE Engineering in Medicine and Biology Society (EMBC). IEEE, pp 783–786 (2017)
4. DeGuglielmo, N., Lobo, C., Moriarty, E.J., Ma, G., Dow, D.E.: Haptic vibrations for hearing impaired to experience aspects of live music. In: Nakano, T. (eds.) Bio-Inspired Information and Communications Technologies. BICT 2021. LNICST, vol. 403, pp. 71–86 . Springer, Cham (2021). https://doi.org/10.1007/978-3-030-92163-7_7
5. Poppinga, B., Magnusson, C., Pielot, M., Rassmus-Gröhn, K.: TouchOver map: audio-tactile exploration of interactive maps. In: Presented at the Proceedings of the 13th International Conference on Human Computer Interaction with Mobile Devices and Services, pp. 545–550 (2011)
6. Muender, T., Bonfert, M., Reinschluessel, A.V., Malaka, R., Döring, T.: Haptic fidelity framework: defining the factors of realistic haptic feedback for virtual reality. In: Presented at the CHI Conference on Human Factors in Computing Systems, pp. 1–17 (2022)
7. Tennison, J.L., Uesbeck, P.M., Giudice, N.A., Stefik, A., Smith, D.W., Gorlewicz, J.L.: Establishing vibration-based tactile line profiles for use in multimodal graphics. ACM Trans. Appl. Percept. (TAP). **17**(2), 1–14 (2020)

8. Porquis, L.B., Finocchietti, S., Zini, G., Cappagli, G., Gori, M., Baud-Bovy, G.: ABBI: a wearable device for improving spatial cognition in visually-impaired children. In: Presented at the 2017 IEEE Biomedical Circuits and Systems Conference (BioCAS), pp 1–4. IEEE (2017)
9. Shiizu, Y., Hirahara, Y., Yanashima, K., Magatani, K.: The development of a white cane which navigates the visually impaired. In: Presented at the 2007 29th Annual International Conference of the IEEE Engineering in Medicine and Biology Society, pp 5005–5008. IEEE (2007)
10. Elmannai, W., Elleithy, K.: Sensor-based assistive devices for visually-impaired people: current status, challenges, and future directions. Sensors **17**(3), 565 (2017)

Development of Capacitive Sensors to Detect and Quantify Fluids in the Adult Diaper

Muhammad Tanweer[1](✉), Raimo Sepponen[2], Ihsan Oguz Tanzer[1], and Kari A. Halonen[1]

[1] Department of Electronics and Nanoengineering, Aalto University, Espoo, Finland
{muhammad.tanweer,oguz.tanzer,kari.halonen}@aalto.fi
[2] Department of Automation and Electrical Engineering, Aalto University, Espoo, Finland
raimo.sepponen@aalto.fi

Abstract. In recent years the rapid technological development in printable electronics technology has made it possible to develop sensors and circuits for wearable medical devices using flexible substrate. In this paper a novel implementation of flexible capacitive sensors to detect and quantify human excreta inside adult diaper is explored. The flexible capacitive sensors are implemented in co-planar geometry by using conductive strips, paint and conductive fabric which are easily available in the market. The developed sensors are used to detect and quantify the fluid and concentration of electrolytes in-vitro (adult diaper and glass jar). An impedance analyzer is used to perform measurements from the sensor and collect necessary specifications in developing suitable printable sensors for a diaper.

Keywords: Wearable Medical Devices · Flexible Capacitive Sensors · Smart Diaper · Printed Sensors · Printed Electronics · Human Fluid Detection & Quantification

1 Introduction

There has been a rapid rise in the elderly population during few decades especially in Europe and Asia [1]. The percentage of elderly people in Europe in the total population is one of the highest, and recorded as the highest in Asia among global population [2]. The recent pandemic disease (COVID- 19) adversely impacted the elderly health worldwide [3]. Recent advances has made it possible to monitor and record physiological parameters using wearable medical devices [4]. Healthcare providers are now adapting to advanced and smart medical solutions provided by the wearable medical devices featured with the application of internet of things (IoT) to reduce the human resource need in medical care. A recent market research report has estimated the rise in global wearable medical device market size to grow from USD 27.2 billion in 2022 to USD 196.6 billion in 2030 [5].

The use of smart adult diapers elderly-care centres is expected to increase because the urine provides enormous biological data which is very helpful in clinical diagnosis

Y. Chen et al. (Eds.): BICT 2023, LNICST 512, pp. 237–245, 2023.
https://doi.org/10.1007/978-3-031-43135-7_23

of various diseases which is otherwise not feasible with the use of conventional diapers [6]. Information about urination frequency together with the urine quantity can be predictor of not only urological diseases like bladder inflammation and kidney function but also various other diseases such as prostatic hyperplasia, urinary track infection (UTI), cognitive deficit, arterial hypertension, diabetes mellitus, depression and many others when combined with other parameters [7].

Recently various fluid detection studies are performed by deploying capacitive sensors using conventional techniques [8, 9]. The advancement of printed electronics technology has brought the possibility to have low cost electronics production on flexible substrates using various biodegradable and environmentally friendly materials to enhance sustainability [10]. In this study, the capacitive sensors in coplanar geometry are developed on a flexible substrate by using printable and lowcost material for wearable usage. The developed sensors are used for detection of liquid and quantification of the volume of liquid inside adult diapers. In Sect. 2 the materials and methods used for the development of flexible sensors are discussed. Section 3 elaborates the measurement results recorded using developed sensors in different scenarios. Finally the Sect. 4 concludes the research work and discusses the possible applications for future developments.

2 Materials and Methods

The materials used in this research work are commercially available. Sensors are developed in Aalto University, Department of Electrical Engineering. Following subsections elaborate the material and methods in detail.

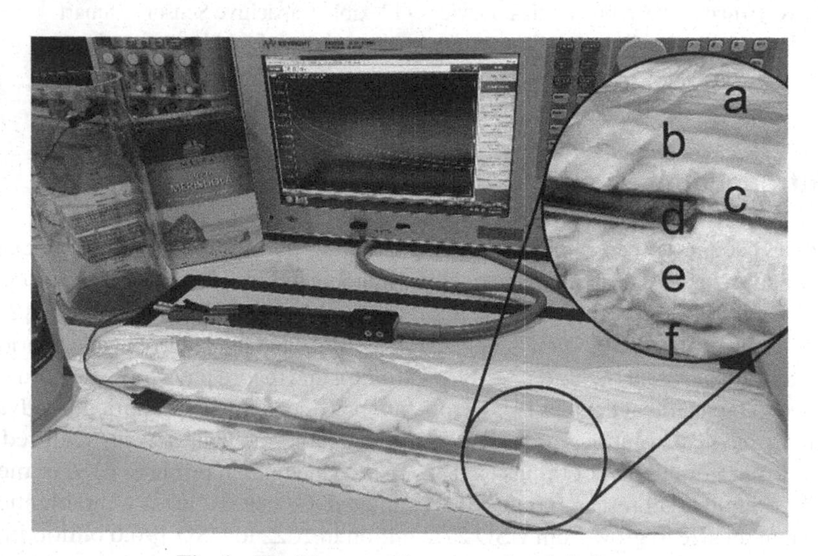

Fig. 1. Flexible capacitive sensor in adult diaper.

2.1 Capacitive Sensors

The capacitance of a conventional parallel plate capacitor is directly proportion to the geometry area (A) of conductors and permittivity ($\varepsilon_r\ \varepsilon_0$) of medium (dielectric) between the plates. The distance between conductor plates is inversely proportional to the capacitance as depicted in Eq. (1). The permittivity varies if there are different material used as dielectric for same geometry of capacitor.

$$C = \varepsilon_r \varepsilon_0 \frac{A}{d} \tag{1}$$

In this study, for the development of capacitive sensors, the conductive strips are rather deployed in coplanar geometry instead of parallel geometry. The capacitance model of sensor with two coplanar flat conductor is given in Eq. (2). Where as Fig. 2(b) presents the geometry of coplanar capacitive sensor where two conductive strips of equal width W and length L are separated by distance S on same plane.

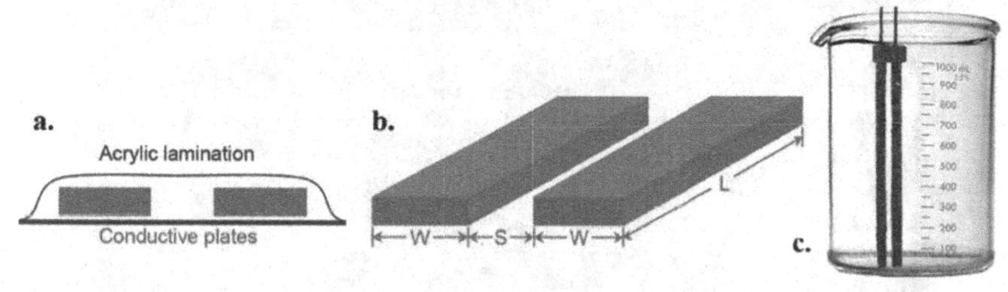

Fig. 2. a) Cross-section of acrylic laminated sensor. b) Geometry of coplanar capacitive sensor with two flat conductive strips. c) Sensor deployed in jar.

The thickness of the conductors is very small and assumed zero here. The flat conductor are assumed in a homogeneous medium with permittivity ($\varepsilon = \varepsilon_r\ \varepsilon_0$) and permeability $\mu = \mu_0$. Where μ_0 is permeability of free space, ε_0 is the permittivity of air and v0 is speed of light [11].

$$C = \frac{\varepsilon_r L \ln\left(-\frac{2}{\sqrt[4]{1-\frac{S^2}{(S+2W)^2}}-1}\left(\sqrt[4]{1-\frac{S^2}{(S+2W)^2}}+1\right)\right)}{377\pi v_0} \tag{2}$$

where $v_0 = \frac{1}{\sqrt{\mu\varepsilon_0}}$.

The coplanar capacitive sensor is developed by using two equal flat strips of various conductive materials having width W = 10 mm and length L = 200 mm. Both strips are separated with a distance of S = 3 mm. Figure 2(a) depicts the cross section of developed flexible coplanar capacitive sensor. These strips are deployed on a flexible substrate and laminated with a thin transparent acrylic sheet to protect the conductors from corrosive chemical reactions while interacting with fluids. The permittivity of acrylic sheet ranges between 2 to 5 which adds it's effect on the capacitance of coplanar sensor. In free space, the capacitance of the sensors for the given materials with acrylic lamination sheet is calculated using Eq. (2) to be between 7.7 pF and 19.3 pF. When the capacitor is immersed in water, it is estimated that the capacitance of the sensor is supposed to rise about 40 fold assuming the permittivity of water to be 80.

In this study different conductive materials are used to develop a flexible capacitive sensor in coplanar geometry as shown in Fig. 3 in order to compare the capacitive behaviour for detection and quantification of liquid in an adult diaper.

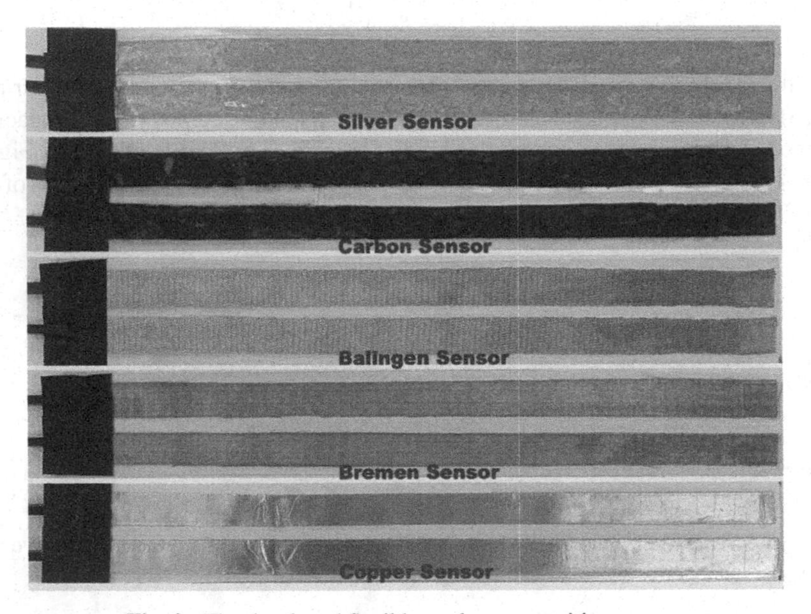

Fig. 3. The developed flexible coplanar capacitive sensors.

Conductive Foil Tape: Copper foil with conductive adhesive (3M, StPaul, Minnesota, USA) having electrical resistance of 0.005 Ω / □is used to develop the flexible capacitive sensor. Named as copper sensor in this article.

Conductive Paint: Off the shelf electrically conductive paints are used on unwoven cotton fabric to contain develop the flexible capacitive sensor.

– Silver based conductive paint (Kemo-Electronic GmbH) having electrical resistance of 0.02 Ω- 0.1 Ω / □is soaked in the fabric to develop the sensor. Named as silver sensor in this article.
– Carbon based conductive paint having electrical resistance of 0.1 Ω – 4 Ω / □is soaked in the fabric to develop the sensor. Named as carbon sensor in this article.

Conductive Fabric: General purpose electrically conductive knitted fabric from the market is used to develop the flexible capacitive sensors.

- 99% pure silver plated polyamide fabric by Bremen having electrical resistance of 0.3 Ω /□is used to develop the sensor named as bremen sensor in this article.
- 99% pure silver plated polyamide fabric by Balingen having electrical resistance of 0.6 Ω /□is used to develop the sensor named as balingen sensor in this article.

2.2 Fluids

The fluids used for this study is off the shelf distilled water by Wurth and tap water from the research lab. Table salt (NaCl) is used to introduce the sodium and chloride electrolytes in the fluid with different concentrations. A 60 mmol/l salt solution is prepared in-rd house to mimic the urine to use in-diaper fluid detection and measurements.

2.3 Adult Diapers

The adult diapers of Caroli brand (by W. Pelz GmbH & Co. KG) are procured for this study work. A cross section of the adult diaper is presented in Fig. 1. The upper hydrophilic layer (a) which absorbs the fluids quickly and maintains the dryness around skin. The distribution layer (b) transfers the fluids evenly to lower absorbent layer. The absorption core is made of fluff and hydrogel (c, e) which is a super absorbent polymer (SAP). A non-woven hydrophobic back-sheet (f) prevents the fluids leakage.

2.4 Lab Measurement Equipment

Laboratory precision scale and beaker is used to prepare and measure the fluids. An impedance analyzer by Keysight Technologies Inc (US) E4990A is used with fixture HP16664A to measure and record the real-time capacitance of sensors. A sweep of 900 kHz on 100 kHz–1 MHz span is run to record the sensor behaviour on wide span for an earlier developed custom front-end electronics interface. The recorded measurements are further analyzed using Matlab to compile the results.

2.5 Measurement Setup

First measurement setup is prepared using a liter jar with flexible capacitive sensor attached vertically to the inner wall of jar as shown in Fig. 2(c) Distilled water, tap water, salt solutions and sugar solutions of various concentrations are used to study the response of capacitive sensor. In the second measurement setup the flexible capacitive sensors are deployed inside absorption core of adult diaper under the distribution layer as depicted in the Fig. 1 to ensure the close contact of sensor with fluids absorbed by hydro-gel. A series of fluid introduction scenarios are performed to study the behaviour of developed sensors inside diaper.

3 Measurements Results

In the first scenario the distilled water is used as fluid for detection and quantification. A quantity of one liter of liquid is poured in jar setup with increment of 100 ml for each measurement. The Fig. 4 presents the fluid quantity detection from all sensors.

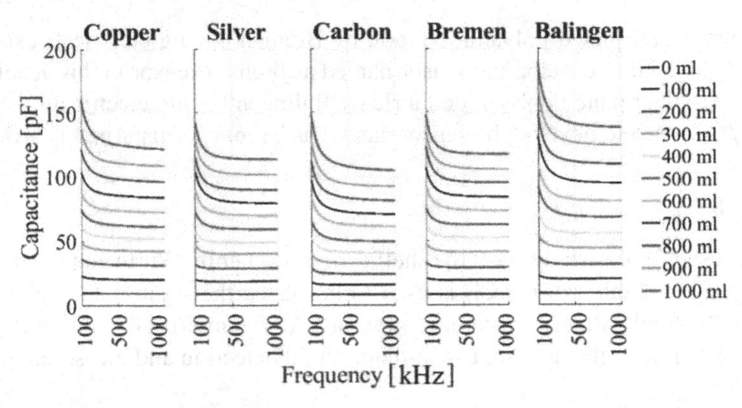

Fig. 4. In jar fluid detection and quantification results compared.

In the second scenario sugar and salt electrolyte concentration in distilled water is measured in the jar using all the sensor types. Figure 5 depicts the comparison of the measured results with silver sensor in three concentration levels of distilled water at level of 500 ml. A change in capacitance is observed with respect to the change in concentration of electrolytes in the liquid affecting the dielectric permittivity. High capacitive behavior of fluid is observed for salt based electrolytes as compared to distilled water. Whereas Fig. 6 shows the capacitive curve of silver sensor for fluid quantification.

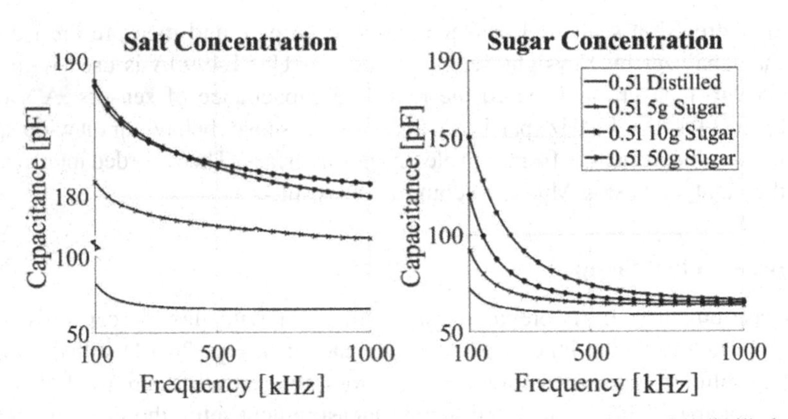

Fig. 5. In jar sensor sugar and salt electrolyte concentration comparison with distilled water.

For sensor measurements inside diaper it is assumed that an adult holds around 400 ml to 600 ml of urine in bladder normally [12]. A pseudo urine is prepared in lab with 60 mmol/l sodium chloride electrolyte concentration. In the third scenario an amount of 600 ml pseudo urine is used to make the diapers wet. Figure 7 shows the measurement results for dry and wet diapers. All sensors have produced promising results for liquid detection.

The fourth scenario is prepared for liquid volume quantification in diaper. For each measurement an amount of 600 ml tap water and pseudo urine is poured into diaper with

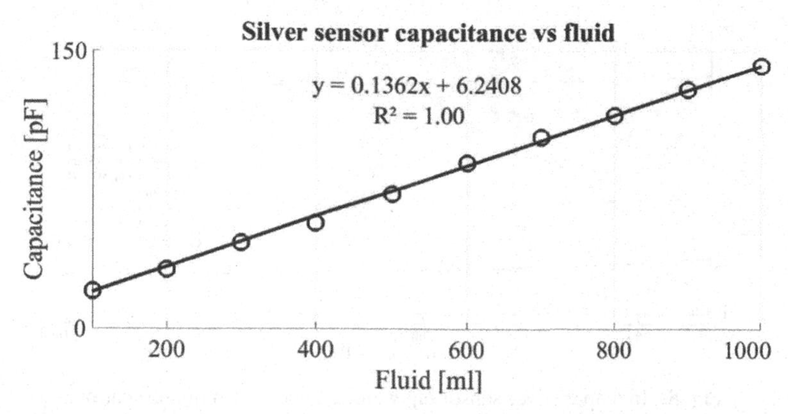

Fig. 6. Capacitive curve of Silver sensor for liquid quantification.

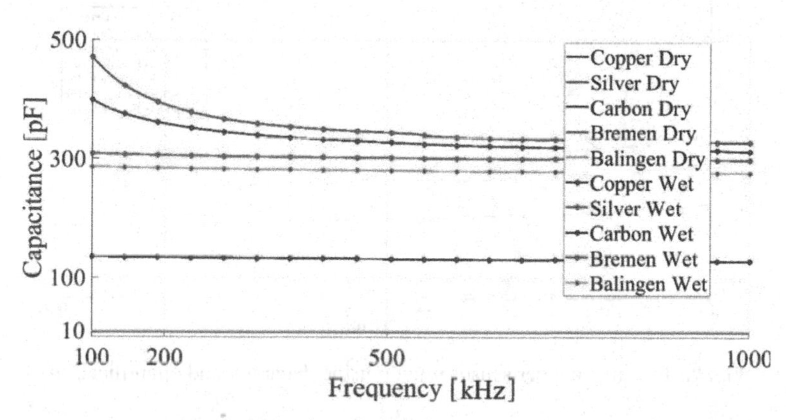

Fig. 7. In diaper sensor dry and wet results compared.

increment of 100 ml at rate of 15 ml/s which is an average urine flow-rate in adults [13] An absorption waiting time of 60 s is introduced before each measurement. Figure 8 and Fig. 9 presents the results of tap water and pseudo urine quantification in silver and copper sensors respectively. It is observed that the increase in capacitance is not linear with respect to liquid volume. We also believe that the liquid volume quantification precision would be enhanced if an artificial intelligence (AI) based algorithm would be used.

In the sixth scenario 600 ml pseudo urine is poured into diaper and sensor capacitance is measured after every 5 min up to 30 min to study the hydro-gel absorption effect on the sensor capacitance. It is observed that there is a slight decrease in capacitance over the time followed by initial rapid reduction in first 5 min. It is because the hydro-gel in edges of diaper takes sometime to uniformly absorb the liquid in overall diaper. Finally an amount of 100ml liquid is poured into diaper after 30 min to observe the effect on sensor capacitance. Figure 10 depicts the gradual capacitance reduction overtime and rapid increase when 100ml liquid is poured after 30 min. The results shows the capability of the sensor to detect intermittent liquid introduction in diaper.

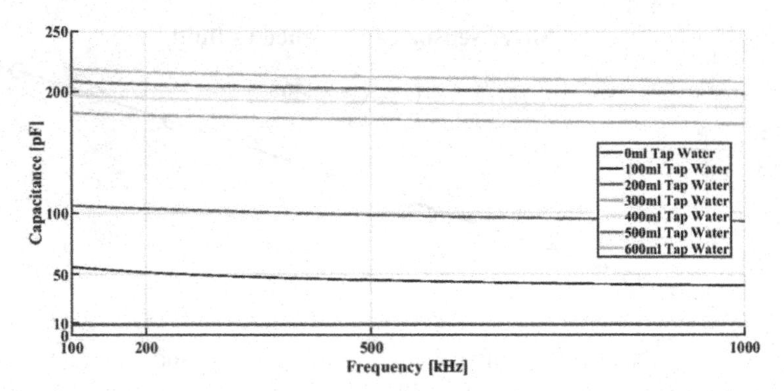

Fig. 8. In diaper silver sensor tap water detection and quantification.

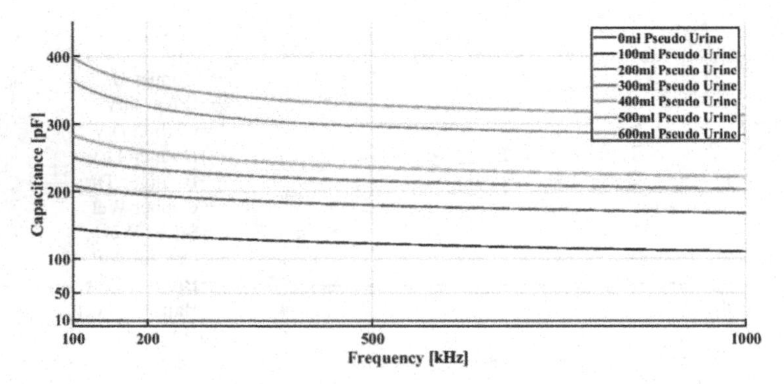

Fig. 9. In diaper copper sensor pseudo urine detection and quantification.

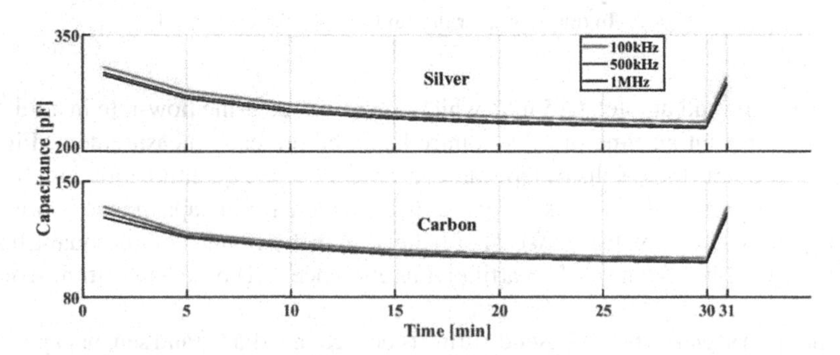

Fig. 10. Hydro-gel absorption trend for all sensors over 30 min time.

4 Conclusion

In this study a series of flexible capacitive sensors are developed using conductive foils, conductive paints and conductive fabrics to detect and quantify the fluids in adult diaper. Several in-vitro measurement scenarios are executed to validate the proof of concept by using impedance analyzer. The measurement results show not only the fluid detection but also a good resolution of fluid volume detection for 100 ml fluid in a diaper. The absorption trend of hydro-gel is also measured for precise volume quantification during frequent urination which brings the possibility to detect bladder inflammation. The frequency sweep results of artificial urine has shown a reduction in dielectric constant for increased salinity over higher frequencies. The proposed implementation of flexible capacitive sensors and front end electronics on flexible substrate using state-of-the-art printable materials along with artificial intelligence based data analysis could enable smart, economical and environmentally friendly adult diapers.

5 Acknowledment

The development of coplanar capacitive sensor is executed at laboratories of Aalto University under Urisens project funded by Business Finland and EHIR funded by Academy of Finland.

References

1. Balachandran, A., et al.: Comparison of population aging in Europe and Asia using a time-consistent and comparative aging measure. J. Aging Health **32**(5–6), 340–351 (2020)
2. United Nations Concise report on strengthening demographic evidence base for the post-2015 development agenda' New York, NY: Department of Economic and Social Affairs (2016). www.un.org
3. Martins, V.J., Gabrielle, V.J.: The effects of COVID-19 among the elderly population: a case for closing the digital divide. Front. Psychiatry **11**(1664–0640) (2020)
4. Tanweer, M., Halonen, K.A.: Development of wearable hardware platform to measure the ECG and EMG with IMU to detect motion artifacts. In: 2019 IEEE 22nd International Symposium on DDECS, pp. 1–4 (2019)
5. Grand View Research. Wearable Medical Device Market Size, Share Trends Analysis Report, 2022–2030 (2022). www.researchandmarkets.com/reports/3640168
6. Li, X., et al.: Smart diaper based on integrated multiplex carbon nanotube-coated electrode array sensors for in situ urine monitoring. ACS Appl. Nano Mater. **5**(4), 4767–4778 (2022)
7. Zhang, Y., et al.: Epidemiology of frequent/urgent urination in older adults in China: a multicenter cross-sectional study. Front. Public Health **9**(669070), 7 (2021)
8. Konno, S., Kim, J., Kazuki, N.: Development of capacitive sensor for diaper absorption volume. Adv. Biomed. Eng. **9**, 106–111 (2020). Released on J-STAGE 23 May 2020
9. Fischer, M., Renzler, M., Ussmueller, T.: Development of a smart bed insert for detection of incontinence and occupation in elder care. IEEE Access **7**, 118498–118508 (2019)
10. Hakola, L., et al.: Sustainable materials and processes for electronics, photonics and diagnostics. In: Proceedings of the Electronics Goes Green 2020+, Sep 2020, pp. 45–52 (2020)
11. EMI Software LLC, Coplanar capacitance calculation toolbox, accessed latest by date: 17.06.2022. www.emisoftware.com/calculator/coplanar-capacitance
12. Bieber, E.J., Sanfilippo, J.S., Horowitz, I.R.: Chapter 28 - Urogynecologic Workup and Testing, linical Gynecology, pp. 375–412. Churchill Livingstone, Philadelphia (2006)
13. Kumar, V., Dhabalia, J.V., Nelivigi, G.G., Punia, M.S., Suryavanshi, M.: Age, gender, and voided volume dependency of peak urinary flow rate and uroflowmetry nomogram in the Indian population. Indian J Urol. **25**(4), 461–466 (2009)

Energy Cyber Attacks to Smart Healthcare Devices: A Testbed

Zainab Alwaisi[1(✉)], Simone Soderi[1,2], and Rocco De Nicola[1,2]

[1] IMT School for Advanced Studies, Lucca, Italy
{zainab.alwaisi,simone.soderi,rocco.denicola}@imtlucca.it
[2] CINI Cybersecurity Laboratory, Roma, Italy

Abstract. The Internet of Things (IoT) has emerged as a subject of intense interest among the research and industrial community as it has significantly impacted human life. The rapid growth of IoT technology has revolutionized human life by inaugurating the concept of smart healthcare, smart devices, smart city, and smart grid. The security of IoT devices has become a serious concern, especially in the healthcare domain, where recent attacks exposed damaging IoT security vulnerabilities. In addition, in IoT networks where the connected devices are vulnerable to attacks such as attacks that affect the resource constraints of healthcare devices, e.g., energy consumption attacks. Therefore, this paper defines the impact of Distributed Denial of Service (DDoS) and Fake Access Points (F-APs) attacks on WiFi smart healthcare devices and investigates in detail how these attacks can be deployed toward victim devices and Access Points (APs). Our work focuses on IoT devices' connectivity and energy consumption when under attack. The main key findings of this paper are as follows: (i) the minimum and maximum attack rate of DDoS attacks that cause service disruptions on the victim side, and (ii) the minimum-the higher effect of energy-consumption Distributed Denial of Service (EC-DDoS) and F-APs attacks on the energy consumption of the smart healthcare devices. Our study reveals the communication protocols, attack rates, payload sizes, and victim devices' ports state as the vital factors in determining the energy consumption of victim devices. These findings facilitate a thorough understanding of IoT devices' potential vulnerabilities within a smart healthcare environment and pave solid foundations for future studies on defense solutions.

Keywords: Internet of Things (IoT) · smart healthcare · security · energy consumption · resource constraints

1 Introduction

The Internet of Things (IoT) refers to real-world objects having communicative and cognitive capabilities using smart devices. IoT is a tremendous communication paradigm where many heterogeneous devices will connect and talk to each other. These communication devices will play an essential role in the life

Y. Chen et al. (Eds.): BICT 2023, LNICST 512, pp. 246–265, 2023.
https://doi.org/10.1007/978-3-031-43135-7_24

of human beings, e.g., healthcare, transportation, and others. IoT is creating a revolutionary impact in technology and people's social life. Over time, IoT devices are overgrowing. According to recent reports from Cisco, the number of connected smart devices over the Internet will escalate to 29.3 billion networked devices by 2023 [19]. As the number and heterogeneity of smart devices are accelerating rapidly, it is becoming challenging to maintain the security of these devices [16]. IoT footprints have been identified in various domains such as manufacturing, agriculture, transportation, electric grid, and healthcare [3]. In an IoT-based healthcare system, security is the primary concern as the data is directly related to human beings [4]. An Intensive Care Unit (ICU) is a hospital's special and critically operational department where specialized treatment is given to patients requiring critical medical care. Usually, patients who are acutely unwell or injured severely and require continuous medical care are admitted to the ICU. The equipment and devices in the ICU play a vital role in keeping the patient alive and healthy. In such a scenario, any communication breakdown due to a cybersecurity breach may cause severe effects on a patient's life and even death in some instances [12].

Moreover, smart healthcare devices typically interact through different wireless communication protocols that allow adversaries to perform different attack types. For example, eavesdropping, creating Fake Access Points (F-APs), Distributed Denial-of-Service (DDoS), and energy-consumption Distributed-Denial-of-Service (EC-DDoS) [15]. Attackers use DDoS attacks to launch malicious traffic to damage target smart healthcare devices by affecting their resources and disconnecting them from the legitimate AP. EC-DDoS attacks lead to increasing the target's energy consumption to destroy it by sending malicious traffic. F-APs attacks force smart healthcare devices to connect to an alternative AP, monitor the transferred packets, and launch malicious attacks to consume more energy. Typically, most IoT devices have limited processing capabilities, and applying advanced security techniques to each device is challenging [2]. Using them in the smart healthcare system may also give unauthorized access to cybercriminals to monitor patients' private data and exploit sensitive information or send attacks to consume more energy and destroy smart devices [20].

Existing studies have performed general static and dynamic analyses and defenses against DDoS and F-AP attacks. However, most of the existing dynamic analyses are conducted in virtual environments, which makes it challenging to accurately measure the compromised devices' resource consumption, especially energy consumption. To address this issue, in this work, we perform our experiments in a controlled environment using real-world devices. With a cost-efficient experimental setup, we can collect sufficient data on the impact of attacks on resource-constrained IoT healthcare devices.

In this paper, we study the effect of DDoS, EC-DDoS, and F-APs attacks on WiFi connectivity and energy consumption of wireless healthcare devices. The results show the significant damage that could be caused by these attacks and draw attention to the urgent need for effective defense solutions. This paper can be used as a framework to test smart healthcare devices' security and create security standards for robust, predictable, and tamper-free operations.

1.1 Motivation and Contribution

Since IoT healthcare devices operate in an interconnected and interdependent environment, new threats constantly emerge. Moreover, as IoT healthcare devices typically use in an unattended environment, intruders may maliciously access these devices. Eavesdropping can access privately-owned information from the communication channel because IoT devices are usually linked through wireless networks. In addition to these security issues, IoT devices cannot afford to incorporate advanced security features because of their limited energy and processing power. Therefore, it is essential to study the effect of malicious attacks on the energy consumption of smart healthcare devices and show their impact, as smart healthcare systems are much more vulnerable and sensitive to their privacy and security. More importantly, considering the massive amount of smart healthcare devices on the market, the impact of energy consumption attacks cannot be neglected. Our main contribution is studying the effect of a practical combination of F-APs and DDoS attacks on smart healthcare devices. The main purpose of choosing DDoS and F-APs attacks as their impact on IoT security is high, so many researchers are working on solutions [8]. Thus, we target the energy consumption of smart devices. In the first step, we design a smart system to measure the current consumption of smart healthcare devices. Also, we build a testbed to monitor the smart devices and capture devices' status, e.g., *On* or *Off*, network traffic, and energy consumption. We identify several critical influential factors, particularly in communication protocols, Attack Rate (AR), payload size, and victim devices' ports status. We study the impact of these factors on the victim devices' resource constraints, such as energy consumption. In the second part of the contribution, the attack continues, and the attacker disconnects the smart devices from the local AP by sending DDoS attacks. At the same time, we study the effect of EC-DDoS on energy consumption by sending malicious attacks to affect the energy resources of smart devices. Also, we implement real-time energy monitoring on real smart devices to register the effect of the attack. In the last part of the contribution, we designed the F-AP to force the smart devices to connect to it once it disconnected from the local AP by DDoS attack. We designed the F-AP to automatically send malicious attacks affecting smart devices' energy consumption.

This paper gives a valuable understanding of the effect of DDoS and F-AP attacks on the energy consumption of smart healthcare devices by presenting a testbed and accurate energy consumption tests. Energy consumption attacks can destroy smart devices and impact patients' lives.

1.2 Organization of the Paper

This paper is organized as follows: Sect. 2 reviews related work about the different types of attacks. Section 3 explain the attack scenario and assumptions. The testbed scenario, data collection, F-APs setup, and the most important influential factors are presented in Sect. 4. Section 5 describes different results regarding network scans, disconnections caused by DDoS attacks, energy consumption measurement, and the effect of EC-DDoS and F-APs on energy consumption. Some concluding remarks can be found in Sect. 6.

2 Related Work

2.1 Fake Access Points Attacks

One of the most challenging security problems for wireless networks is detecting F-AP attacks. This attack is also called the rogue AP attack or the evil twin attack [11].

Detection of rogue AP attacks in the wireless network of a smart healthcare system is an essential aspect of wireless security [21]. A rogue device detection system using various techniques such as site survey, noise checking, MAC address list checking, and wireless traffic analysis has been proposed in [14]. The authors concentrated on detecting internal rogue devices, such as devices connected via a wireless network and used by employees on a corporate network. But this approach cannot be applied to IoT devices due to resource constraints.

Mehndi *et al.* [13] proposed an approach that considers the Mac Address, Service Set Identifier (SSID), and signal strength of the AP to decide whether the AP is rogue or not. In detecting authorized APs, the MAC addresses of all visible APs are matched against a list of authorized APs. Tools such as Ettercap[1], Wireshark[2], and Snort[3] are used for filtering instances where the MAC address is spoofed. While Kilincer, Ertam, and Şengür [6] proposed an automated technique for detecting and preventing F-APs attacks in the network of IoT devices. The proposed experiment uses a Single Board Computer (SBC) and a wireless antenna (ODROID module). The operation was about: 1) creating an F-APs, 2) scanning the surroundings using the SBC and WiFi modules, and 3) detecting fake AP broadcasts. The F-APs have been assigned to an unauthorized Virtual Local Area Network (VLAN). This study [6] is limited and focuses on F-APs attack detection and prevention. However, the data collected about the network and some attacks are still possible without connecting.

2.2 DDoS and EC-DDoS Attacks

DDoS and EC-DDoS attacks are security threats by attackers that enter the WiFi network coverage area and inject many different forged packets. Adversaries use this attack for two purposes, restricting usage of the WiFi bandwidth and preventing licensed users from communicating with the licensed AP to paralyze or reduce the WiFi network's performance [17].

Different studies have focused on the impact of DDoS attacks on Web servers when compromised IoT devices launch the attack. For example, Kambourakis [5],

[1] https://ettercap.github.io/ettercap/.
[2] http://www.wireshark.org/.
[3] http://www.snort.org/.

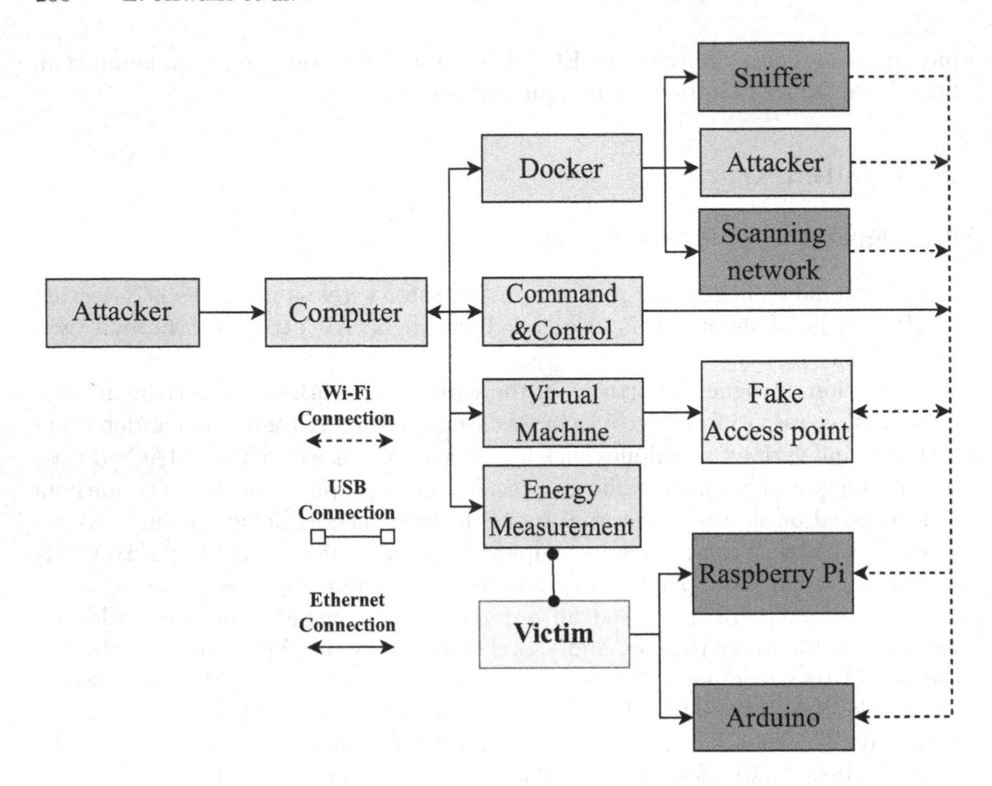

Fig. 1. Testing Environment.

Marzano [10], Tushir [18], and Kolias [7] discussed the outbreak of the Mirai botnet (and its variants), which compromised IoT devices to launch a DDoS attack against data centers. They claim that even naive techniques can be used to take control of such devices and create a massive and highly disruptive army of zombie devices. Liu and Qiu [9] examined de-authentication and disassociation DDoS attacks, where the attacker overwhelms the wireless device through fake de-authentication and disassociation packets. They show that Transmission Control Protocol (TCP) throughput drops and User Datagram Protocol (UDP) packet loss rises by increasing the AR. They have developed a client-device-based queuing model to show that the current IEEE 802.11w standard cannot resolve de-authentication and disassociation at high attack rates. Moreover, there are different approaches to monitoring IoT devices' energy consumption to detect IoT cyberattacks. Tushir *et al.* [18] quantitatively studied the impact of DDoS attacks on smart home IoT devices and their energy consumption. However, they did not present any detection or mitigation solutions.

Despite this, many developed methods exist to detect and prevent DDoS and F-AP attacks in IoT systems. In our approach, we mainly focused on the impact of the combination of DDoS, EC-DDoS, and F-APs attacks on energy consumption, response time, and connectivity of smart healthcare devices.

3 Attack Scenarios and Assumptions

Attack Scenarios. Figure 2 illustrates the potential scenarios where DDoS, EC-DDoS, and F-APs attacks can be applied. First, F-AP is designed to look like the actual AP. In those scenarios, the attacker can set a F-AP to launch different attacks to affect the energy resources of the smart healthcare devices. The F-AP signals could be more vital to the victim than the actual AP. Once disconnected from the actual AP by sending DDoS attacks, the tool forces smart healthcare devices to automatically reconnect to the F-AP, allowing the attacker to intercept all the traffic to that smart healthcare device, such as Man-In-The-Middle (MITM) attacks. Sniffing tools can also be applied to get or edit the information sent or received from the victim's devices. Additionally, the attackers may use F-AP and EC-DDoS attacks to destroy smart healthcare devices by affecting resource usage, e.g., energy consumption.

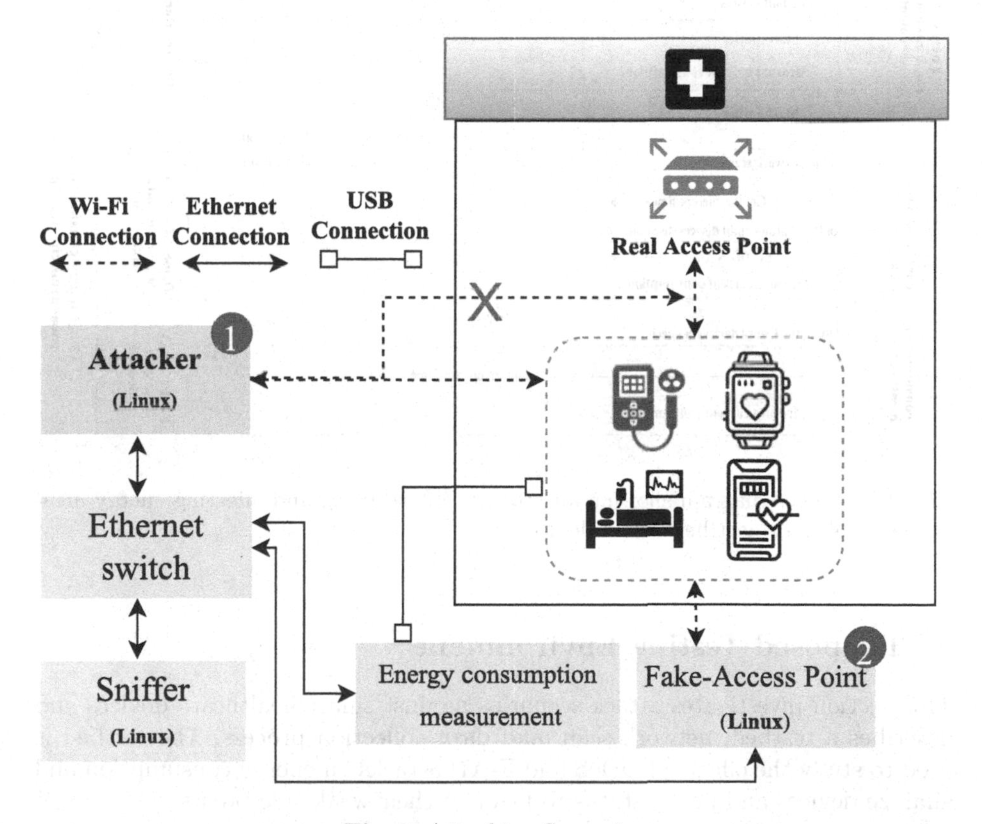

Fig. 2. Attacking Scenarios.

Assumptions. The attacks we consider require the attacker to send DDoS attacks to force the device to disconnect from the legitimate AP and affect its resources. Then the adversary could set up the F-APs attack at different locations. The adversary may set the F-APs at a distance from the victim to avoid being caught. As a result, the smart device is connected to the F-APs created by the attacker. The potential F-APs attack relies on sniffing over the WiFi to capture all packets traveling to and from the monitored smart device. Also, the F-AP is designed to affect the energy consumption of smart devices by automatically sending malicious attacks to connected smart devices.

Therefore, in this paper, we investigate the effect of F-APs and EC-DDoS attacks on smart healthcare devices' energy consumption by implementing them with malware designed to increase the energy consumption of smart healthcare devices and destroy them.

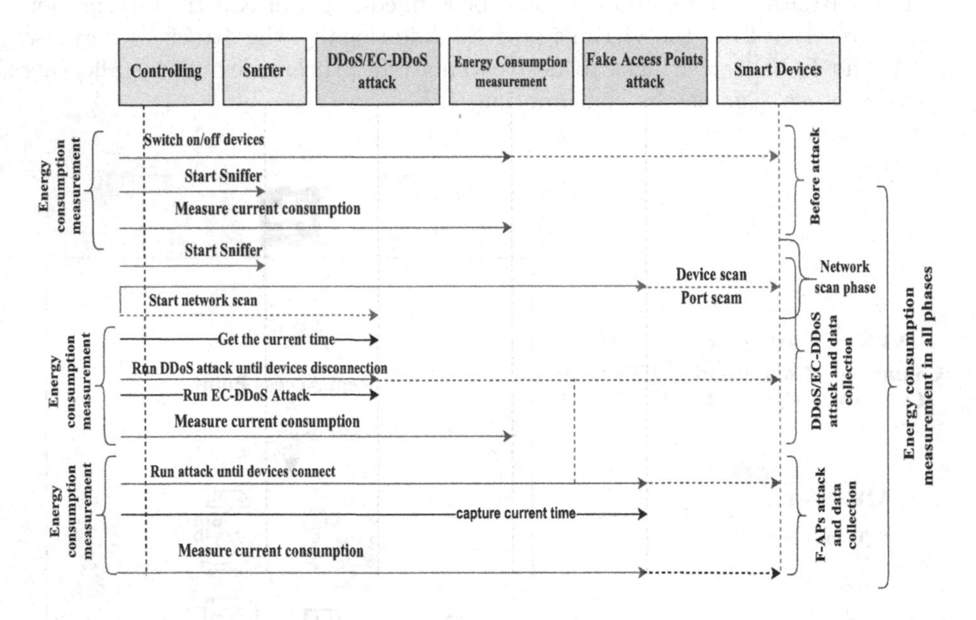

Fig. 3. Sequence diagram showing an attacker intercepting and affecting energy measurement of the smart healthcare devices.

4 Proposed Testing Environment

This section investigates attack scenarios against smart healthcare devices and describes a testbed, network scan, and data collection process. The testbed is used to study the effects of DDoS and F-AP attacks on energy consumption and analyze devices and ports' status to identify their weak side points.

4.1 Experiment Setup

There are two types of attacks on IoT devices: internal and external. Internal attacks happen when the adversary has access to the local network; this is possible by hacking the WiFi or gaining access to IoT devices. For example, attackers may gain access to the device by launching internal attacks to access a local Linux-based device remotely or by sending packets from outside the network for the external attack. For instance, if the attacker can force smart healthcare devices to disconnect from the actual AP and connect to the F-APs, then the adversary can send packets to the device outside the local network.

The testbed contains different smart healthcare devices, e.g., Arduino and Raspberry Pi, as proof of concept. Furthermore, three Linux-based images were created using Docker, as shown in Fig. 1. These images involve 1) an attacker sending malicious packets to the victim devices, 2) a sniffer for capturing WiFi traffic, and 3) a control system to scan the network and get different information about the port and devices' status.

Moreover, we created the F-APs using a TP-Link TLWN772N USB adapter and a Linux-based software system. Then, we designed a smart meter to measure the energy consumption of smart healthcare devices using a non-invasive current sensor [1] with an Arduino and some other resistors. Also, we used different software tools for attacking data generation and collection. On the adversary side, we used Nmap[4] to launch a network scan and identify devices' status, such as *online* or *offline*, IP address, and MAC address. Then, we ran a TCP/UDP port scan on the victim devices to identify port status (open/closed, filtered/not filtered, and others). Furthermore, we used hping3[5] to generate DDoS and EC-DDoS attacks by adjusting the AR, source IP address, destination IP address, payload, attack type, flags of TCP sessions (SYN, ACK, FIN, push, or urgent), and port types. In addition, we used hostapd (host access point daemon) and dnsmasq with a TP-Link TLWN772N USB adapter to create the F-APs. Hostapd is a user-space daemon software enabling a network interface card to act as an AP and authentication server. Dnsmasq is a lightweight, easy-to-configure DNS forwarder designed to provide DNS (and, optionally DHCP and TFTP) services to a small-scale network. The TP-Link TLWN772N is a USB adapter that acts as a F-AP.

We used tshark to evaluate the impact of EC-DDoS and F-APs attacks on the resource constraints of smart healthcare devices and capture WiFi traffic. Figures 2 and 4 show the attacking scenarios. Different stages are used to run our experiment. In the first stage, we measure the energy consumption once the device is turned *On*. Another measurement is used when the device has connected to the AP. Then, we run a network scan to capture the port and device status. Once we ensure that the device is connected to the Internet, we send DDoS and EC-DDoS attacks for two purposes: first, to consume more energy, and second, to disconnect the device from the local AP. Then, we run energy consumption measurements to calculate the energy consumption of the devices and study the devices' behaviors under DDoS and EC-DDoS attacks. Next, we run the F-APs attack on smart devices. We first check whether the devices are already disconnected from the local AP. We then force the smart devices to connect to the F-APs and finally start measuring the energy consumption of

[4] https://nmap.org/.

[5] https://www.kali.org/tools/hping3/.

smart healthcare devices. Also, we study the behavior of smart devices in terms of energy consumption and connectivity. The F-AP works as a MITM attack. The F-APs used for different purposes: 1) monitoring the devices, 2) sending malicious packets to consume more energy, and 3) affecting the CPU usage of the smart healthcare devices.

Algorithm 1 the affecting of F-APs and DDoS attacks on energy consumption

```
1: procedure SmartDevice(a)                          ▷ consume more energy of (a).
2:     Sniff air for network scanning
3:     E1:energy consumption before attack
4:     E2:energy consumption after attack
5:     if a = connected then                          ▷ a is connected to the AP
6:         Calculate energy consumption
7:         Disconnect using DDoS attack
8:     else if a = disconnected then                  ▷ from the actual AP
9:         Send DDoS attack
10:        Calculate energy consumption after attack
11:    if E1(a) < E2(a) then
12:        Connect the device to F-APs
13:        Send malicious attack to consume more energy
14:    else if E1(a) < E2(a) then                     ▷ energy consumption after attack
15:        Send another packets of DDoS attack
16:        Consume more energy and disconnect the devices
17:    else
18:        Try to consume more energy
19:        Calculate energy consumption, AR, survival duration (SD), and thresh-
   old(AR)
20:    while E1 ≠ 0 do                                ▷ Consume more energy to destroy the device
21:        E1(a) ← E2(a)
```

4.2 Collecting Data

We built a smart healthcare test environment by deploying different smart devices, as shown in Fig. 4. Data is aggregated from various smart devices for analysis purposes. Moreover, the aggregated data is categorized into behavioral and network data. Behavioral data refers to the status of the smart devices, like *On* or *Off*, and the device's readings. Network data refers to smart healthcare devices' TCP and UDP packet data. We integrate this data to learn typical behaviors in a smart healthcare environment.

Figure 3 shows different phases of our attack scenario; in the first phase, we control the smart devices by switching them *On* or *Off*; this is essential to calculate the energy consumption of the smart devices before launching any attack. In the second phase, we measure the energy consumption of the smart devices for at least 30 minutes to calculate the average energy consumption before launching

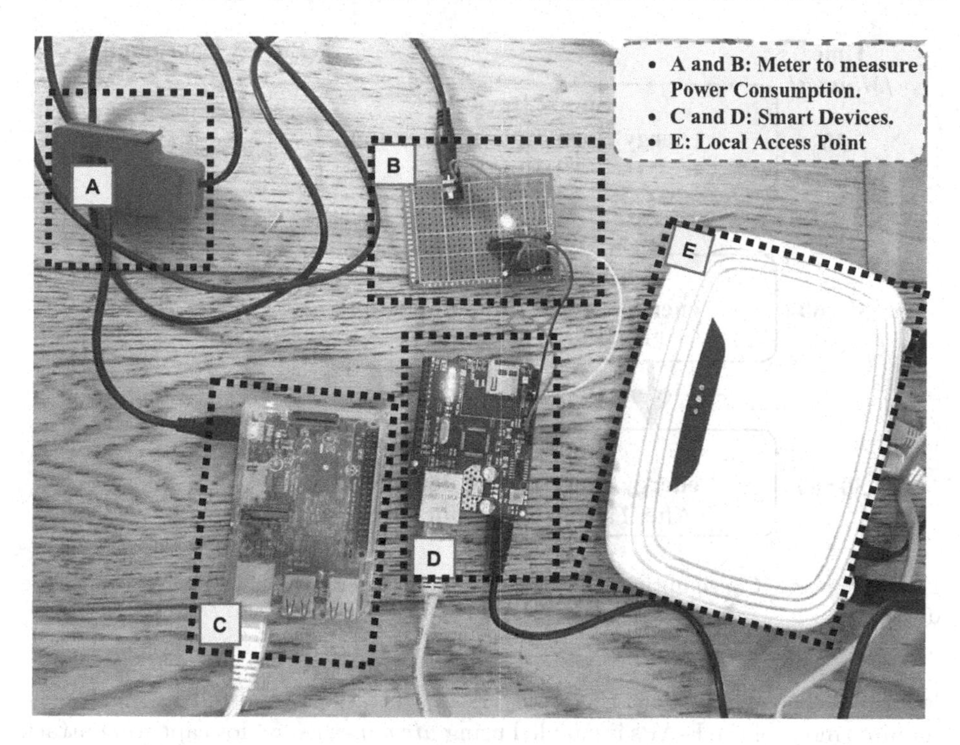

Fig. 4. Proof of Concept for Wireless Network smart healthcare devices.

any attack; then, we start sniffing the network to get information about ports and devices' status. In the final phase, we launch DDoS and EC-DDoS attacks to impact smart devices' connectivity and energy consumption. Simultaneously, the energy consumption of appliances is measured to show the impact of this attack on the energy consumption of smart healthcare devices. After that, once the devices are disconnected from the legitimate AP using DDoS attacks, we calculate AR, Survival Duration (SD), and the threshold of the AR. Next, we force the devices to connect to the F-APs, where the fourth phase will start. We start monitoring and collecting information about the smart devices using F-APs facilities in this phase. We then launch malicious attacks through the F-APs to consume more energy and study the behaviors of smart devices' energy consumption under F-APs attacks. Through that, we achieve the main purpose of destroying smart healthcare devices by using energy consumption attacks caused by F-AP and EC-DDoS attacks as shown in Fig. 5.

4.3 Setting up a Fake Access Point

We used hostapd (host access point daemon) and dnsmasq with a TP-Link TLWN772N USB adapter to create the F-APs attack, broadcast a fake signal, and capture the victim's packets. The F-AP configures the same SSID, Basic Service Set Identifier (BSSID), broadcast channel, and security settings as the

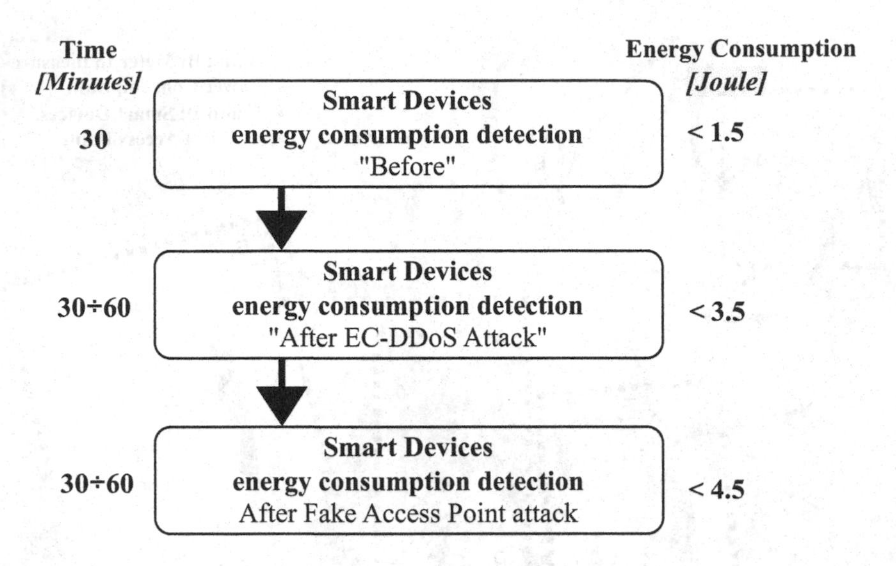

Fig. 5. Energy Consumption affects before and after attacking smart healthcare devices.

legitimate AP. The scenario with the F-APs attack is reported in Fig. 2. The monitor mode of the F-APs is enabled using *airmon-ng start* for capturing attack injections or packets from and to the smart healthcare devices. The F-APs work as a MITM attack to capture packets transferred between the smart devices and the server. Moreover, the F-AP is designed to send malicious attacks to consume more energy of the connected smart healthcare devices[6].

4.4 Determining the Weak Side

In this section, we focus mainly on the impact of disconnections and power consumption, considering victim devices that differ in their hardware, e.g., CPU, WiFi chip, and memory. Therefore they respond differently to a given attack. Arduino and Raspberry Pi are considered.

During an attack, the port state and communication protocol can significantly influence victims' responses. Accordingly, we launch three attacks: TCP-SYN, UDP, and Internet Control Message Protocol (ICMP) echo request attacks. In the case of UDP and TCP-SYN attacks, we launch attacks against ports with different states, e.g., open, closed, filtered, and open-filtered. With ICMP attacks, do not specify the port number. Moreover, being smart IoT devices resource-constrained, their behavior may differ dramatically according to the payload size of attack packets. As a result, we set the payload of attack packets to 0 B, i.e., no payload (NP), and 1500 B, i.e., high payload (HP). Although it has been recognized that the impact of DDoS, EC-DDoS attacks are directly related to AR. Section 5.2 analysis the impact of the AR on service disruption

[6] https://github.com/developerZA/EnergyConsumptionAttack.git.

by calculating the minimum AR that causes the devices to be disconnected from the legitimate AP. The impact of EC-DDoS and F-APs attacks on the energy consumption of smart healthcare devices has also been studied in Sect. 5.

5 Experimental Results and Analysis

In this section, we describe an experimental workplace to test the effect of DDoS, EC-DDoS, and F-APs attacks on the energy consumption of smart healthcare devices. This experiment focuses on collecting incoming malicious attacks and the usage statistics of a victim device and analyzing the attack effects on the victim devices in terms of energy. The network scan of the smart devices is used to obtain the status of the ports and then to determine the weak side of smart devices by calculating their AR and SD. Moreover, we study the effect of EC-DDoS and F-APs attacks on energy consumption.

5.1 Network Scan

The network scan gathers information about the victim's smart devices, such as *online* or *offline* status, IP address, and MAC address. The port scan permits an attacker to discover the status of TCP and UDP ports. The possible states of the used ports are: open, closed, filtered, and open-filtered. Table 1 reports the status of ports for the devices used in our testbed.

Table 1. Network scan result in terms of port status for TCP and UDP protocols.

Device	TCP scanned ports	UDP scanned ports
Raspberry Pi	3 open, 998 open-filtered, 65389 filtered and 0 closed ports	4 open and 700 open-filtered and 0 closed ports
Arduino	1 open, 22 filtered, 1000 open-filtered and 0 closed ports	1000 open-filtered ports

5.2 Attack Rate and DDoS Attacks

We consider threshold AR as the minimum AR measured in Packets Per Second (PPS) that disconnects the victim device from the AP. The SD is the time duration between the start of an attack and the device disconnection caused by the attack. We set the maximum attack duration between 8 and 30 minutes. We have launched ICMP and TCP-SYN/UDP attacks on the victim devices' open, filtered, and closed ports to collect their threshold AR and SD. The AR applied to the Raspberry Pi is between 500 to 10,000 PPS for both NP and PH. In contrast, the AR sent to the Arduino for NP attacks is between 100 to 800 PPS, as the threshold AR is 800 PPS. We did not use a PH attack against the

Arduino because it disconnects with minimal AR. Table 2 reports the average SD in minutes for the smart devices. We can see that the Arduino device disconnects in all cases with different attacks. Instead, the Raspberry Pi disconnects only with low packets.

Table 2. Survival Duration (SD) caused by DDoS attack.

Survival Duration	Raspberry Pi		Arduino	
	NP [Minutes]	*PH [Minutes]*	*NP [Minutes]*	*PH [Minutes]*
SD (ICMP)	7.58	none	3.6	3.13
SD (TCP)	6.2	none	3.3	2.44
SD (UDP)	7.8	none	3.8	2.44

The Raspberry Pi survives with a higher AR than the Arduino, with 20 k packets at NP. The AR of the Arduino is 800 packets at NP and 200 packets at PH. Looking at the tshark files, we can see that the Raspberry Pi broadcasts probe requests and sends de-authentication packets to the legitimate AP. The main difference between the smart devices is that the Raspberry Pi has more powerful hardware than the Arduino.

Through the experiment, when we calculate the received AR by the victim devices, we find that the victim devices rarely receive the actual AR sent by the attacker. For example, we sent about 15 k packets to the open ports of the Raspberry Pi; the received packets were about 14544 packets. Also, we can notice that the increase in the average packet rates sent by the attacker causes an approximately logarithmic increase in the received packets by the victim.

5.3 Energy Measurement and EC-DDoS Attacks

We developed a smart circuit using a non-invasive current sensor, as shown in Fig. 6 to measure the current consumption of smart healthcare devices. This smart circuit samples voltage, ampere, watt, and current per second. The current consumption values for each smart healthcare device are stored in the database (DB). In our experiment, we use the Joule (J) values to calculate the energy consumption of smart devices.

To calculate the energy consumption of the devices versus the incoming attack reception rate of the victim devices, we need to collect the data from both the sensors and the tshark data. Therefore, all data relevant to this experiment is stored automatically in the DB.

During packet collection, the attacks are sent using the same TCP, UDP, and ICMP flood commands. Using the topology depicted by Fig. 2, the malicious TCP, UDP, and ICMP traffic are separately sent to the victim device, while all usage statistics in terms of energy consumption and connectivity are recorded on the victim device. Each attack is simulated for a duration of 1 second for a total of 30 minutes, and all usage statistics are recorded for the same duration.

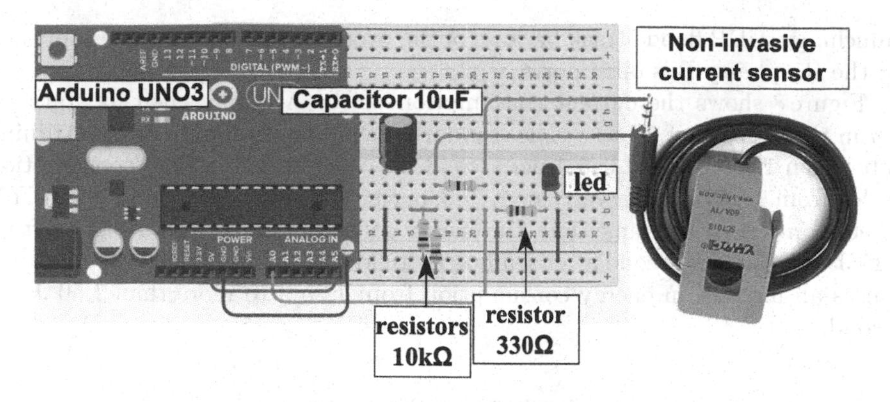

Fig. 6. Circuit for measuring current consumption.

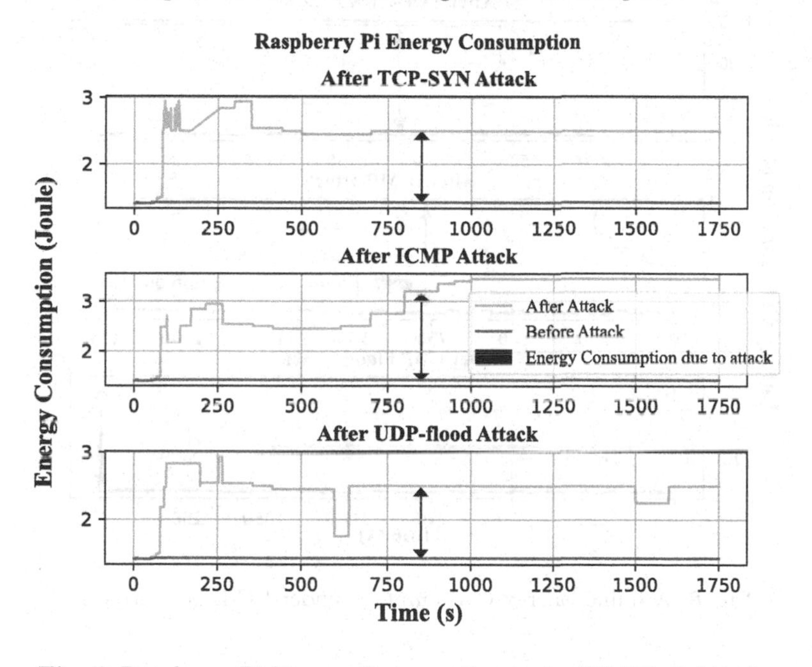

Fig. 7. Raspberry Pi Energy Consumption under EC-DDoS Attack.

Figure 7 shows the device's energy consumption when its status is *On* in the absence of attacks on that device. The standard energy consumption of the Raspberry Pi is between 1.410 J and 1.420 J per second. However, the current consumption varies from 1.410 J to more than 3.3 J per second after launching TCP-SYN attacks on open ports of the Raspberry Pi. In contrast, we can notice that the energy consumption increases to more than 3.60 J per second after launching ICMP attacks on open ports of the Raspberry Pi. Additionally, the energy consumption fluctuates between 1.4 J and 3.50 J per second after launching a UDP flood attack because of the overload that might have happened on the Raspberry Pi's open ports.

Figure 8 shows the current consumption of the Arduino when its status is *On* in the absence of attacks. The standard energy consumption of the Arduino is between 1.060 J and 1.065 J per second. In contrast, the energy consumption varies from 1.065 J to more than 1.75 J per second after launching TCP-SYN attacks on NP. At the same time, the energy consumption increases slightly from 1.15 J to 1.25 J per second after sending an ICMP attack. The UDP flood attack causes an increase in energy consumption from 1.25 J to more than 1.50 J per second.

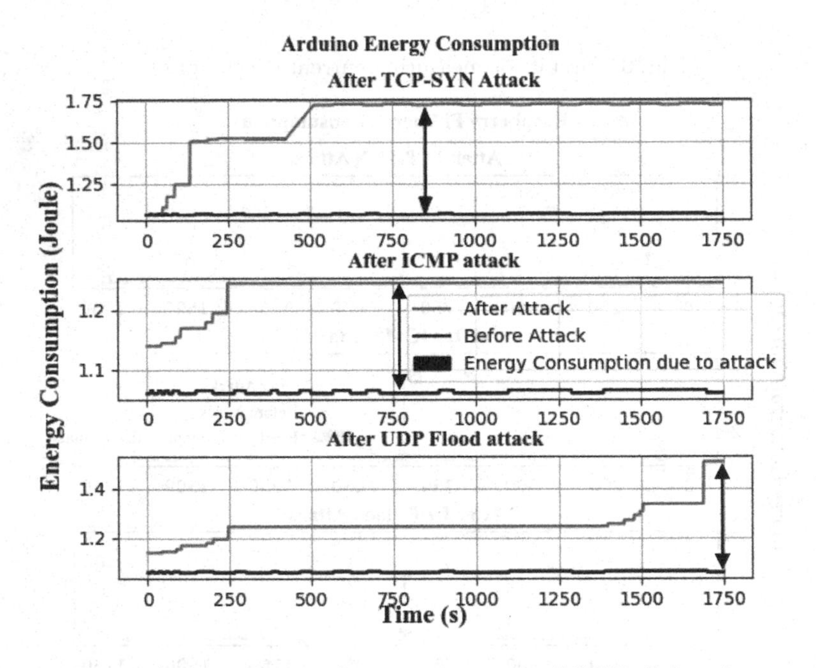

Fig. 8. Arduino Energy Consumption under EC-DDoS Attacks.

Note that the EC-DDoS attack rates sent by the attacker are below the threshold of DDoS attack rates that cause disconnection on smart healthcare devices. In the next section, we study the smart devices' energy consumption behavior under F-APs attacks.

Algorithm 2 Fake Access Points Attack

```
 1: procedure SmartDevice, F-APs(a,b)            ▷ consume more energy of (a).
 2:     Sniff air for network scanning
 3:     Measure energy consumption of SH
 4:     if a ⊆ b then
 5:         if a = connected then                ▷ a is connected to F-APs
 6:             Sniff air for network scanning
 7:             Send malicious packets
 8:             Calculate energy consumption after attack
 9:         else if a = NotConnected then
10:             Try to reconnect it to F-APs
11:     while a ⊈ b do                           ▷ if there are no new devices
12:         Sniff air for finding new devices
```

5.4 Energy Consumption and F-APs Attacks

Once the devices are disconnected from the legitimate AP, the F-APs attack takes over its responsibility to consume more energy and monitor them.

The signal of the F-APs is more vital to the victim's smart devices than the legitimate AP. When the devices are disconnected, the signal from the F-APs will be sent to the smart devices to force them to connect to affect their energy resources. Afterward, the monitoring mode of the F-APs will be enabled to monitor packets transferred from and to the smart devices. At this stage, the sniffer is essential to launch further attacks on the target device and collect information about it, such as IP and port status. The F-AP is designed to be more flexible in sending malicious packets automatically to affect the energy resources once the smart devices are connected.

The required time for the Raspberry Pi to connect to the F-APs is between 3 and 5 minutes. While the Arduino takes 7 to 10 minutes, sometimes we force the Arduino to connect to the F-APs.

Figure 9 shows how the energy consumption of the Raspberry Pi changed after connecting it to the F-APs; the malicious packets were randomly selected and sent to the Raspberry Pi. The energy consumption increases to more than 4.00 J per second. At the same time, the energy consumption of the Arduino increases slightly to reach more than 2.00 J per second after connecting it to the F-AP. Therefore, we can conclude that the F-APs attack successfully affects smart healthcare devices' energy consumption.

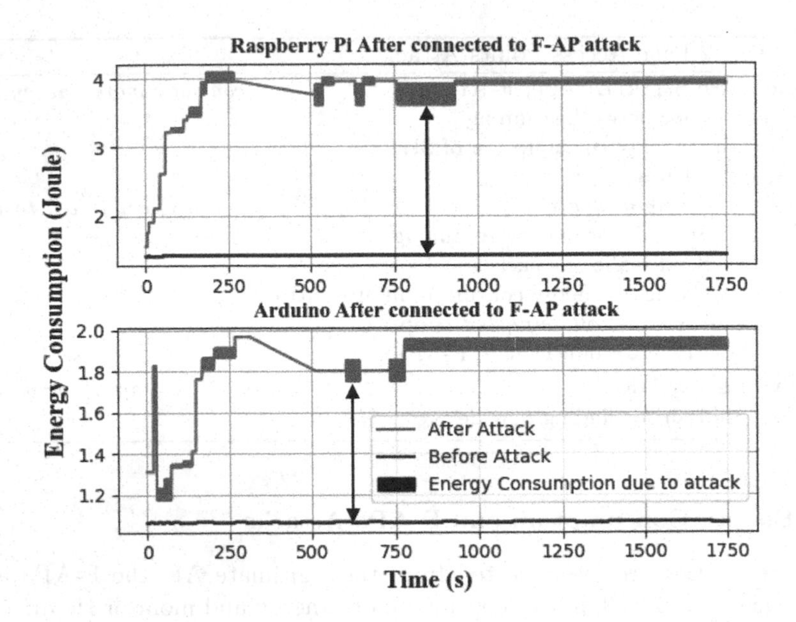

Fig. 9. Raspberry Pi and Arduino Energy Consumption under F-APs Attack.

5.5 Results and Analysis

In our experiment, we studied the effect of DDoS, EC-DDoS, and F-APs attacks against the Raspberry Pi and Arduino for about 30 to 60 minutes and measured the energy consumption. During such attacks, the smart devices continuously receive the packets and spend resources processing these packets.

Our analysis shows that effective DDoS attacks can be launched at NP if the victim replies to ICMP packets. ICMP or TCP-SYN/UDP attacks could be used on open and closed ports. However, to launch EC-DDoS attacks that cost the victim device's maximum energy without being disconnected from the legitimate AP, the attacker can launch a PH TCP-SYN attack against open ports or ICMP attacks if the device responds to ICMP packets.

Moreover, to force the smart healthcare devices to connect to the F-AP attack, the signals of the latter should appear stronger to the victim than the legitimate APs. The attacker launches malicious attacks through the F-AP to induce maximal energy consumption without being disconnected by considering the threshold of the AR. Figure 10 shows the overall infection of EC-DDoS and F-APs attacks on both devices (Arduino and Raspberry Pi); as it can be seen, the energy consumption of the Raspberry Pi device varies from 1.42 J to be more than 3 J per second. At the same time, the energy consumption of the Arduino varies from 1.06 J per second to more than 2 J per second. It is observed that DDoS, EC-DDoS, and F-APs attacks significantly impact the energy consumption of IoT devices. When an IoT device is flooded with TCP, UDP, and ICMP packets, there are significant increases in energy usage, which might destroy the IoT devices in the end. This study offers a better understanding of energy con-

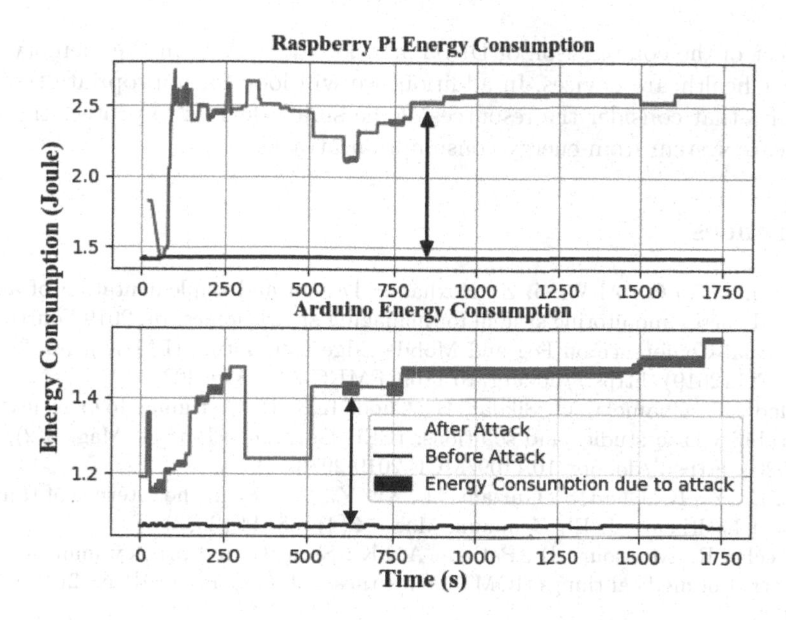

Fig. 10. Raspberry Pi and Arduino Energy Consumption under Attacks where the F-APs affect 45% of the energy consumption of the Raspberry Pi and the Arduino, while the affection of EC-DDoS attack is about 55%.

sumption attacks caused by the combination of F-AP, DDoS, and EC-DDoS attacks on the smart healthcare system. The analysis of such resource consumption will benefit the deep understanding of DDoS and F-APs attacks' impact on resource-constrained smart healthcare environments and facilitated future research on lightweight defense mechanisms against such attacks.

6 Conclusions and Future Work

This paper studied the impact of EC-DDoS and F-APs attacks on the resource usage of different smart healthcare devices and, more specifically, on energy consumption. We first used Docker images to collect data, scan the smart devices' networks, and sniff the network. Then, we calculated the AR, SD, and threshold of the AR on the victim side. The main purpose of the calculation is to study the effect of DDoS attacks on the connectivity of smart healthcare devices. We also studied other influential factors such as ports, device state, attack type (i.e., protocols used), and AR. We then analyzed the impact of DDoS, EC-DDoS, and F-APs attacks on the energy consumption of smart devices. Specifically, we designed the F-APs attack to affect the energy resources of the smart devices by automatically sending malicious attacks to the connected smart healthcare devices. Through our work, we offer a better understanding of the effect of DDoS, EC-DDoS, and F-APs attacks on smart healthcare devices' energy consumption and connectivity within a wireless network. In the future, we will also study the effect of the combination of DDoS attacks and F-APs on the memory usage of smart healthcare devices. In addition, we will look for appropriate tools and strategies that consider the resources of the smart devices to protect the smart healthcare system from energy consumption attacks.

References

1. Agyeman, M.O., Al-Waisi, Z., Hoxha, I.: Design and implementation of an IoT-based energy monitoring system for managing smart homes. In: 2019 Fourth International Conference on Fog and Mobile Edge Computing (FMEC), pp. 253–258. FMEC (2019). https://doi.org/10.1109/FMEC.2019.8795363
2. Alladi, T., Chamola, V., Sikdar, B., Choo, K.K.R.: Consumer IoT: security vulnerability case studies and solutions. IEEE Consum. Electron. Mag. **9**(2), 17–25 (2020). https://doi.org/10.1109/MCE.2019.2953740
3. He, D., Ye, R., Chan, S., Guizani, M., Xu, Y.: Privacy in the internet of things for smart healthcare. IEEE Commun. Mag. **56**(4), 38–44 (2018)
4. Hireche, R., Mansouri, H., Pathan, A.S.K.: Security and privacy management in internet of medical things (IOMT): a synthesis. J. Cybersecur. Priv. **2**(3), 640–661 (2022)
5. Kambourakis, G., Kolias, C., Stavrou, A.: The MIRAI botnet and the IoT zombie armies. In: MILCOM 2017–2017 IEEE Military Communications Conference (MILCOM), pp. 267–272. IEEE, USA (2017). https://doi.org/10.1109/MILCOM.2017.8170867
6. Kilincer, F., Ertam, F., Sengur, A.: Automated fake access point attack detection and prevention system with IoT devices. Balkan J. Electr. Comput. Eng. **8**, 50–56 (2020). https://doi.org/10.17694/bajece.634104
7. Kolias, C., Kambourakis, G., Stavrou, A., Voas, J.: DDoS in the IoT: mirai and other botnets. Computer **50**(07), 80–84 (2017). https://doi.org/10.1109/MC.2017.201
8. Kumari, P., Jain, A.K.: A comprehensive study of DDoS attacks over IoT network and their countermeasures. Comput. Secur. **127**, 103096 (2023)
9. Liu, C., Qiu, J.: Performance study of 802.11w for preventing dos attacks on wireless local area networks. Wirel. Pers. Commun. **95**(2), 1031–1053 (2017). https://doi.org/10.1007/s11277-016-3812-9
10. Marzano, A., et al.: The evolution of Bashlite and Mirai IoT botnets. In: 2018 IEEE Symposium on Computers and Communications (ISCC), pp. 00813–00818. IEEE, Brazil (2018). https://doi.org/10.1109/ISCC.2018.8538636
11. Metwally, E.A., Haikal, N.A., Soliman, H.H.: Detecting semantic social engineering attack in the context of information security. In: Magdi, D.A., Helmy, Y.K., Mamdouh, M., Joshi, A. (eds.) Digital Transformation Technology. LNNS, vol. 224, pp. 43–65. Springer, Singapore (2022). https://doi.org/10.1007/978-981-16-2275-5_3
12. Morgan, V., Birtus, M., Zauskova, A.: Medical internet of things-based healthcare systems, wearable biometric sensors, and personalized clinical care in remotely monitoring and caring for confirmed or suspected covid-19 patients. Am. J. Med. Res. **8**(1), 81–91 (2021)
13. Samra, M., Mengi, M., Sharma, S., Gondhi, N.K.: Detection and mitigation of rogue access point **1**, 195–198 (2015)
14. Saruhan, I.H.: Detecting and preventing rogue deviceson the network. SANS Institute (2007)

15. Shahid, J., Ahmad, R., Kiani, A.K., Ahmad, T., Saeed, S., Almuhaideb, A.M.: Data protection and privacy of the internet of healthcare things (IOHTS). Appl. Sci. **12**(4), 1927 (2022)
16. Shouran, Z., Ashari, A., Kuntoro, T.: Internet of things (IoT) of smart home: privacy and security. Int. J. Comput. Appl. **182**(39), 3–8 (2019)
17. Tang, Z., et al.: Exploiting wireless received signal strength indicators to detect evil-twin attacks in smart homes. Mob. Inf. Syst. **2017**, 1248578:1–1248578:14 (2017). https://doi.org/10.1155/2017/1248578
18. Tushir, B., Dalal, Y., Dezfouli, B., Liu, Y.: A quantitative study of DDoS and e-DDoS attacks on WiFi smart home devices. IEEE Internet Things J. **8**(8), 6282–6292 (2021). https://doi.org/10.1109/JIOT.2020.3026023
19. (VNI), C.V.N.I.: Cisco visual networking index (VNI) (2021). https://shorturl.at/abkSU
20. Yaqoob, I., et al.: The rise of ransomware and emerging security challenges in the internet of things. Comput. Netw. **129**, 444–458 (2017). shorturl.at/imzM4, https://shorturl.at/acG26, special Issue on 5G Wireless Networks for IoT and Body Sensors
21. Zhang, Z., Yu, T., Ma, X., Guan, Y., Moll, P., Zhang, L.: Sovereign: self-contained smart home with data-centric network and security. IEEE Internet Things J. **9**(15), 13808–13822 (2022)

Ensembles of Heuristics and Computational Optimisation in Highly Flexible Manufacturing System

Rotimi Ogunsakin[1(✉)], Nikolay Mehandjiev[1], and Cesar Marin[2]

[1] Alliance Manchester Business School, Booth Street East, Manchester M13 9SS, UK
rotimi.ogunsakin@manchester.ac.uk
[2] Information Catalyst for Enterprise Ltd., Haslington, Crewe CW15QR, UK

Abstract. The objective of a Flexible Manufacturing System (FMS) is to respond faster to changes in products and demands with minimum changeover cost. However, layout changes in FMS are not automatic and required human intervention. Therefore, when requirements for layout changes are frequent, such as in a dynamic production environment, like mass personalisation production environments, layout reconfiguration becomes expensive and unrealistic. In this paper, we relax this core assumption of static FMS layout and introduce a decentralised approach to the design and coordination of manufacturing systems' entities, whereby both products and production machines are mobile and autonomous. We apply three different optimisation methods, of which two are ensembles of computational and heuristics optimisation approaches based on Gradient Descent and Ant Colony Optimisation (ACO), to optimise mobile machines locations under non-deterministic manufacturing conditions as obtainable in a mass personalisation context. These approaches enable mobile production machines to coordinate and autonomously adjust their location and layout in real-time to minimise the cost of material flow between production machines. The proposed approach offers a promising outlook on the design and coordination of manufacturing systems under unpredictable manufacturing conditions.

Keywords: nature-inspired optimisation · ensemble learning · flexible manufacturing system · real-time optimisation · mass personalisation

1 Introduction

Mass personalisation seeks to introduce individualised production into manufacturing systems, which will require the requirement for constant reconfiguration of assembly systems and layouts to manufacture at mass production cost [13,18]. Unfortunately, current assembly systems are limited in their reconfigurability capacities when additional processes and changes in process sequence or processing time are required at short notice, such as in minutes or hours. This is due to the physical constraints presented by the fixed transfer system of conveyor systems and fixed production machines locations [13].

Y. Chen et al. (Eds.): BICT 2023, LNICST 512, pp. 266–279, 2023.
https://doi.org/10.1007/978-3-031-43135-7_26

Indeed, finding innovative ways to reduce throughput times and increase product varieties within manufacturing firms is of long-standing interest within the operations research literature. Optimisation techniques such as in Assembly Line Balancing Problem (ALBP) [3,21], and Facility Layout Problem (FLP) [12] have also been suggested. However, the underlying assumption of these studies is that the production machines, workstations and employees remain fixed in their current factory layout. Up to date, virtually no studies have explored how throughput times can be improved by reconfiguring production lines and moving production machines during production (in real-time) to meet rapid changes in supply and demand in the context of mass personalisation [1].

To address this research gap in the manufacturing and operations research literature, we proffer a method for optimising production machine location and layout in real-time during production, depending on the mix of order inflow, using an ensemble of computational and heuristic means. This approach is analogous to performing assembly line balancing and facility layout in real-time and in a mass personalisation context, where order arrivals are non-deterministic, product mixes are stochastic, and real-time reconfiguration is required. The term ensemble is used loosely in this paper to refer to a combination of multiple optimisation approaches [30].

To achieve this, we explored the implication of relaxing the core assumption of static production machines and the use of fixed transfer lines for part movement by designing a decentralised and distributed manufacturing system, which we have labelled as the Lot Size-of-1 Manufacturing System (S1MS). In S1MS, we introduce mobile production machines and autonomous material handling systems to replace conveyor systems (see Fig. 1). We created a digital-object of the physical manufacturing system, called the virtual S1MS (V-S1MS), which connects and informs the real S1MS existing in the real world. The optimisation process is carried out in the V-S1MS, which informs the real S1MS in the real-world. This architectural change improves flexibility in material handling, minimises the cost of material flow by eliminating bypassing and backtracking, and minimises congestion that may occur when material flow is constrained by rigid conveyor movement.

Three different versions of S1MS were implemented using three distinct approaches to coordinate and autonomously organise and optimise the locations of the mobile production machines in real-time during manufacturing. Two of the approaches, referred to as the Gradient-Decent for Autonomous Resource Positioning (G-DARP), and Sensing Radius and Cluster Analysis for Autonomous Resource Positioning (SR-CAARP) are based on a combination of an iterative optimisation algorithm referred to as the gradient descent (GD) algorithm [14,20] and ant-based heuristic optimisation algorithm. The GD uses only local data and not global data to avoid computational overload and overfitting. The third approach, Ant Colony Optimisation for Autonomous Resource Positioning (ACO-ARP) is based on a nature-inspired algorithm referred to as the Ant Colony Optimisation Algorithm (ACO) [8,9].

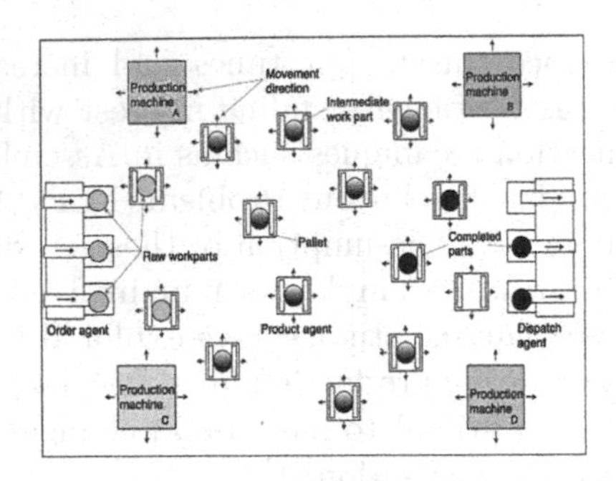

(a) Initial S1M's layout.

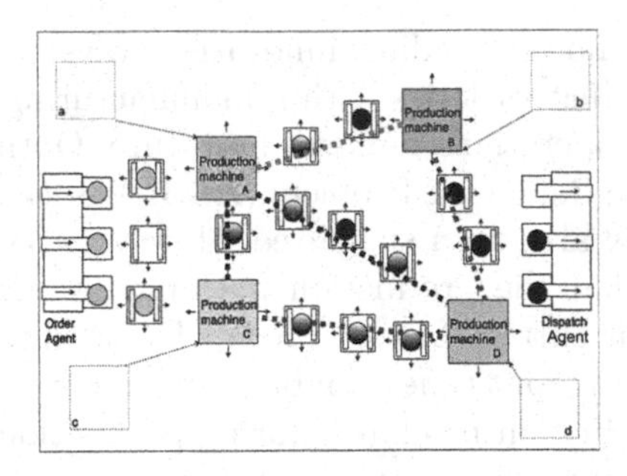

(b) Final S1M's layout.

Fig. 1. Layout of Lot Size-of-1 Manufacturing System (S1MS). (a) Initial layout of S1MS during production. Production machines, which are aggregations of required resources for performing specific manufacturing processes on products, and with the capability to move during the manufacturing process depending on the mix of order in-flow. Product agents are aggregations of mechanisms relating to product handling, transportation, and routing, which can be implemented as mobile robots/AGV/AIV with integrated pallets for holding parts that are undergoing the production process. The product agents are capable of independently coordinating their production process by figuring out an optimal route based on order-mix through the use of a stigmergic coordination mechanism. (b) The layout of S1MS after a period, production machines A, B, C, and D have moved from their previous positions a, b, c, d to a new position. Product agents have also autonomously figured-out optimal routes for production based on the existing product mix using virtual pheromones. The coordination of the movement of the production machine is achieved using the optimisation techniques proposed in this paper.

We determined the viability of these three approaches using computer simulation. We explored how the introduction of mobile production machines in the different implementations of S1MS affects the throughput achieved in a mass personalisation context, compared with a base manufacturing system with stationary production machines. This system is referred to as the baseline system (BS). The throughput of the three approaches and that of the BS for processing personalised orders with a "lot size of one" were compared. This is achieved by measuring the average production rates and average cycle-time per unit during production. The performance of the three S1MS implementations was compared to determine which was best and in what scenarios the different approaches were most applicable.

Finally, we present a high-level description of the three approaches and corresponding simulation results. The output of the simulation shows that a combination of computational and heuristic approaches outperforms a pure heuristics approach, While the heuristic approach outperforms the base system. We conclude that a combination of computational optimisation using only locally available variables and heuristics optimisation has the potential to address the slow convergence of heuristics optimisation techniques.

2 Manufacturing Systems with Characteristics for Dynamic Production Environment

Manufacturing companies are investigating smart, flexible and adaptive manufacturing systems capable of autonomous self-adaptation, self-reconfiguration, and capability for lot size-of-1 manufacturing [22]. Examples of such systems include Flexible Manufacturing Systems (FMS), Reconfigurable Manufacturing Systems (RMS), Holonic Manufacturing Systems (HMS), and Evolvable Assembly Systems (EAS).

Flexible Manufacturing System is an integrated system of machine modules and material handling equipment under computer control for the automatic processing of palletised parts [6,23]. The objective of FMS is to respond faster to changes in products and demands by manufacturing several types of parts cost-effectively; within pre-defined part families that can change over time; with minimum changeover cost; on the same system and at the required volume and quality [10].

Reconfigurable Manufacturing System, on the other hand, is designed to enable rapid change in hardware and software components for quick response to sudden market changes by adjusting its functionality and production capacity [16]. RMS proposes a manufacturing system where machine components and material handling units can be added, removed, modified, or interchanged as needed to respond quickly to changes in requirements, demands, and functionality [17].

Holonic Manufacturing System is inspired by Arthur Koestler's holons concept [15]. Holons are autonomous, self-reliant units with a degree of independence, such that contingencies can be handled without being instructed by higher authority; and simultaneously subjected to control from single or multiple higher authorities [4,30]. This implies that holons can exist in complex systems like manufacturing systems as both a whole and a part simultaneously. The "whole" property ensures the stability of forms in the system, while the "part" property signifies intermediate forms and ensures stability for higher forms. The holons concept comparatively provides more flexibility for manufacturing systems through decentralisation, and aggregation [5,11].

The evolvable assembly system (EAS) is a holistic system approach to enhance the capability of the manufacturing system to respond to rapid changes in product demand, market and processes. The EAS architecture comprises four phases which are: the reconfiguration phase, operation phase, monitor phase, and adaptation phase [6]. These four phases provide the capabilities for EAS components to adapt to changing conditions of operations. Examples of EAS implementation are Instantly Deployable Evolvable Assembly Systems (IDEAS) and Smart Manufacturing and Reconfigurable Technologies (SMART), which have attracted several European Projects. An example is the IDEA Project sponsored by the EU with participating companies and institutions such as FESTO, and the University of Nottingham [28].

3 Optimimisation Approaches in S1MS

3.1 Gradient-Descent Approach

We created a digital simulation of the physical manufacturing system, called the virtual S1MS (V-S1MS), which connects and informs the real S1MS existing in the real world. We present the structure and the layout optimisation process of the V-S1MS in this study, hence corresponding production entities are virtual. We modelled the shop floor as a two-dimensional plane. We assume Euclidean distance as a distance measure instead of rectilinear distance; this is because the movement of materials by the AIVs/AGVs and the movement of production machines are not constrained by any physical means. We then approximate the space of possible location h of production machines as a linear function of the location of AIVs/AGVs. Then we approximate the optimal location H, referred to as the global-hotspot as a function of h.

Therefore, we define a cost function $J_H(\theta_0, \theta_1)$, which evaluates the closeness of $h_\theta(x)$ to the desire AIV.

$$J_H(\theta_0, \theta_1) = \frac{1}{2m} \sum_{i=1}^{m} (H_\theta(x^{(i)}) - y^{(i)})^2 = \frac{1}{2m} \sum_{i=1}^{m} \left(\frac{1}{n} \sum_{i=1}^{n} (h_\theta(x^{(i)}) - y^{(i)} \right)^2 \quad (1)$$

To minimise the cost function $J_H(\theta_0, \theta_1)$, we use the gradient descent algorithm, which searches for potential *hotspots* and eventually arrives at a global-hotspot by starting with some initial-guess for θ and repeatedly changes θ to minimize $J_H(\theta_0, \theta_1)$, until it cannot be minimised further.

The gradient descent algorithm is expressed as follows:

$$repeat \left\{ \theta_j = \theta_j - \alpha \frac{\partial}{\partial \theta_j} J(\theta_0, \theta_1) \colon (For\ j = 0\ and\ j = 1) \right\} \quad (2)$$

The final algorithm for finding the global hotspot is expressed as follows:

$$H_\theta(x) = \frac{1}{m} \beta \sum_{i=1}^{m} \left((\theta_0 - \alpha \frac{1}{n} \sum_{i=1}^{n} (h_\theta(x^{(i)}) - y^{(i)})) + (\theta_1 - \alpha \frac{1}{n} \sum_{i=1}^{n} (h_\theta(x^{(i)}) - y^{(i)})) \right) \quad (3)$$

where β is the location bias, α is the learning rate, θ_0 and θ_1 are the intercept and slope of the line joining the different possible *hotspots* $h_\theta(x)$. $H_\theta(x)$ is the *global-hotspot*, which is the optimal location for the production machine, based on the location x, y of AIV with matching operation task.

3.2 Nature-Inspired Approach

A nature-inspired approach referred to as stigmergy is used as the coordination mechanism for coordinating the movement and re-positioning of production machines in real-time to minimise the cost of material flow and the distance between production machines without overlapping. The type of stigmergic coordination used is called *ant-based algorithm*. It is based on the stigmergic coordination found in an ant colony. This is generally referred to as Ant Colony Optimisation (ACO) algorithm or simply Ant Algorithm [8,9]. This algorithm has been extensively used as an approach to solving optimisation problems in manufacturing, operations research and supply chain [2,7,12,19,24,25].

However, ACO in the context of S1MS requires an abstraction within the problem domain. This will allow for seamless integration of the ACO algorithm into S1MS for effective coordination of production processes. To achieve this, the production environment is modelled as a Direct Acyclic Graph G and the ACO algorithm is used to find a feasible minimum cost path between two nodes over the graph $G = (C, L, W)$, where C is the node, L is the edge, and W is the weight associated with the edges, and feasibility is defined with respect to a set of constraints Ω. The minimum cost path in the context of S1MS implies the optimal positioning of production machines for the production of varied product mix with minimum cycle time. Details of the algorithm are given in [29].

3.3 The Three Optimisation Approaches in S1MS

Gradient Descent for Autonomous Resource Positioning (G-DARP)

In Fig. 2(a), mobile production machines, referred to as processing stations, move toward the global-hotspot *H*, which is equidistant from the product (the AGV + loaded parts for operations) with product-ids 1 & 2. This is because of the number of products selected for optimisation, which is the node-count = 2. Therefore, the spot that is optimum for production machine discovery, which is the global-hotspot, will be a location between the two products. Figure 2(b) shows that as the processing station approaches the global-hotspot H, it will be able to execute production tasks for other products with matching production processes (depicted as those with the same colour as the production machine)

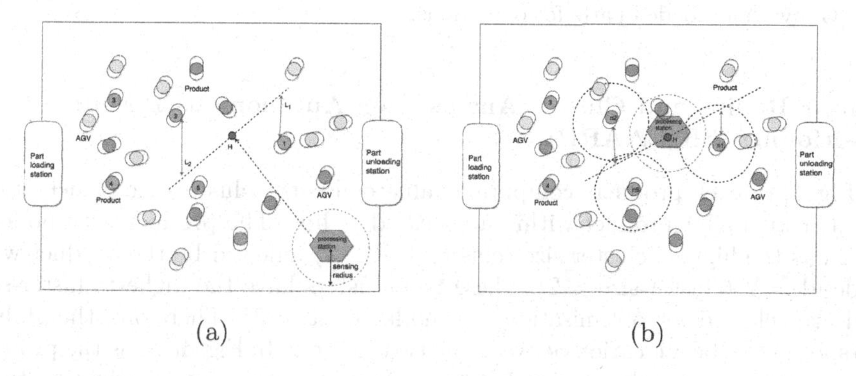

Fig. 2. V-S1MS using G-DARP for navigation and autonomous positioning of production machines (node-count = 2): product → AGV + loaded parts for operations, AGV → AGV without loaded parts for operations.

and are within its sensing-radius (for example, product with node-id = $n1, n2$ and $n5$). The products that have already located the hotspots use pheromones to lead other products to the hotspots. The final global hotspot is passed to the S1MS in the real world.

In Fig. 3, the processing station moves toward the global-hotspot H, which is equidistant from the products with product-id 1 to 5, but with a bias towards products with a smaller distance from each other. Therefore, computing a global-hotspot using the G-DARP algorithm is biased towards the centre of the floor-space between the closest products. Thus, the G-DARP algorithm tends to over-fit when node-count is high and thus providing a sub-optimal solution. It also tends to under-fit and provides a less optimal solution when node-count is low. However, it is intuitive to expect a better solution as the number of node-count increases, hence the poor performance is assumed to be a weakness of the algorithm and thus, a new algorithm that overcomes this deficiency is implemented and juxtaposed.

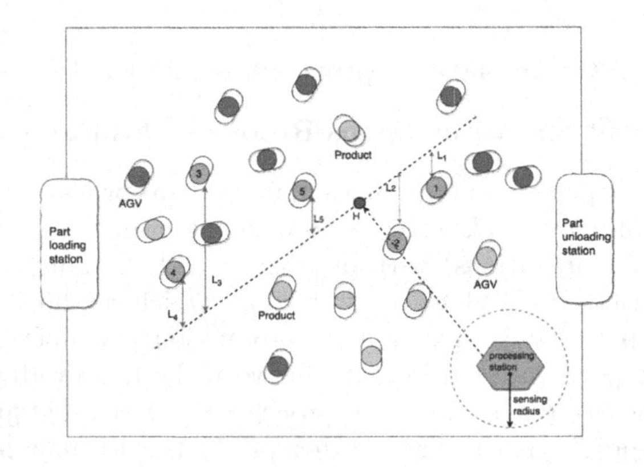

Fig. 3. V-S1MS using G-DARP for navigation and autonomous positioning of production machines (node-count > 2): product → AGV + loaded parts for operations, AGV → AGV without loaded parts for operations.

Sensor Radius and Cluster Analysis for Autonomous Resource Positioning (SR-CAARP)

In Fig. 4(a), each products compute a value called the cluster-size, which is the number of similar product within a specified radius. The product with node-id = X has the highest cluster size (cluster-size = 3) followed by the product with node-id = Y (cluster-size = 2). These two clusters have the highest cluster-size and are selected for optimisation (i.e. node-count = 2). Therefore, the global-hotspot H will be a location between the two clusters. In Fig. 4(b), as the production machine approaches the global-hotspot H, it can execute production tasks for products within the clusters that are with matching production operations (depicted as those with the same colour as the production machine) and are

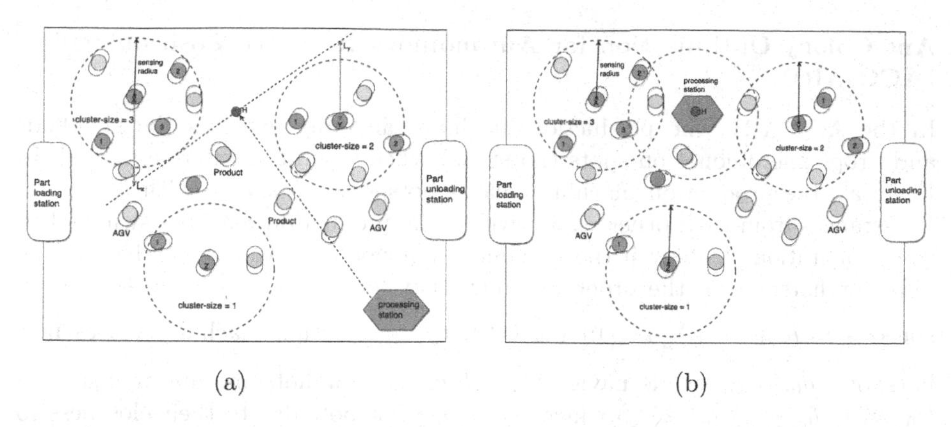

Fig. 4. V-S1MS using SR-CAARP for navigation and autonomous positioning of production machines (node-count = 2): product → AGV + loaded parts for operations, AGV → AGV without loaded parts for operations.

within its sensing-radius (for example, product-agent with node-id = $n1, n2$ and $n3$).

In Fig. 5(a), the products with node-id = X, Y, Z are selected for optimisation (node-count = 3). Therefore, the global-hotspot H will be a location between the three clusters. In Fig. 5(b), as the production machine approaches the global-hotspot H, it can execute production tasks for products within the clusters that are with matching operations requests (depicted as those with the same colour as the production machine) and are within its sensing-radius (for example, product-agent with node-id = $n1, n2$ and $n3$). In both cases, the products that have already located the hotspots use pheromones to lead other products to the hotspots. The final global hotspot is passed to the S1MS in the real world.

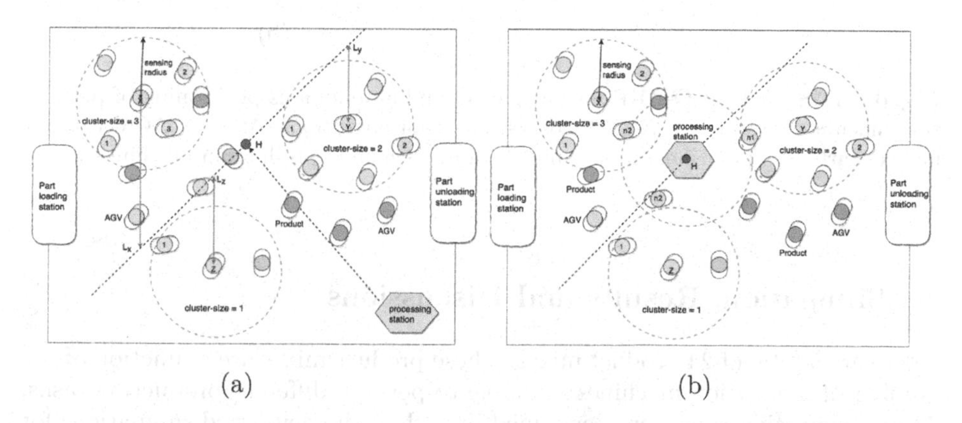

Fig. 5. V-S1MS using SR-CAARP for navigation and autonomous positioning of production machines (node-count = 3): product → AGV + loaded parts for operations, AGV → AGV without loaded parts for operations.

Ant Colony Optimisation for Autonomous Resource Positioning (ACO-ARP)

In the ACO-ARP, the production machine randomly explores the shopfloor and drops pheromones on spots (hotspots) where operations were executed. In Fig. 6(a), the production machine creates hotspots a, b and c, with l_1, l_2 and l_3 distances from each other respectively. The production machine switched to the exploitation strategy if the maximum hotspots are reached, and visits the different hotspots in the order in which they are created ($a \rightarrow b \rightarrow c$) with speed $s = log(\beta \cdot \frac{1}{n} \sum_{i=1}^{n} l_i)$. In Fig. 6(b), the production machine creates new hotspots (a_0, b_0, c_0) while navigating a, b, c, the new hotspots are at distances $(l_{0_1} \ll l_1; l_{0_2} \ll l_2; l_{0_3} \ll l_3)$ which are better hotspots due to their closeness to each other. At this instance, the speed $s = log(\beta \cdot \frac{1}{n} \sum_{i=1}^{n} l_i) \approx 0$ for the production machine and thus it remains stationary at a distance close to the hotspots, which becomes the global hotspot GH. The final global hotspot is passed to the S1MS in the real world.

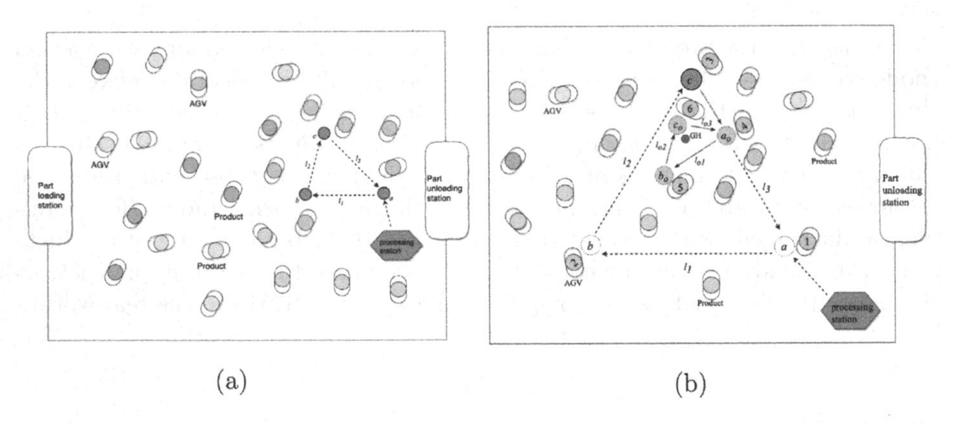

Fig. 6. S1MS using ACO-ARP for navigation and autonomous positioning of production machines: product $\rightarrow$ AGV + loaded parts for operations, AGV $\rightarrow$ AGV without loaded parts for operations, processing stations $\rightarrow$ mobile production machines

4 Simulation, Results and Discussions

There are a total of 24 product mixes. These product mixes are a function of the number of production machines available to perform different production tasks. Thirty simulation runs were performed in each of the developed simulations for 140,000 simulation steps each. The distribution of product mix in the system was skewed at an interval of 20,000 simulation steps during the simulation to further observe the behaviour of S1MS during unpredictable changes in product mix.

This distribution is estimated based on the probability of selecting a product mix with a particular machining sequence among the 24 possible product mixes. This probability is referred to as mix probability. The distribution of the different order types within the product mix is shown in Fig. 7.

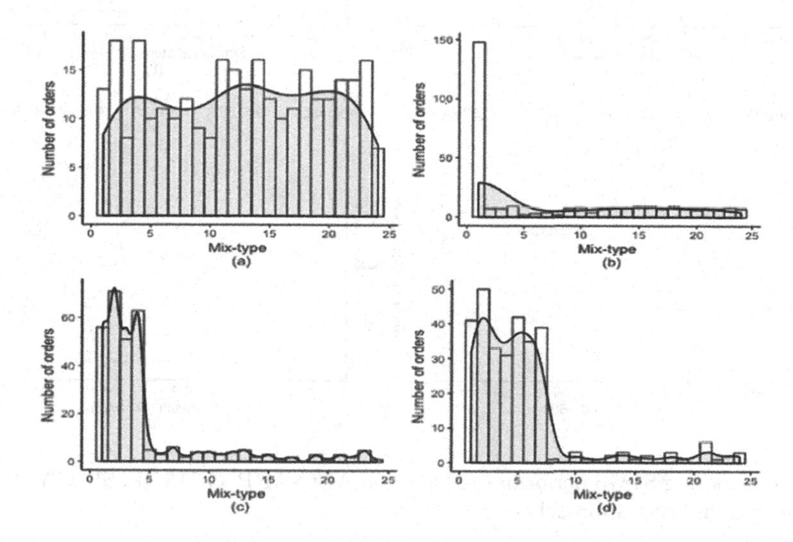

Fig. 7. Distribution of products within the product mix during simulation: (a) the default scenario, (b) scenario 1, (c) scenario 2, and (d) scenario 3

The following parameters were measured in each of the simulation setups to investigate differences in performance in the different approaches:

1. *Production rate*: This is the average number of products produced per 1k simulation steps during the simulation runs (140,000 simulation steps in total).
2. *Average cycle-time per product unit*: This is the average time (measured in simulation steps) taken to manufacture each product, i.e. the time each product spent in the production system.

G-DARP, SR-CAARP, ACO-ARP were compared with the BS using the production rate and average cycle-time per unit. The outcome of these two measures is referred to as performance afterwards. Four separate experiments were designed for this purpose, as shown in Fig. 7. The performance and behaviour of the different systems during production and unexpected changes in the product mix were compared (see Fig. 8).

Figure 8 showed that the production rate and average cycle-time when the total number of work-in-progress $N = 50$. It was observed that the three systems with mobile production machines, namely ACO-ARP, G-DARP and SR-CAARP, outperformed the BS. This is as a result of the capability of the three approaches to effectively minimise the cost of material flow and distances between production machines in real-time.

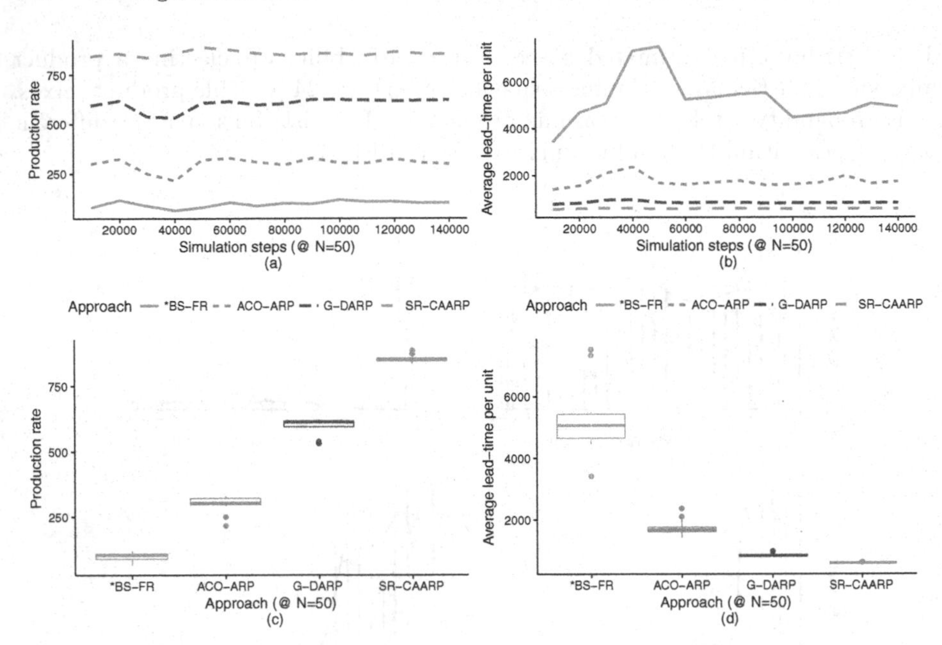

Fig. 8. Comparing S1MS performance based on ACO-ARP, G-DARP, SR-CAARP and baseline system (BS) approaches

The SR-CAARP approach, which is the optimised version of the G-DARP has the highest performance followed by the G-DARP, i.e. it has the highest production rate (see Fig. 8(a) and Fig. 8(c)) and lowest average cycle-time per unit (see Fig. 8(b) and Fig. 8(d)). Both of these approaches employed a local interaction mechanism complement with the ant-inspired algorithm for coordinating and autonomously positioning production machines to minimise the cost of material flow and distances between the production machines. The ACO-ARP comes just below the two ensemble approaches in performance, but its method for coordination and autonomous positioning of production machines is based on local interaction and heuristics optimisation algorithm, and, therefore, far less computationally expensive. The baseline system with fixed production machines records the least performance.

5 Conclusion

The three different approaches in S1MS were effective in seamlessly coordinating the movement and positioning of production machines during manufacturing. However, the ensemble approaches, G-DARP and SC-CAARP, performed better than the ACO-ARP. This is because the ensemble approach explores clusters of locally available information, such as product position, cluster size, and sensing radius, to optimise the location of production machines, while it uses the pheromones to speed up convergence. However, as the complexity and volume of this information grow as the value of N increases, slightly more computational resources may be required for cluster computation and analysis. The nature-inspired approach (ACO-ARP) used only locally available information and followed simple but effective rules to achieve seamless coordination of the movement and positioning of production machines.

Overall, the ACO-ARP has a computation advantage, but it does not outweigh the poor performance compared to the ensemble approaches. However, the ACO-ARP can serve as a backup in situations when local information required for computation suddenly becomes unavailable, such as network failure.

References

1. Andersen, A.-L., Brunoe, T.D., Nielsen, K., Rösiö, C.: Towards a generic design method for reconfigurable manufacturing systems: analysis and synthesis of current design methods and evaluation of supportive tools. J. Manuf. Syst. **42**, 179–195 (2017). https://doi.org/10.1016/j.jmsy.2016.11.006
2. Bautista, J., Pereira, J.: Ant algorithms for a time and space constrained assembly line balancing problem. Eur. J. Oper. Res. **177**(3), 2016–2032 (2007). https://doi.org/10.1016/j.ejor.2005.12.017
3. Becker, C., Scholl, A.: Balancing assembly lines with variable parallel workplaces: problem definition and effective solution procedure. Eur. J. Oper. Res. **199**(2), 359–374 (2009). https://doi.org/10.1016/j.ejor.2008.11.051
4. Botti, V., Giret, A.: Holonic manufacturing systems. In: Bowen, H.K., Robert, S., Carin-Isabel, K. (eds.) A multi-Agent Methodology for Holonic Manufacturing. New Balance Athletic Shoe, Inc., Harvard Business School Case, 606-094, April 2006. https://doi.org/10.1007/978-1-84800-310-1. Revised June 2008
5. Brusaferri, A., Ballarino, A., Carpanzano, E.: Distributed intelligent automation solutions for self-adaptive manufacturing plants. In: Ortiz, Á., Franco, R.D., Gasquet, P.G. (eds.) BASYS 2010. IAICT, vol. 322, pp. 205–213. Springer, Heidelberg (2010). https://doi.org/10.1007/978-3-642-14341-0_24
6. Chaplin, J.C., et al.: Evolvable assembly systems: a distributed architecture for intelligent manufacturing. IFAC-PapersOnLine **28**(3), 2065–2070 (2015). https://doi.org/10.1016/j.ifacol.2015.06.393
7. De Santis, R., Montanari, R., Vignali, G., Bottani, E.: An adapted ant colony optimization algorithm for the minimization of the travel distance of pickers in manual warehouses. Eur. J. Oper. Res. **267**(1), 120–137 (2018). https://doi.org/10.1016/j.ejor.2017.11.017
8. Dorigo, M., Blum, C.: Ant colony optimization theory: a survey. Theoret. Comput. Sci. **344**(2–3), 243–278 (2005). https://doi.org/10.1016/j.tcs.2005.05.020
9. Dorigo, M., Di Caro, G.: Ant colony optimization: a new meta-heuristic. In: Proceedings of the 1999 Congress on Evolutionary Computation-CEC99 (Cat. No. 99TH8406), vol. 2, pp. 1470–1477 (1999). https://doi.org/10.1109/CEC.1999.782657
10. El Maraghy, H.A.: Flexible and reconfigurable manufacturing systems paradigms. Flex. Serv. Manuf. J. **17**, 261–276 (2006). https://doi.org/10.1007/s10696-006-9028-7

11. Frayret, J.-M., D'Amours, S., Montreuil, B.: Coordination and control in distributed and agent-based manufacturing systems. Prod. Plann. Control **15**(1), 42–54 (2004). https://doi.org/10.1080/09537280410001658344
12. Hani, Y., Amodeo, L., Yalaoui, F., Chen, H.: Ant colony optimization for solving an industrial layout problem. Eur. J. Oper. Res. **183**(2), 633–642 (2007). https://doi.org/10.1016/j.ejor.2006.10.032
13. Huettemann, G., Gaffry, C., Schmitt, R.H.: Adaptation of reconfigurable manufacturing systems for industrial assembly - review of flexibility paradigms, concepts, and outlook. Procedia CIRP **52**, 112–117 (2016). https://doi.org/10.1016/j.procir.2016.07.021
14. Kiefer, J., Wolfowitz, J.: Stochastic estimation of the maximum of a regression function. Ann. Math. Stat. **23**(3), 462–466 (1952). https://doi.org/10.1214/aoms/1177729392
15. Koestler, A.: The Ghost in the Machine. Arkana Books, London (1989)
16. Koren, Y., Shpitalni, M.: Design of reconfigurable manufacturing systems. J. Manuf. Syst. **29**(4), 130–141 (2010). https://doi.org/10.1016/j.jmsy.2011.01.001
17. Koren, Y., et al.: Reconfigurable manufacturing systems. CIRP Ann. Manuf. Technol. **48**(2), 527–540 (1999)
18. Pine, B.J.: Mass Customization - The New Frontier in Business Competition. Harvard Business School Press (1993). ISBN 0-87584-372-7
19. Rajendran, C., Ziegler, H.: Ant-colony algorithms for permutation flowshop scheduling to minimize makespan/total flowtime of jobs. Eur. J. Oper. Res. **155**(2), 426–438 (2004). https://doi.org/10.1016/S0377-2217(02)00908-6
20. Robbins, H., Monro, S.: A stochastic approximation method. Ann. Math. Stat. **22**(3), 400–407 (1951)
21. Samadhi, T.M., Hoang, K.: Shared computer-integrated manufacturing for various types of production environment. Int. J. Oper. Prod. Manag. **15**(5), 95–108 (1995). https://doi.org/10.1108/01443579510083695
22. Sanderson, D., Chaplin, J.C., De Silva, L., Holmes, P., Ratchev, S.: Smart manufacturing and reconfigurable technologies: towards an integrated environment for evolvable assembly systems. In: Proceedings - IEEE 1st International Workshops on Foundations and Applications of Self-Systems. FAS-W 2016, pp. 263–264 (2016). https://doi.org/10.1109/FAS-W.2016.61
23. Sethi, A.K., Sethi, S.P.: Flexibility in manufacturing: a survey. Int. J. Flex. Manuf. Syst. **2**(4), 289–328 (1990). https://doi.org/10.1007/BF00186471
24. Shishvan, M.S., Sattarvand, J.: Long term production planning of open pit mines by ant colony optimization. Eur. J. Oper. Res. **240**(3), 825–836 (2014). https://doi.org/10.1016/j.ejor.2014.07.040
25. Silva, C.A., Sousa, J.M.C., Runkler, T.A., Sá da Costa, J.M.G.: Distributed supply chain management using ant colony optimization. Eur. J. Oper. Res. **199**(2), 349–358 (2009). https://doi.org/10.1016/j.ejor.2008.11.021
26. Valckenaers, P., Van Brussel, H.: Holonic manufacturing execution systems. CIRP Ann. Manuf. Technol. **54**(1), 427–432 (2005). https://doi.org/10.1016/S0007-8506(07)60137-1
27. Valckenaers, P., Hadeli, G., Saint, B., Verstraete, P., Van Brussel, H.: MAS coordination and control based on stigmergy. Comput. Ind. **58**(7), 621–629 (2007). https://doi.org/10.1016/j.compind.2007.05.003
28. Weirather, J., Institutsleitung, M.D., Muenchen, T.U.: Instantly deployable evolvable assembly systems. Project Final Rep. **49**, 1–69 (2015)

29. Ogunsakin, R., Marin, C.A., Mehandjiev, N.: Towards engineering manufacturing systems for mass personalisation: a stigmergic approach. Int. J. Comput. Integr. Manuf. **34**(4), 341–369 (2021)
30. Valentini, G., Masulli, F.: Ensembles of learning machines. In: Marinaro, M., Tagliaferri, R. (eds.) WIRN 2002. LNCS, vol. 2486, pp. 3–20. Springer, Heidelberg (2002). https://doi.org/10.1007/3-540-45808-5_1

A Intelligent Nanorobots Fish Swarm Strategy for Tumor Targeting

ShanChao Wen[2], Yue Sun[1,2](✉), SiYang Chen[1], and Yifan Chen[2,3]

[1] School of mechanical and electrical engineering, Chengdu University of Technology, Chengdu, China
sunyuestc90@126.com
[2] School of Life Science and Technology, University of Electronic Science and Technology of China, Chengdu, China
[3] Putian University, Putian, China

Abstract. This paper proposes a nanorobots fish swarm algorithm (NFSA) for tumor targeting. The alterations in the tumor microenvironment caused by tumor growth produce the biological gradient field (BGF), which is regulated by the adjacent tortuous and dense capillary network. NFSA is used to measure tumor-targeting efficiency in comparison to the benchmarks of Brute-force and the conventional gradient descent algorithm. Our goal is to increase the efficiency of targeting tumors in the early stages by using existing swarm intelligence algorithms to manipulate nanorobot swarms (NS) through magnetic fields. The extracorporeal observation system sensed the motion of NS under the influence of a BGF and then estimated the gradient of BGF. The invasive percolation algorithm models the vascular network to evaluate the performance of searching strategies. We also apply the exponential evolution step mechanism to boost the tumor-targeting efficiency of NFSA. The results show that NFSA has higher overall tumor targeting efficiency and a fast convergence property than previous algorithms. We hope that the NS in a multi-agent system could pave the way for challenges in tumor targeting.

Keywords: Tumor targeting · Nanorobots swarm · Biological gradient field · Vascular network · Swarm intelligence

1 Introduction

The high cancer mortality rate is primarily due to the diffusion and migration of malignant tumors [1]. Early therapy, for instance, can cut breast cancer mortality by 39%. As a result, early detection and diagnosis of malignancies can significantly improve the cancer cure rate. Since conventional medical imaging tools, such as Magnetic Resonance Imaging (MRI) and Computed Tomography (CT), are limited in accuracy and resolution, early tumor diagnosis is challenging. Consequently, magnetic nanoparticles (MNPs) play a critical role in medical imaging

Y. Chen et al. (Eds.): BICT 2023, LNICST 512, pp. 280–291, 2023.
https://doi.org/10.1007/978-3-031-43135-7_27

diagnostics today [2]. Under the guidance of a regulated magnetic field, MNPs as contrast agents may discriminate malignant tumors from healthy tissue.

Due to the high demand for oxygen and nutrients during malignant tumor growth, the capillaries growth surrounding the tumor is abnormal, forming and tortuous and high-density interconnected network. As a consequence, NS can penetrate the tumor region by passively permeating capillaries [3]. It is difficult to eradicate the NS since the tumor region lacks a regular lymphatic system. Accurate medical imaging and target drug delivery treatment can be accomplished once NS is enriched in the tumor's area.

Nanorobots in medical imaging provides a solution for early tumor detection and precise drug targeting. Based on tumor detection, Chen et al. [4] proposed an *in vivo* computing system, which achieves considerable aggregation of nanorobots in the tumor area with the shortest possible physiological path without previous knowledge of tumor location.

Because of the complicated human vascular network and biological milieu, an NS could easily be split into numerous smaller clusters of NS while propagation in the vascular network. Moreover, the tumor detection efficiency of a single NS is lower than that of multiple NS. As a result, a swarm intelligence-based multi-agent system for tumor targeting is required. A sequential evolutionary strategy is proposed in the [5], which is influenced by traditional wireless communication "time division multiplexing (TDM)" and creates a strict and realistic framework for tumor targeting in multi-NS management. Since the challenges of extracorporeal operating system in manipulating multiple NS separately, thus, this paper assumes all the NS under a uniform magnetic field control.

The biological population evolutionary algorithm is inspired by behaviors of the biological population in nature [6], which provides a solution for the swarm intelligent optimization. This paper proposes a multiple NS fish swarm strategy for tumor targeting, using sequential evolutionary methods and an artificial fish swarm algorithm.

The rest of this paper is organized as follows: The biological gradient field is illustrated in the Sect. 2. The vascular network modeling is established in the Sect. 3 using the invasive percolation algorithm. Moreover, the Sect. 4 describes the NFSA optimization approach in detail. The comparison of brute-force, conventional gradient algorithm, and NFSA are analyzed in the Sect. 5. We conclude in the Sect. 6 section with a proposal for future direction.

2 Biological Gradient Fields

2.1 The Generation of BGFs

Tumor lesions produce multi-property alterations to the tumor microenvironment, and the biological and chemical features of the near-site and far-site tumors are vastly different [7]. For example, in the tumor microenvironment, the distribution of oxygen density, pH, glucose, and other nutrients (such as growth factors and hormones) may be uneven or insufficient. These homologous tissue heterogeneities caused by malignancy can be thought of as BGF.

Through the tumor microenvironment's enhanced permeability and retention effect (EPR), NS with contrast agents can reach the area around the tumor via passive transport. Through the EPR effect of the tumor micro-environment, NS can be continuously enriched in tumor lesions [8].

2.2 BGF in Vivo Computing

NS does not directly sense the value of BGF, but it serves as a computational agent by measuring and estimating the changes in the features of NS [9]. For instance, the pH-responsive Oregon Green 488 (OG) conjugates with a trypsin-cleavable peptide as PH changes, acting as a biosensor to detect the dynamic of PH profile [10]. The extracorporeal operating system continually detect NS towards the tumor, depending on the BGF value. Since there are currently no commonly accepted quantitative models of BGF, this paper adopts the Sphere objective function with global minimization values to characterize BGF.

For the convenience of calculation, the BGF value is normalized to 0–1. The tumor search parameter space P is set to -5mm $\leq$ (x,y) $\leq$ 5 mm. The expression is as follows:
Landscape :

$$U_T(x,y) = \begin{cases} 1 & \sqrt{x^2+y^2} \leq 0.3 \; and \; (x,y) \in P \\ -(x^2+y^2+10)/85+1 & \sqrt{x^2+y^2} > 0.3 \; and \; (x,y) \in P \\ 0 & (x,y) \notin P \end{cases} \quad (1)$$

where space P denotes a discrete vascular network space. The Sphere function, smooth and uniform, is regarded as the most basic objective function. The circle with a radius of 0.3 mm in the central position is considered a tumor, and the objective function value of this region corresponds to the landscape's global maximum value 1, also the optimal global solution.

3 Vascular Network Around Tumor Micro-Environment

3.1 Growth Mechanism of Vascular Network Around Tumor Micro-Environment

Human tissues are supplied with oxygen and nutrients via the vascular network system. Capillaries, which connect arteries and veins, require more oxygen and nutrients than normal tissues due to tumor cells' erratic proliferation; hence their capillaries will suffer pathological structural changes. As a result, changes in the status of surrounding tissues can be reflected in the geometry of blood vessels. The otherwise normal vascular network may proliferate by budding or overlapping when the tumor diameter reaches 1–2 mm. The vascular networks near the tumor degrade significantly as the tumor grows. The blood vessels surrounding the tumor, on the other hand, will proliferate rapidly (Fig. 1).

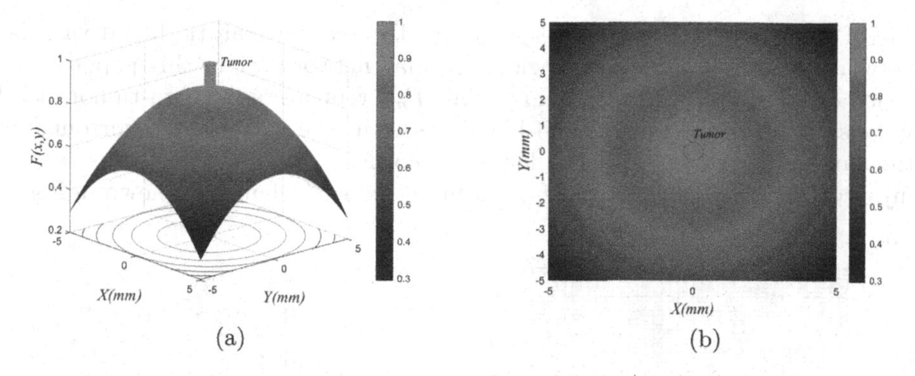

Fig. 1. The diagram of the Sphere objective function: (a) the geometry of the Sphere function and (b) its contour curve

Pathological vascularity is one of the hallmarks of a potential tumor. The normal vascular network undergoes a series of tumor-specific changes in response to the demand for nutrients and oxygen for tumor growth, including vascular regeneration (formation of new blood vessels), vascular remodeling (integration of existing vessels with tumor vessels), and vascular degeneration. Furthermore, the permeability of these vascular networks is improved, and the diameter of their endothelial and extracellular pores is increased, allowing nanoparticles to pass through the pore and reach the tumor lesion. In summary, malignant tumor vasculature is characterized by the following two features:

- **1) The density of blood vessels in the paratumoral area is greater than in normal tissue.**
  2) The centeral area of tumor suffered vascular degeneration.

3.2 Modeling of Vascular Network Periphery of the Tumor

We assume that the capillary network around normal tissues is latticed and regularized, and the vascular network's growth direction is parallel to the coordinate axis. The distance between capillaries determines the capillaries density. The expected increase in capillaries density is usually observed in the periphery area of the tumor, whereas the morphology of the vasculature in the tumor center is characterized by decreased density [11]. The intrusive percolation algorithm presents the characteristics of blood vessels with tumors, which is used to construct the near-tumor capillary network [12].

In the intrusion percolation model, each lattice point is randomly assigned a uniformly distributed intensity value. Then from the set position as the starting point, the vessel grows to the position with the smallest lattice intensity among all the lattice nodes adjacent to the current network. The procedure is repeated until the desired level of lattice occupancy is achieved. The lattice occupation is used to simulate the fractal dimension of the pathological vascular network of the tumor. The fractal dimensions 1.6, 1.8, and 1.9 correspond to 40%, 60%, and

80%of the lattice occupation, respectively. To create a realistic blood vascular network, the size of the pathological vascular network set in this paper is the space P_m of $-1\,\text{mm} \leq (\text{x,y}) \leq 1\,\text{mm}$, while P_{m1} represents the para-tumor blood vessels. P_{m2} is expressed as the blood vessel in the central area surrounding the tumor.$P_{m1}, P_{m2} \in P_m$. The lower-left corner of the blood vessel is the NS injection area, the range of which is $-5\,\text{mm} \leq (\text{x,y}) \leq -4\,\text{mm}$, as shown in Fig. 2,

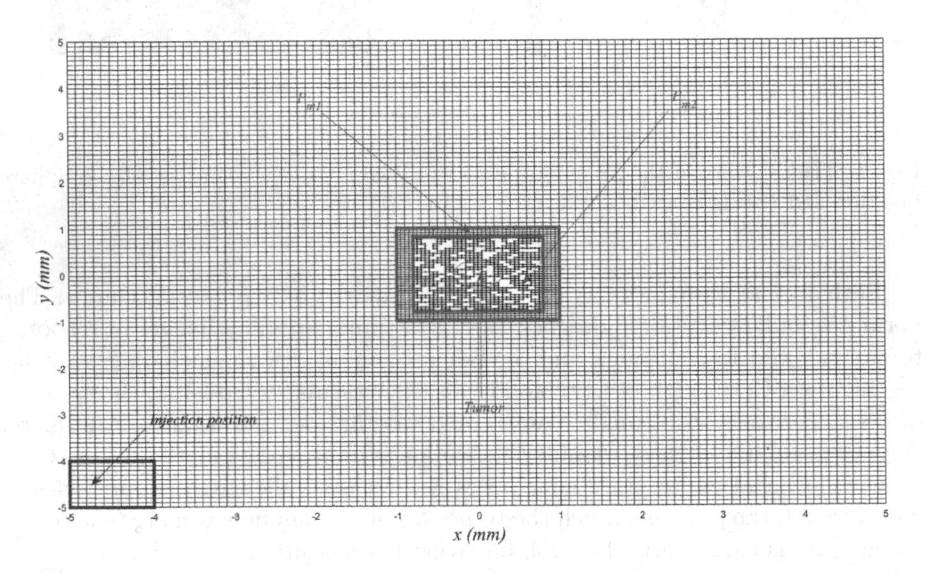

Fig. 2. The vascular network is modeled by invasive percolation technique. P_{m1} represents the blood vessels adjacent to the tumor, P_{m2} represents the blood vessels in the central area of the tumor, and the lower left corner of the blood vessel is the NS injection area.

4 Computing Method of Nano Fish Swarm Algorithm

Although the uniform magnetic field generated by the Helmholtz coil can regulate the movement direction of NS, the best orientations of multiple NS are frequently incongruent. As a result, while the magnetic field may cause one NS to move in the optimal direction, the magnetic field will cause other NS to move in the non-optimal direction. Based on this, this paper adopts the control method of time-division multiplexing proposed in [5]. The critical parts of the model are divided into the following points.

4.1 Time Division Multiplexing Control Strategy

The whole framework of the time-division multipurpose control model is divided into three modes: FC (Forced Control), DC (Directional Control), and IT (ImagingTracking). In IT mode, Multiple NS move through blood viscous force without the intervention of a magnetic field, and BGF can be calculated by an extracorporeal operating system based on NS's performance; In DC mode, The magnetic field controls the NS in a specific order. However, in a cycle, the magnetic field only cares about whether the movement direction of one NS is optimal and ignores the other NS. In FC mode, the motion of NS is similar to that in DC mode. A cluster of NS moves in the same direction as the one NS under the uniform magnetic field control, which may result in a non-optimal direction. The rest of NS, in this case, is also called in forced control.

- 1) **Initialization**: at the initial time $t_{DC,1}^{(1)}$, nano-particle swarm NS_1,NS_2,..., NS_N enters the injection area, randomly allocates each position in the injection area, and uses the vector $\overrightarrow{x_1}(t_{DC,1}^{(1)})$, $\overrightarrow{x_2}(t_{DC,1}^{(1)})$, $\overrightarrow{x_3}(t_{DC,1}^{(1)})$,..., $\overrightarrow{x_N}(t_{DC,1}^{(1)})$ represent its initial position, where the superscript indicates the number of times of current evolution for all NS; when all NS completes a FC\DC-IT movement, it is regarded as evolutionary completion and enters the next generation; subscript is the number of times of cycles in current generation.
- 2) **Directional control (DC)**: NS_1 is used as a representative to depict the motion process since all NS are controlled in the same way. The initial position of NS_1 is $\overrightarrow{x_1}(t_{DC,1}^{(1)})$, and after a DC control, the position changes $\overrightarrow{x_1}(t_{DC,1}^{(1)}) \rightarrow \overrightarrow{x_1}(t_{IT,1}^{(1)})$, where $t_{IT,1}^{(1)} = t_{DC,1}^{(1)} + T_{FC/DC,1}^{(1)}$, $T_{FC/DC,1}^{(1)}$is the control time, the control time is not necessarily fixed, and this parameter can be adjusted to meet the needs of the situation. Under the vascular network shown in Fig. 2, the specific location of the NS_1 is calculated as follows:

$$\overrightarrow{x_1}(t_{IT,1}^{(1)}) = \overrightarrow{x_1}(t_{DC,1}^{(1)}) + d_1(t_{DC,1}^{(1)})\overrightarrow{u}_{\angle\varphi(t_{DC,1}^{(1)})} + \overrightarrow{e_1}(t_{DC,1}^{(1)}) \tag{2}$$

$$\left\| d_1(t_{DC,1}^{(1)}) \right\|_1 = v_{ma} \cdot T_{DC/FC,1}^{(1)} \tag{3}$$

$$\left\| d_1(t_{DC,1}^{(1)}) \right\|_1 = d_1(t_{DC,1}^{(1)}) \cos\angle\varphi(t_{DC,1}^{(1)}) + d_1(t_{DC,1}^{(1)}) \sin\angle\varphi(t_{DC,1}^{(1)}) \tag{4}$$

  where $\|\bullet\|_1$ indicates l_1 norm, $\angle\varphi(t_{DC,1}^{(1)})$ depends on the optimization direction needed by the situation, $\overrightarrow{u}_{\angle\varphi(t_{DC,1}^{(1)})}$ is the unit vector in the optimization direction, between 0 and $\pi/2$, since NS could not perform the opposite direction of blood flow. The discretization error generated by the discretization of the vascular network, to ensure that the destination is on the vascular node, is $e_1(t_{DC,1}^{(1)})$, and the magnetic field control speed is v_{ma}.
- 3) **Imaging tracking (IT)**: when a FC\DC movement occurs, the extracorporeal control system needs to re-track all the NS to calculate the motion direction of the next optimization. At this stage, the NS walk randomly along the blood flow velocity to sense the BGF. After the completion of the IT, the position change of the NS_1 is $\overrightarrow{x_1}(t_{IT,1}^{(1)}) \rightarrow \overrightarrow{x_1}(t_{DC,2}^{(1)})$, where $t_{DC,2}^{(1)} = t_{IT,1}^{(1)} + T_{IT,1}^{(1)}$ traveling $T_{IT,1}^{(1)}$ is the walking time, which is similar to that of $T_{FC/DC,1}^{(1)}$. The specific calculation method is as follows:

$$\overrightarrow{x_1}(t_{DC,2}^{(1)}) = \overrightarrow{x_1}(t_{IT,1}^{(1)}) + (x(t_{IT,1}^{(1)}), y(t_{IT,1}^{(1)})) \tag{5}$$

$$\left\|d_1(t_{IT,1}^{(1)})\right\|_1 = v_{vessel} \cdot T_{IT,1}^{(1)} + e(t_{IT,1}^{(1)}) \tag{6}$$

$$\left\|d_1(t_{IT,1}^{(1)})\right\|_1 = x(t_{IT,1}^{(1)}) + y(t_{IT,1}^{(1)}) \tag{7}$$

where v_{vessel} is the blood flow velocity of the vascular network, and the blood flow velocity of blood vessels in different regions is different, and the error compensation of $e(t_{IT,1}^{(1)})$ node is to ensure that the movement destination is on the vessel node. $\left\|d_1(t_{IT,1}^{(1)})\right\|_1$ is the distance of NS flowing with blood in it mode. Since NS moves up or right in the vascular network under the influence of blood viscous force. The value of $x(t_{IT,1}^{(1)}), y(t_{IT,1}^{(1)})$ is equally distributed under the constraint of (7).

- 4) **Forced control (FC)**: this mode can be regarded as the DC control stage of other NS, at this time, due to the global force of the magnetic field, the NS_1 will be forced to move. After the completion of the FC, the position of the NS1 changes to $\vec{x_1}(t_{DC,2}^{(1)}) \rightarrow \vec{x_1}(t_{IT,2}^{(1)})$, $t_{IT,2}^{(1)} = t_{DC,2}^{(1)} + T_{FC/DC,2}^{(1)}$, $T_{FC/DC,2}^{(1)}$ is the control time of other NS, and the specific position calculation rule of NS is similar to that of the DC mode.

4.2 Nanorobot Fish Swarm Algorithm

NFSA is an algorithm that combines an artificial fish swarming algorithm with a time-division multiplexing control strategy. The artificial fish swarm algorithm is proposed based on the swarm behavior of fish, where the clustering and foraging behavior in artificial fish swarm algorithm (AFSA) can be combined with the swarm intelligence control of NS [13]. In order to ensure the targeting efficiency of the whole population, NFSA adopts a piecewise iterative method in the DC phase based on the time-division multiplexing strategy and the specific process is shown in Fig. 3 In DC mode, there are two control types: foraging (DC_c) and clustering (DC_b). In foraging behavior, the sequential control strategy is adopted, and NS_1, NS_2... NS_N adjusts its orientation sequentially. Under foraging behavior, The direction with the fastest fitness increase is given precedence by NS. The weakest-first evolutionary strategy is adopted under clustering behavior. Individuals with the worst BGF adaptation are given priority in moving to the center of m NS with the best adaption. This design is because some NS with the best BGF fitness tend to enter the tumor center first, which has more convoluted blood capillaries. The movement of these NS is more likely to be restricted by vascular space, which is favorable for allowing NS with low BGF fitness to catch up and approach tumor lesions as much as possible, enhancing the targeting rate of all NS. The duration T_{DC_c} of foraging behavior tends to influence the improvement of the best NS fitness, while the duration T_{DC_b} of clustering behavior tends to influence the improvement of the fitness of all NS. How to balance T_{DC_c} and T_{DC_b} duration to maximize the targeting efficiency

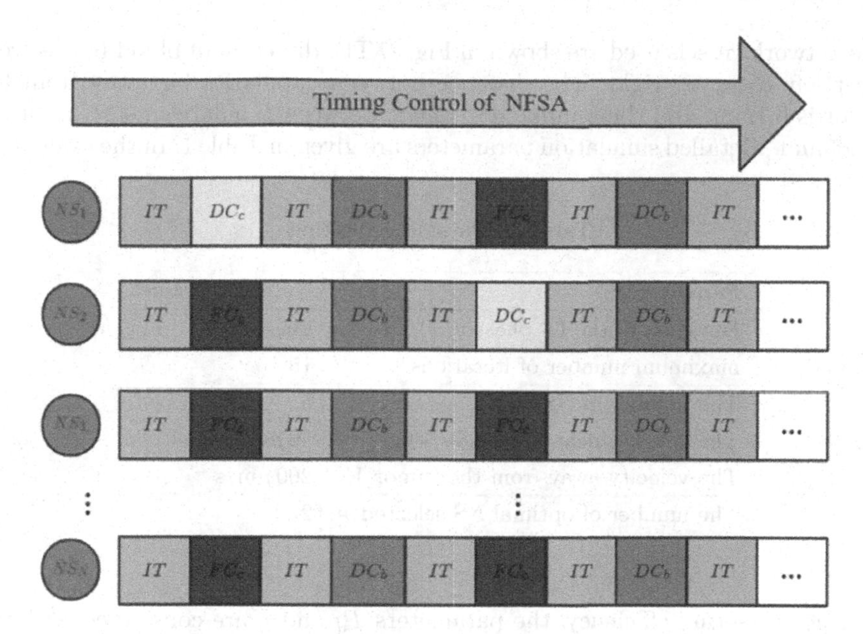

Fig. 3. Timing control logic of NFSA

needs deeper level related research. Taking NS1 as an example, the specific calculation model of clustering behavior (DC_b) and foraging behavior (DC_c) is as follows:

Foraging behavior

$$\overrightarrow{u}_{\angle\varphi(t_{DC,1}^{(1)})} = \frac{grad(f(\overrightarrow{x_1}(t_{DC,1}^{(1)})))}{\left|grad(f(\overrightarrow{x_1}(t_{DC,1}^{(1)})))\right|} \tag{8}$$

Clustering behavior

$$\overrightarrow{u}_{\angle\varphi(t_{DC,1}^{(1)})} = \frac{\frac{\sum_{j\in V_m}\overrightarrow{x_j}(t_{DC,1}^{(1)})}{m} - \overrightarrow{x_1}(t_{DC,1}^{(1)}))}{\left|\frac{\sum_{j\in V_m}\overrightarrow{x_j}(t_{DC,1}^{(1)})}{m} - \overrightarrow{x_1}(t_{DC,1}^{(1)}))\right|} \tag{9}$$

where V_m is the set of m NS with the best fitness. Until they reach the tumor, all ns are subjected to continual iterative control in the manner indicated in Fig. 3.

5 Comparison and Performance Analysis of Algorithms

5.1 Simulation Parameter Setting

Numerical simulations are performed in MatLab to compare the performance of NFSA, conventional GD algorithm, and Brute-force search for tumor targeting.

The network vessels used are shown in Fig. 2. The direction of blood flow is from lower left to upper right. The distance between capillaries far away from the tumor is 0.1mm, and the minimum distance between capillaries near the tumor is 0.05mm. Detailed simulation parameters are given in Table 1. In the evaluation

Table 1. Parameter Settings

Parameters	Numerical values
Duration of the IT phase	4 s
maximum number of iterations	15
The number of NS	5
The velocity near the tumor V_{Pm}	50 μm/s
The velocity away from the tumor V	200 μm/s
The number of optimal NS selected m	2

of tumor-targeting efficiency, the parameters P_d and η are considered within a limited number of iterations, whereas P_d represents the ratio of the number of successful tumor detection to the total number of simulation experiments, and η denotes the ratio of the total number of computational agents that detected the tumor to the total number of NS. The random direction of range $0 - \frac{\pi}{2}$ is employed in the DC phase, taking into account the influence of Brute-force search as a reference group and blood flow velocity. In addition, since the tumor diameter is too small, the change step of the exponential evolution mechanism is also adopted in the DC/FC stage.

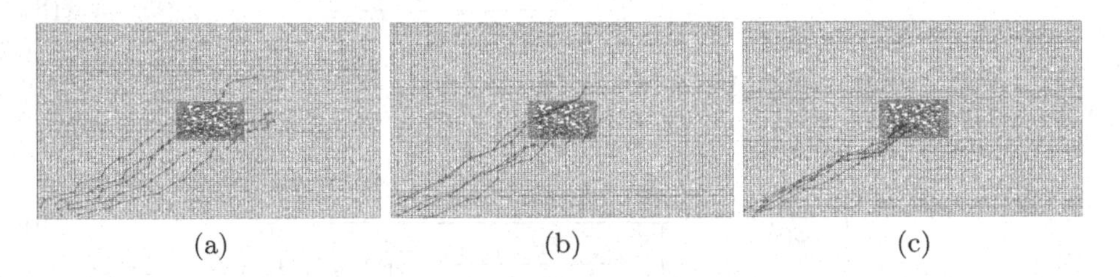

(a) (b) (c)

Fig. 4. Multiple NS paths with three strategies in one simulation: (a) Multiple NS paths using Brute-force search; (b) Multiple NS paths using traditional GD algorithm in TDM Framework; (c) Multiple NS paths using AFSA

5.2 Simulation Results

Figure 4 shows the NS path trajectory performed in one simulation using the three algorithms. Figure 5 shows the corresponding average fitness, the average

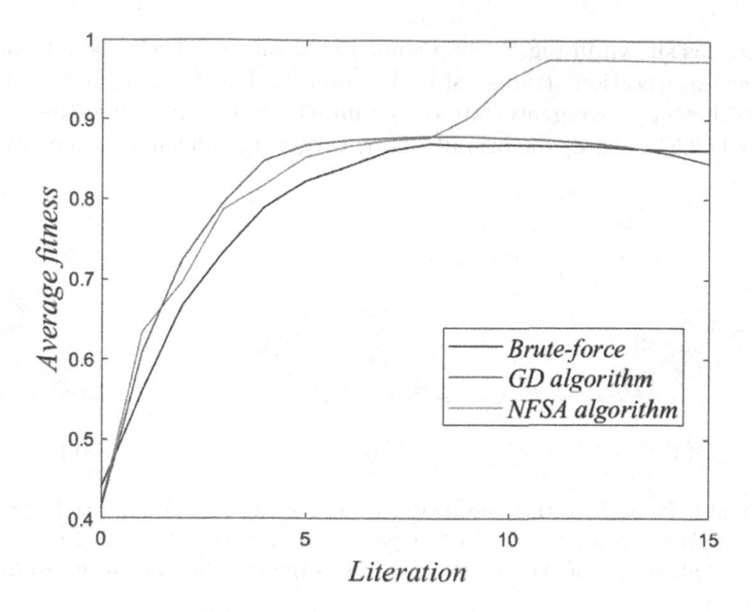

Fig. 5. The change of the average fitness of multiple NS as the number of iterations increases under different strategies

value of BGF, with increasing number of iterations. Figure 6 presents the histogram of the distribution of the number of NS that can reach the central region of the tumor using the three algorithms in 100 experiments, and the path of Brute-force search is scattered since the direction of magnetic field control in its DC phase is random. The overall targeting efficiency of the NFSA-based swarm intelligence system is significantly higher than that of the GD method and Brute-force search, as shown in Fig. 6. The NS in the DC phase uses the GD algorithm to move in the same direction as its gradient, whereas the other NS in the FC phase is offset due to the global action of the magnetic field force. As a result, the total targeting efficiency of multi-NS systems using the time-division multiplexing technique and the GD algorithm is average. In the multiple NS system based on the NFSA algorithm, when the optimal NS is close to the tumor area or the tumor center area. The movement of optimal NS is restricted to capillary degradation around the tumor. Simultaneously, the NS with the lowest fitness is chosen to approach some of the best NS. Even though the global magnetic field impacts it, the most optimal NS status is unaffected, and the overall NS targeting rate can be improved.

From Fig. 5, Although the average fitness of NS employing Brute-force search improves over time as a result of blood flow, its targeting rate remains low. The fitness will rapidly decline once they pass through the tumor. In contrast, the convergence speed of multiple NS systems using the time-division multiplexing strategy and GD algorithm is the fastest since BGF usually changes fastest in the gradient direction. However, unable to overcome the effect of the global force of the magnetic field, the average fitness decreases rapidly after passing through the tumor, as shown in Fig. 5. After some iterations in NFSA, the advantages of the piecewise iteration strategy start to emerge. The foraging behavior (DC_c) ensures a better convergence rate of the multi-NS system, while the clustering behavior (DC_b) ensures the overall tumor targeting efficiency of the system.

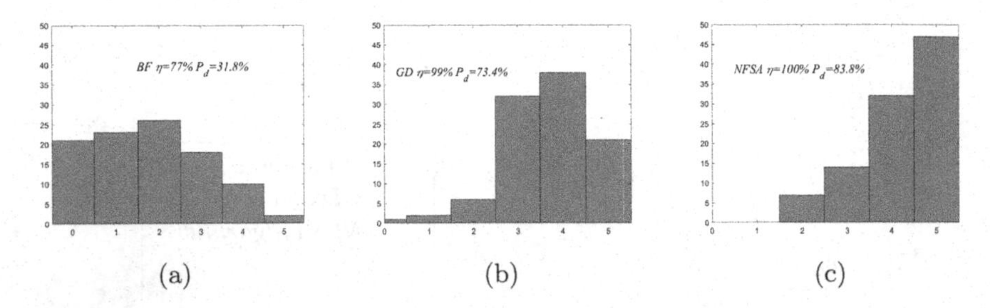

Fig. 6. Multi-NS paths with three strategies under one simulation: (a) Targeting efficiency using Brute-force search of 100 experiments; (b) Targeting efficiency using traditional GD algorithm of 100 experiments; (c) Targeting efficiency using AFSA of 100 experiments

6 Conclusion

In this paper, NFSA is proposed in the framework of TDM strategy and artificial fish swarm algorithm, which provides an effective multi-agent system for tumor targeting. The vascular network periphery of the tumor is tortuous and dense interconnected, which is a considerable challenge to controlling the NS swarm. The simulation results demonstrate that the swarm intelligence system based on NFSA has considerable search convergence speed and more optimized overall tumor targeting efficiency than the traditional GD algorithm and Brute-force search. We may apply more evolutionary algorithms and more complex fractal blood vessels to provide more practical and effective tumor targeting programs in future work.

References

1. Smith, R., et al.: Cancer screening in the United States, 2019: a review of current American cancer society guidelines and current issues in cancer screening. CA Cancer J. Clin. **69**, 184–210 (2019)
2. Sun, Y., Qing, Y., Chen, Y.: In vivo computing for smart tumor targeting in taxicab-geometry vasculature. IEEE Trans. NanoBiosci. **21**, 445–453 (2022)
3. Harney, A.S., et al.: Real-time imaging reveals local, transient vascular permeability, and tumor cell intravasation stimulated by tie2hi macrophage-derived VEGFA. Cancer Disc. **5**(9), 932–943 (2015). ISSN 2159-8274
4. Chen, Y., et al.: Biosensing-by-learning direct targeting strategy for enhanced tumor sensitization. IEEE Trans. NanoBiosci. **18**(3), 498–509 (2019)

5. Shi, S., et al.: Exponential evolution mechanism for in vivo computation. Swarm Evol. Comput. **65**, 100931 (2021). ISSN 2210–6502
6. Simon, D.: Evolutionary optimization algorithms: biologically-inspired and population-based approaches to computer intelligence (2013)
7. Ali, M., Chen, Y., Cree, M.J.: Autonomous *In Vivo* computation in internet of nano bio things. IEEE Internet Things J. **9**(8), 6134–6147 (2022)
8. Ao, H., et al.: Enhanced tumor accumulation and therapeutic efficacy of liposomal drugs through over-threshold dosing. J. Nanobiotechnol. **20**, 137 (2022). https://doi.org/10.1186/s12951-022-01349-1
9. Bejarano, L., Jordão, M.J.C., Joyce, J.A.: Therapeutic targeting of the tumor microenvironment. Cancer Disc. **11**(4), 933–959 (2021). ISSN 2159–8274
10. Sun, Y., Bian, H., Chen, Y.: A photolysis-assist molecular communication for tumor biosensing. Sensors **22**(7), 2495 (2022)
11. Paul, R.: Flow-correlated dilution of a regular network leads to a percolating network during tumor-induced angiogenesis. Eur. Phys. J. E Soft Matter **30**(1), 101–114 (2009). https://doi.org/10.1140/epje/i2009-10513-8. ISSN 1292–8941
12. Shi, S., et al.: Nanorobots-assisted tumor sensitization and targeting for multifocal tumor, pp. 362–365 (2020)
13. Liu, Y., et al.: Parameter identification of collaborative robot based on improved artificial fish swarm algorithm. In: 2020 International Conference on High Performance Big Data and Intelligent Systems (HPBD IS), pp. 1–7 (2020)

Author Index

Y. Chen et al. (Eds.): BICT 2023, LNICST 512, pp. 293–294, 2023.
https://doi.org/10.1007/978-3-031-43135-7

Printed in the United States
by Baker & Taylor Publisher Services